SOLUBILITY DATA SERIES

Editor-in-Chief

A. S. KERTES
Institute of Chemistry
The Hebrew University
Jerusalem, Israel

EDITORIAL BOARD

INTERNATIONAL UNION OF PURE AND APPLIED CHEMISTRY

ANALYTICAL CHEMISTRY DIVISION
COMMISSION ON EQUILIBRIUM DATA
SUBCOMMITTEE ON SOLUBILITY DATA

SOLUBILITY DATA SERIES

Volume 2

KRYPTON, XENON AND RADON — *Gas Solubilities*

SOLUBILITY DATA SERIES

Volume 2

KRYPTON, XENON AND RADON — *Gas Solubilities*

Volume Editor

H. LAWRENCE CLEVER

Chemistry Department,
Emory University,
Atlanta, GA, USA

Evaluators

RUBIN BATTINO
Wright State University,
Dayton, OH, USA

WILLIAM GERRARD
Polytechnic of
North London,
London, UK

COLIN L. YOUNG
University of
Melbourne,
Parkville, Victoria,
Australia

Compilers

ARDIS L. CRAMER
Emory University

M. ELIZABETH DERRICK
Valdosta State College

SUSAN A. JOHNSON
Emory University

TRINA KITTREDGE
Emory University

PATRICK L. LONG
Emory University

PERGAMON PRESS

OXFORD · NEW YORK · TORONTO · SYDNEY · PARIS · FRANKFURT

U.K.	Pergamon Press Ltd., Headington Hill Hall, Oxford OX3 0BW, England
U.S.A.	Pergamon Press Inc., Maxwell House, Fairview Park, Elmsford, New York 10523, U.S.A.
CANADA	Pergamon of Canada, Suite 104, 150 Consumers Road, Willowdale, Ontario M2J 1P9, Canada
AUSTRALIA	Pergamon Press (Aust.) Pty. Ltd., P.O. Box 544, Potts Point, N.S.W. 2011, Australia
FRANCE	Pergamon Press SARL, 24 rue des Ecoles, 75240 Paris, Cedex 05, France
FEDERAL REPUBLIC OF GERMANY	Pergamon Press GmbH, 6242 Kronberg-Taunus, Pferdstrasse 1, Federal Republic of Germany

First edition 1979

British Library Cataloguing in Publication Data

Krypton, xenon and radon. - (International Union of Pure and Applied Chemistry. IUPAC solubility data series; vol.2).
1. Krypton - Solubility - Tables
2. Xenon - Solubility - Tables   3. Radon - Solubility - Tables
I. Clever, H Lawrence   II. Series
546'.754'5420212   QD181.K6   79-40145
ISBN 0-08-022352-4

In order to make this volume available as economically and as rapidly as possible the author's typescript has been reproduced in its original form. This method unfortunately has its typographical limitations but it is hoped that they in no way distract the reader.

Printed in Great Britain by A. Wheaton & Co. Ltd., Exeter

CONTENTS

FOREWORD

*If the knowledge is
undigested or simply wrong,
more is not better*

How to communicate and disseminate numerical data effectively in chemical science and technology has been a problem of serious and growing concern to IUPAC, the International Union of Pure and Applied Chemistry, for the last two decades. The steadily expanding volume of numerical information, the formulation of new interdisciplinary areas in which chemistry is a partner, and the links between these and existing traditional subdisciplines in chemistry, along with an increasing number of users, have been considered as urgent aspects of the information problem in general, and of the numerical data problem in particular.

Among the several numerical data projects initiated and operated by various IUPAC commissions, the *Solubility Data Project* is probably one of the most ambitious ones. It is concerned with preparing a comprehensive critical compilation of data on solubilities in all physical systems, of gases, liquids and solids. Both the basic and applied branches of almost all scientific disciplines require a knowledge of solubilities as a function of solvent, temperature and pressure. Solubility data are basic to the fundamental understanding of processes relevant to agronomy, biology, chemistry, geology and oceanography, medicine and pharmacology, and metallurgy and materials science. Knowledge of solubility is very frequently of great importance to such diverse practical applications as drug dosage and drug solubility in biological fluids, anesthesiology, corrosion by dissolution of metals, properties of glasses, ceramics, concretes and coatings, phase relations in the formation of minerals and alloys, the deposits of minerals and radioactive fission products from ocean waters, the composition of ground waters, and the requirements of oxygen and other gases in life support systems.

The widespread relevance of solubility data to many branches and disciplines of science, medicine, technology and engineering, and the difficulty of recovering solubility data from the literature, lead to the proliferation of published data in an ever increasing number of scientific and technical primary sources. The sheer volume of data has overcome the capacity of the classical secondary and tertiary services to respond effectively.

While the proportion of secondary services of the review article type is generally increasing due to the rapid growth of all forms of primary literature, the review articles become more limited in scope, more specialized. The disturbing phenomenon is that in some disciplines, certainly in chemistry, authors are reluctant to treat even those limited-in-scope reviews exhaustively. There is a trend to preselect the literature, sometimes under the pretext of reducing it to manageable size. The crucial problem with such preselection - as far as numerical data are concerned - is that there is no indication as to whether the material was excluded by design or by a less than thorough literature search. We are equally concerned that most current secondary sources, critical in character as they may be, give scant attention to numerical data.

On the other hand, tertiary sources - handbooks, reference books, and other tabulated and graphical compilations - as they exist today, are comprehensive but, as a rule, uncritical. They usually attempt to cover whole disciplines, thus obviously are superficial in treatment. Since they command a wide market, we believe that their service to advancement of science is at least questionable. Additionally, the change which is taking place in the generation of new and diversified numerical data, and the rate at which this is done, is not reflected in an increased third-level service. The emergence of new tertiary literature sources does not parallel the shift that has occurred in the primary literature.

With the status of current secondary and tertiary services being as briefly stated above, the innovative approach of the *Solubility Data Project* is that its compilation and critical evaluation work involve consolidation and reprocessing services when both activities are based on intellectual and scholarly reworking of information from primary sources. It comprises compact compilation, rationalization and simplification, and the fitting of isolated numerical data into a critically evaluated general framework.

The *Solubility Data Project* has developed a mechanism which involves a number of innovations in exploiting the literature fully, and which contains new elements of a more imaginative approach for transfer of reliable information from primary to secondary/tertiary sources. *The fundamental trend of the Solubility Data Project is toward integration of secondary and tertiary services with the objective of producing in-depth critical analysis and evaluation which are characteristic to secondary services, in a scope as broad as conventional tertiary services.*

Fundamental to the philosophy of the project is the recognition that the basic element of strength is the active participation of career scientists in it. Consolidating primary data, producing a truly critically-evaluated set of numerical data, and synthesizing data in a meaningful relationship are demands considered worthy of the efforts of top scientists. Career scientists, who themselves contribute to science by their involvement in active scientific research, are the backbone of the project. The scholarly work is commissioned to recognized authorities, involving a process of careful selection in the best tradition of IUPAC. This selection in turn is the key to the quality of the output. These top experts are expected to view their specific topics dispassionately, paying equal attention to their own contributions and to those of their peers. They digest literature data into a coherent story by weeding out what is wrong from what is believed to be right. To fulfill this task, the evaluator must cover *all* relevant open literature. No reference is excluded by design and every effort is made to detect every bit of relevant primary source. Poor quality or wrong data are mentioned and explicitly disqualified as such. In fact, it is only when the reliable data are presented alongside the unreliable data that proper justice can be done. The user is bound to have incomparably more confidence in a succinct evaluative commentary and a comprehensive review with a complete bibliography to both good and poor data.

It is the standard practice that any given solute-solvent system consists of two essential parts: I. Critical Evaluation and Recommended Values, and II. Compiled Data Sheets.

The Critical Evaluation part gives the following information:
(i) a verbal text of evaluation which discusses the numerical solubility information appearing in the primary sources located in the literature. The evaluation text concerns primarily the quality of data after consideration of the purity of the materials and their characterization, the experimental method employed and the uncertainties in control of physical parameters, the reproducibility of the data, the agreement of the worker's results on accepted test systems with standard values, and finally, the fitting of data, with suitable statistical tests, to mathematical functions;
(ii) a set of recommended numerical data. Whenever possible, the set of recommended data includes weighted average and standard deviations, and a set of smoothing equations derived from the experimental data endorsed by the evaluator;
(iii) a graphical plot of recommended data.

The compilation part consists of data sheets of the best experimental data in the primary literature. Generally speaking, such independent data sheets are given only to the best and endorsed data covering the known range of experimental parameters. Data sheets based on primary sources where the data are of a lower precision are given only when no better data are available. Experimental data with a precision poorer than considered acceptable are reproduced in the form of data sheets when they are the only known data for a particular system. Such data are considered to be still suitable for some applications, and their presence in the compilation should alert researchers to areas that need more work.

The typical data sheet carries the following information:
(i) components - definition of the system - their names, formulas and Chemical Abstracts registry numbers;
(ii) reference to the primary source where the numerical information is reported. In cases when the primary source is a less common periodical or a report document, published though of limited availability, abstract references are also given;
(iii) experimental variables;
(iv) identification of the compiler;
(v) experimental values as they appear in the primary source. Whenever available, the data may be given both in tabular and graphical form. If auxiliary information is available, the experimental data are converted also to SI units by the compiler.

Under the general heading of Auxiliary Information, the essential experimental details are summarized:

(vi) experimental method used for the generation of data;

(vii) type of apparatus and procedure employed;

(viii) source and purity of materials;

(ix) estimated error;

(x) references relevant to the generation of experimental data as cited in the primary source.

This new approach to numerical data presentation, developed during our four years of existence, has been strongly influenced by the diversity of background of those whom we are supposed to serve. We thus deemed it right to preface the evaluation/compilation sheets in each volume with a detailed discussion of the principles of the accurate determination of relevant solubility data and related thermodynamic information.

Finally, the role of education is more than corollary to the efforts we are seeking. The scientific standards advocated here are necessary to strengthen science and technology, and should be regarded as a major effort in the training and formation of the next generation of scientists and engineers. Specifically, we believe that there is going to be an impact of our project on scientific-communication practices. The quality of consolidation adopted by this program offers down-to-earth guidelines, concrete examples which are bound to make primary publication services more responsive than ever before to the needs of users. The self-regulatory message to scientists of 15 years ago to refrain from unnecessary publication has not achieved much. The literature is still, in 1978, cluttered with poor-quality articles. The Weinberg report (in "Reader in Science Information", Eds. J. Sherrod and A. Hodina, Microcard Editions Books, Indian Head, Inc., 1973, p. 292) states that "admonition to authors to restrain themselves from premature, unnecessary publication can have little effect unless the climate of the entire technical and scholarly community encourages restraint..." We think that projects of this kind translate the climate into operational terms by exerting pressure on authors to avoid submitting low-grade material. The type of our output, we hope, will encourage attention to quality as authors will increasingly realize that their work will not be suited for permanent retrievability unless it meets the standards adopted in this project. It should help to dispel confusion in the minds of many authors of what represents a permanently useful bit of information of an archival value, and what does not.

If we succeed in that aim, even partially, we have then done our share in protecting the scientific community from unwanted and irrelevant, wrong numerical information.

A. S. Kertes

December, 1978

PREFACE

The users of this volume will find (1) the available experimental solubility data of krypton, xenon, and radon gas in liquids as reported in the scientific literature, (2) tables of smoothed mole fraction solubility data for the systems which were studied over a temperature interval and (3) tables of either tentative or recommended solubility data when two or more laboratories reported solubility data over the same range of temperature and pressure for a system. Users have the option of using the experimental values, either directly, or in their own smoothing equations, or of using the smoothed values prepared by the compilers and evaluators. The goal was to cover the literature thoroughly enough so that the user need not do a detailed literature search for krypton, xenon, and radon solubility data prior to 1978.

Some words of explanation are required with respect to units, corrections, smoothing equations, auxiliary data and data sources, nomenclature, and other points. The experimental data are presented in the units found in the original paper. In addition the original data are often converted to other units, especially mole fraction. Temperatures have been converted to Kelvin. In evaluations of solubility data, S.I. units are used.

Only in the past 10 to 15 years have experimental methods for the determination of the solubility of gases in liquids developed to the point where 0.5 percent or better accuracy is attained. Only a small fraction of the literature's gas solubility data are accurate to 0.5 percent. The corrections for non-ideal gas behavior and for expansion of the liquid phase on dissolution of the gas are small and well within the normal experimental error. Thus such corrections were not made for the krypton and xenon gas solubility data at low pressure. For radon gas solubility data the real gas volume of $22,290 \ cm^3 \ mol^{-1}$ at standard conditions was used in the calculation of mole fraction.

Small, often unknown partial pressures, of radioactive krypton, xenon and radon are frequently used in the measurement of Ostwald coefficients. Bunsen coefficients and mole fraction gas solubilities at 101.325 kPa (1 atm) were calculated from the Ostwald coefficient, assuming that the Ostwald coefficient was independent of pressure. This is a reasonable often made assumption. However, it is an assumption, and the Bunsen coefficients and mole fraction solubilities calculated from Ostwald coefficients measured at low unknown partial pressures of radioactive gas should be accepted with caution.

The lack of high accuracy is also the reason that only a two-constant equation is used to smooth and evaluate most of the gas solubility data. A Gibbs energy of solution equation linear in temperature is used

$$\Delta G°/J \ mol^{-1} = - \ RT \ ln \ X_1 = A + BT$$

or in alternate form

$$ln \ X_1 = - \ \Delta G°/RT = - \ (A/R)/T - (B/R)$$

where A is $\Delta H°$, B is $-\Delta S°$, X_1 is the mole fraction solubility at a gas partial pressure of 101.325 kPa (1 atm), and R is 8.31433 J K^{-1} mol^{-1}. The constants A and B require five digits to reproduce the mole fraction solubility to three significant figures. Although the constants are given to five digits it is not intended to imply that the values of the changes in enthalpy and entropy of solution are significant to more than two or three digits.

An inconsistency, which we believe is justified, is found with respect to the solubility data in water. Much time and effort was expended in evaluating the solubility data of each gas in water. A recommended equation and table of values are presented. However, for systems which contain water and other solvent compounds such as electrolytes or water miscible polar organic compounds, the experimental gas solubility in water from that paper is given, even when it is at variance with our recommended values. These data of sometimes poorer quality are presented because the author's ratio of gas solubility in water to solubility in the aqueous solution may be more accurate than the solubility itself. This may be especially true of some of the solubility data in aqueous electrolyte solutions.

Solvent density data were often required in making solubility unit conversions. The density data were not directly referenced. The main sources of density data are

   Circular 461 of the U.S. National Bureau of Standards
   American Petroleum Research Project 44 Publications
   The International Critical Tables, Volume III (E.W. Washburn, Editor)
      McGraw-Hill Co., 1931
   Snow Table, Pure and Applied Chemistry 1976, 45, 1-9
   Thermodynamic Properties of Aliphatic Alcohols, R. C. Wilhoit and
      B. J. Zwolinski, J. Phys. Chem. Ref. Data 1973, 2, Supplement No. 1
   Organic Solvents, J. A. Riddick and W. B. Bunger (Technique of Chemistry, Volume II, A. Weissberger, Editor) Wiley-Interscience, New
      York, 1970, 3rd Ed.

The solubility data are supplemented with partial molal volume and calorimetric enthalpy of solution data when they are available.

Chemical Abstracts recommended names and registry numbers were used throughout. Common names are cross referenced to Chemical Abstract recommended names in the index.

The Editor would appreciate users calling errors and omissions to his attention.

The Editor gratefully acknowledges the advice and comments of members of the IUPAC Commission on Equilibrium Data and the Subcommittee on Solubility Data; the cooperation and hard work of the Evaluators and compilers; and the untiring efforts of the typists Peggy Tyler, Carolyn Dowie, and Lesley Flanagan.

Acknowledgment is made to the Donors of the Petroleum Research Fund, administered by the American Chemical Society, for partial support of the compilation and evaluation of the gas solubility data.

H. Lawrence Clever

Atlanta, GA
December, 1978

THE SOLUBILITY OF GASES IN LIQUIDS

C. L. Young, R. Battino, and H. L. Clever

INTRODUCTION

The Solubility Data Project aims to make a comprehensive search of the literature for data on the solubility of gases, liquids and solids in liquids. Data of suitable accuracy are compiled into data sheets set out in a uniform format. The data for each system are evaluated and where data of sufficient accuracy are available values recommended and in some cases a smoothing equation suggested to represent the variation of solubility with pressure and/or temperature. A text giving an evaluation and recommended values and the compiled data sheets are published on consecutive pages.

DEFINITION OF GAS SOLUBILITY

The distinction between vapor-liquid equilibria and the solubility of gases in liquids is arbitrary. It is generally accepted that the equilibrium set up at 300K between a typical gas such as argon and a liquid such as water is gas-liquid solubility whereas the equilibrium set up between hexane and cyclohexane at 350K is an example of vapor-liquid equilibrium. However, the distinction between gas-liquid solubility and vapor-liquid equilibrium is often not so clear. The equilibria set up between methane and propane above the critical temperature of methane and below the critical temperature of propane may be classed as vapor-liquid equilibrium or as gas-liquid solubility depending on the particular range of pressure considered and the particular worker concerned.

The difficulty partly stems from our inability to rigorously distinguish between a gas, a vapor, and a liquid,which has been discussed in numerous textbooks. We have taken a fairly liberal view in these volumes and have included systems which may be regarded, by some workers, as vapor-liquid equilibria.

UNITS AND QUANTITIES

The solubility of gases in liquids is of interest to a wide range of scientific and technological disciplines and not solely to chemistry. Therefore a variety of ways for reporting gas solubility have been used in the primary literature and inevitably sometimes, because of insufficient available information, it has been necessary to use several quantities in the compiled tables. Where possible, the gas solubility has been quoted as a mole fraction of the gaseous component in the liquid phase. The units of pressure used are bar, pascal, millimeters of mercury and atmosphere. Temperatures are reported in Kelvin.

EVALUATION AND COMPILATION

The solubility of comparatively few systems is known with sufficient accuracy to enable a set of recommended values to be presented. This is true both of the measurement near atmospheric pressure and at high pressures. Although a considerable number of systems have been studied by at least two workers, the range of pressures and/or temperatures is often sufficiently different to make meaningful comparison impossible.

Occasionally, it is not clear why two groups of workers obtained very different sets of results at the same temperature and pressure, although both sets of results were obtained by reliable methods and are internally consistent. In such cases, sometimes an incorrect assessment has been given. There are several examples where two or more sets of data have been classified as tentative although the sets are mutually inconsistent.

Many high pressure solubility data have been published in a smoothed form. Such data are particularly difficult to evaluate, and unless specifically discussed by the authors, the estimated error on such values can only be regarded as an "informed guess".

Many of the high pressure solubility data have been obtained in a more general study of high pressure vapor-liquid equilibrium. In such cases a note is included to indicate that additional vapor-liquid equilibrium data are given in the source. Since the evaluation is for the compiled data, it is possible that the solubility data are given a classification which is better than that which would be given for the complete vapor-liquid data (or vice versa). For example, it is difficult to determine coexisting liquid and vapor compositions near the critical point of a mixture using some widely used experimental techniques which yield accurate high pressure solubility data. For example, conventional methods of analysis may give results with an expected error which would be regarded as sufficiently small for vapor-liquid equilibrium data but an order of magnitude too large for acceptable high pressure gas-liquid solubility.

It is occasionally possible to evaluate data on mixtures of a given substance with a member of a homologous series by considering all the available data for the given substance with other members of the homologous series. In this study the use of such a technique has been very limited.

The estimated error is often omitted in the original article and sometimes the errors quoted do not cover all the variables. In order to increase the usefulness of the compiled tables estimated errors have been included even when absent from the original article. If the error on *any* variable has been inserted by the compiler this has been noted.

PURITY OF MATERIALS

The purity of materials has been quoted in the compiled tables where given in the original publication. The solubility is usually more sensitive to impurities in the gaseous component than to liquid impurities in the liquid component. However, the most important impurities are traces of a gas dissolved in the liquid. Inadequate degassing of the absorbing liquid is probably the most often overlooked serious source of error in gas solubility measurements.

APPARATUS AND PROCEDURES

In the compiled tables brief mention is made of the apparatus and procedure. There are several reviews on experimental methods of determining gas solubilities and these are given in References 1-7.

METHODS OF EXPRESSING GAS SOLUBILITIES

Because gas solubilities are important for many different scientific and engineering problems, they have been expressed in a great many ways:

The Mole Fraction, X(g)

The mole fraction solubility for a binary system is given by:

$$X(g) = \frac{n(g)}{n(g) + n(l)}$$

$$= \frac{W(g)/M(g)}{\{W(g)/M(g)\} + \{W(l)/M(l)\}}$$

here n is the number of moles of a substance (an *amount* of substance), W is the mass of a substance, and M is the molecular mass. To be unambiguous, the partial pressure of the gas (or the total pressure) and the temperature of measurement must be specified.

The Weight Per Cent Solubility, wt%

For a binary system this is given by

$$wt\% = 100 \ W(g)/\{W(g) + W(l)\}$$

where W is the weight of substance. As in the case of mole fraction, the pressure (partial or total) and the temperature must be specified. The weight per cent solubility is related to the mole fraction solubility by

$$X(g) = \frac{\{wt\%/M(g)\}}{\{wt\%/M(g)\} + \{(100 - wt\%)/M(l)\}}$$

The Weight Solubility, C_W

The weight solubility is the number of moles of dissolved gas per gram of solvent when the partial pressure of gas is 1 atmosphere. The weight solubility is related to the mole fraction solubility at one atmosphere partial pressure by

$$X(g) \text{ (partial pressure 1 atm)} = \frac{C_W M(l)}{1 + C_W M(l)}$$

where M(l) is the molecular weight of the solvent.

The Moles Per Unit Volume Solubility, n

Often for multicomponent systems the density of the liquid mixture is not known and the solubility is quoted as moles of gas per unit volume of liquid mixture. This is related to the mole fraction solubility by

$$X = \frac{n \, v^o(l)}{1 + n \, v^o(l)}$$

where $v^o(l)$ is the molar volume of the liquid component.

The Bunsen Coefficient, α

The Bunsen coefficient is defined as the volume of gas reduced to 273.15K and 1 atmosphere pressure which is absorbed by unit volume of solvent (at the temperature of measurement) under a partial pressure of 1 atmosphere. If ideal gas behavior and Henry's law is assumed to be obeyed,

$$\alpha = \frac{V(g)}{V(l)} \frac{273.15}{T}$$

where V(g) is the volume of gas absorbed and V(l) is the original (starting) volume of absorbing solvent. The mole fraction solubility X is related to the Bunsen coefficient by

$$X \text{ (1 atm)} = \frac{\alpha}{\alpha + \dfrac{273.15}{T} \dfrac{v^o(g)}{v^o(l)}}$$

where $v^o(g)$ and $v^o(l)$ are the molar volumes of gas and solvent at a pressure of one atmosphere. If the gas is ideal,

$$X = \frac{\alpha}{\alpha + \dfrac{273.15R}{v^o(l)}}$$

Real gases do not follow the ideal gas law and it is important to establish the real gas law used for calculating α in the original publication and to make the necessary adjustments when calculating the mole fraction solubility.

The Kuenen Coefficient, S

This is the volume of gas, reduced to 273.15K and 1 atmosphere pressure, dissolved at a partial pressure of gas of 1 atmosphere by 1 gram of solvent.

The Ostwald Coefficient, L

The Ostwald coefficient, L, is defined as the ratio of the volume of gas absorbed to the volume of the absorbing liquid, all measured at the same temperature:

$$L = \frac{V(g)}{V(l)}$$

If the gas is ideal and Henry's Law is applicable, the Ostwald coefficient is independent of the partial pressure of the gas. It is necessary, in practice, to state the temperature and total pressure for which the Ostwald coefficient is measured. The mole fraction solubility, X, is related to the Ostwald coefficient by

$$X = \left[\frac{RT}{P(g)\ L\ v^{o}(l)} + 1 \right]^{-1}$$

where P is the partial pressure of gas. The mole fraction solubility will be at a partial pressure of P(g).

The Absorption Coefficient, β

There are several "absorption coefficients", the most commonly used one being defined as the volume of gas, reduced to 273.15K and 1 atmosphere, absorbed per unit volume of liquid when the <u>total</u> pressure is 1 atmosphere. β is related to the Bunsen coefficient by

$$\beta = \alpha(1-P(l))$$

where P(l) is the partial pressure of the liquid in atmosphere.

The Henry's Law Constant

A generally used formulation of Henry's Law may be expressed as

$$P(g) = K_H X$$

where K_H is the Henry's Law constant and X the mole fraction solubility. Other formulations are

$$P(g) = K_2 C(l)$$

or

$$C(g) = K_c C(l)$$

where K_2 and K_c are constants, C the concentration, and (l) and (g) refer to the liquid and gas phases. Unfortunately, K_H, K_2 and K_c are all sometimes referred to as Henry's Law constants. Henry's Law is a limiting law but can sometimes be used for converting solubility data from the experimental pressure to a partial gas pressure of 1 atmosphere, provided the mole fraction of the gas in the liquid is small, and that the difference in pressures is small. Great caution must be exercised in using Henry's Law.

The Mole Ratio, N

The mole ratio, N, is defined by

$$N = n(g)/n(l)$$

Table 1 contains a presentation of the most commonly used inter-conversions not already discussed.

For gas solubilities greater than about 0.01 mole fraction at a partial pressure of 1 atmosphere there are several additional factors which must be taken into account to unambiguously report gas solubilities. Solution densities or the partial molar volume of gases must be known. Corrections should be made for the possible non-ideality of the gas or the non-applicability of Henry's Law.

TABLE 1 Interconversion of parameters used for reporting solubility

$$L = \alpha(T/273.15)$$

$$C_w = \alpha/v_o\rho$$

$$K_H = \frac{17.033 \times 10^6 \rho_{soln}}{\alpha\, M(1)} + 760$$

$$L = C_w\, v_{t,gas}\rho$$

where v_o is the molal volume of the gas in $cm^3 mol^{-1}$ at 0°C, ρ the density of the solvent at the temperature of the measurement, ρ_{soln} the density of the solution at the temperature of the measurement, and $v_{t,gas}$ the molal volume of the gas ($cm^3 mol^{-1}$) at the temperature of the measurement.

SALT EFFECTS

The effect of a dissolved salt in the solvent on the solubility of a gas is often studied. The activity coefficient of a dissolved gas is a function of the concentration of all solute species (see ref. 8). At a given temperature and pressure the logarithm of the dissolved gas activity coefficient can be represented by a power series in C_s, the electrolyte concentration, and C_i, the nonelectrolyte solute gas concentration

$$\log f_i = \sum_{m,n} k_{mn}\, C_s^n\, C_i^m$$

It is usually assumed that only the linear terms are important for low C_s and C_i values when there is negligible chemical interaction between solute species.

$$\log f_i = k_s C_s + k_i C_i$$

where k_s is the salt effect parameter and k_i is the solute-solute gas interaction parameter. The dissolved gas activity is the same in the pure solvent and a salt solution in that solvent for a given partial pressure and temperature

$$a_i = f_i S_i = f_i^O S_i^O \text{ and } f_i = f_i^O\, \frac{S_i^O}{S_i}$$

where S_i and S_i^O are the gas solubility in the salt solution and in the pure solvent, respectively, and the f's are the corresponding activity coefficients. It follows that $\log \dfrac{f_i}{f_i^O} = \log \dfrac{S_i^O}{S_i} = k_s C_s + k_i(S_i - S_i^O)$. When the

quantity $(S_i - S_i^O)$ is small the second term is negligible even though k_s and k_i may be of similar magnitude. This is generally the case for gas solubilities and the equation reduces to

$$\log \frac{f_i}{f_i^O} = \log \frac{S_i^O}{S_i} = k_s C_s$$

which is the form of the empirical Setschenow equation in use since the 1880's. A salt that increases the activity coefficient of the dissolved gas is said to salt-out and a salt that decreases the activity coefficient of the dissolved gas is said to salt-in.

Although salt effect studies have been carried out for many years, there appears to be no common agreement of the units for either the gas solubility or the salt concentration. Both molar (mol dm^{-3}) and molal (mol kg^{-1}) are used for the salt concentration. The gas solubility ratio S_i^O/S_i is given as Bunsen coefficient ratio and Ostwald coefficient ratio, which would be the same as a molar ratio; Kueunen coefficient ratio, volume dissolved in 1 g or 1 kg of solvent which would be a molal ratio; and mole fraction ratio. Recent theoretical treatments use salt concentration in mol dm^{-3} and $S_i^{\,o}/S_i$ ratio as mole fraction ratio with each salt ion acting as a mole. Evaluations which compare the results of several workers are made in the units most compatible with present theory.

TEMPERATURE DEPENDENCE OF GAS SOLUBILITY

In a few cases it has been found possible to fit the mole fraction solubility at various temperatures using an equation of the form

$$\ln x = A + B / (T/100K) + C \ln (T/100K) + DT/100K$$

It is then possible to write the thermodynamic functions $\overline{\Delta G}_1^{\,o}$, $\overline{\Delta H}_1^{\,o}$, $\overline{\Delta S}_1^{\,o}$ and $\overline{\Delta C}^{\,o}_{p_1}$ for the transfer of the gas from the vapor phase at 101,325 Pa partial pressure to the (hypothetical) solution phase of unit mole fraction as:

$$\overline{\Delta G}_1^{\,o} = -RAT - 100\ RB - RCT \ln (T/100) - RDT^2/100$$

$$\overline{\Delta S}_1^{\,o} = RA + RC \ln (T/100) + RC + 2\ RDT/100$$

$$\overline{\Delta H}_1^{\,o} = -100\ RB + RCT + RDT^2/100$$

$$\overline{\Delta C}^{\,o}_{p_1} = RC + 2\ RDT/100$$

In cases where there are solubilities at only a few temperatures it is convenient to use the simpler equations

$$\overline{\Delta G}_1^{\,o} = - RT \ln x = A + BT$$

in which case $A = \overline{\Delta H}_1^{\,o}$ and $-B = \overline{\Delta S}_1^{\,o}$.

REFERENCES

1. Battino, R.; Clever, H. L. Chem.Rev. 1966, 66, 395.
2. Clever, H. L.; Battino, R. in Solutions and Solubilities, Ed. M. R. J. Dack, J. Wiley & Sons, New York, 1975, Chapter 7.
3. Hildebrand, J. H.; Prausnitz, J. M.; Scott, R. L. Regular and Related Solutions, Van Nostrand Reinhold, New York, 1970, Chapter 8.
4. Markham, A. E.; Kobe, K. A. Chem. Rev. 1941, 63, 449.
5. Wilhelm, E.; Battino, R. Chem. Rev. 1973, 73, 1.
6. Wilhelm, E.; Battino, R.; Wilcock, R. J. Chem. Rev. 1977, 77, 219.
7. Kertes, A. S.; Levy, O.; Markovits, G. Y. in Experimental Thermochemistry Vol. II, Ed. B. Vodar and B. LeNaindre, Butterworth, London, 1974, Chapter 15.
8. Long, F. A.; McDevit, W. F. Chem. Rev. 1952, 51, 119.

APPENDIX I. Conversion Factors k and k^{-1}

Non-SI Unit	k 1 (non-SI Unit) = k (SI Unit)	k^{-1} 1 (SI Unit) = k^{-1} (non-SI Unit)
LENGTH		SI Unit, m
Å (angstrom)	1×10^{-10} (*)	1×10^{10} (*)
cm (centimeter)	1×10^{-2} (*)	1×10^{2} (*)
in (inch)	254×10^{-4} (*)	$3\ 937\ 008 \times 10^{-5}$
ft (foot)	$3\ 048 \times 10^{-4}$ (*)	$3\ 280\ 840 \times 10^{-6}$
AREA		SI Unit, m^2
cm^2	1×10^{-4} (*)	1×10^{4} (*)
in^2	$64\ 516 \times 10^{-8}$ (*)	$1\ 550\ 003 \times 10^{-3}$
ft^2	$9\ 290\ 304 \times 10^{-8}$ (*)	$1\ 076\ 391 \times 10^{-5}$
VOLUME		SI Unit, m^3
cm^3	1×10^{-6} (*)	1×10^{6} (*)
in^3	$16\ 387\ 064 \times 10^{-12}$ (*)	$6\ 102\ 374 \times 10^{-2}$
ft^3	$2\ 831\ 685 \times 10^{-8}$	$3\ 531\ 467 \times 10^{-5}$
l (litre)	1×10^{-3} (*)	1×10^{3} (*)
UKgal (UK gallon)	$45\ 461 \times 10^{-7}$	$21\ 997 \times 10^{-2}$
USgal (US gallon)	$37\ 854 \times 10^{-7}$	$26\ 417 \times 10^{-2}$
MASS		SI Unit, kg
g (gram)	1×10^{-3} (*)	1×10^{3} (*)
t (tonne)	1×10^{3} (*)	1×10^{-3} (*)
lb (pound)	$45\ 359\ 237 \times 10^{-8}$ (*)	$2\ 204\ 623 \times 10^{-6}$
DENSITY		SI Unit, $kg\ m^{-3}$
$g\ cm^{-3}$	1×10^{3} (*)	1×10^{-3} (*)
$g\ l^{-1}$	1 (*)	1 (*)
$lb\ in^{-3}$	$2\ 767\ 991 \times 10^{-2}$	$3\ 612\ 728 \times 10^{-11}$
$lb\ ft^{-3}$	$1\ 601\ 847 \times 10^{-5}$	$6\ 242\ 795 \times 10^{-8}$
$lb\ UKgal^{-1}$	$99\ 776 \times 10^{-3}$	$100\ 224 \times 10^{-7}$
$lb\ USgal^{-1}$	$1\ 198\ 264 \times 10^{-4}$	$8\ 345\ 406 \times 10^{-9}$
PRESSURE		SI Unit, Pa (pascal, $kg\ m^{-1}s^{-2}$)
$dyn\ cm^{-2}$	1×10^{-1} (*)	1×10 (*)
at ($kgf\ cm^{-2}$)	$980\ 665 \times 10^{-1}$ (*)	$1\ 019\ 716 \times 10^{-11}$
atm (atmosphere)	$101\ 325$ (*)	$9\ 869\ 233 \times 10^{-12}$
bar	1×10^{5} (*)	1×10^{-5} (*)
$lbf\ in^{-2}$ (p.s.i.)	$6\ 894\ 757 \times 10^{-3}$	$1\ 450\ 377 \times 10^{-10}$
$lbf\ ft^{-2}$	$47\ 880 \times 10^{-3}$	$20\ 886 \times 10^{-6}$
inHg (inch of mercury)	$3\ 386\ 388 \times 10^{-3}$	$2\ 952\ 999 \times 10^{-10}$
mmHg (millimeter of mercury, torr)	$1\ 333\ 224 \times 10^{-4}$	$7\ 500\ 617 \times 10^{-9}$

APPENDIX I. Conversion Factors k and k^{-1}

Non-SI Unit	k 1 (non-SI Unit) = k (SI Unit)		k^{-1} 1 (SI Unit) = k^{-1} (non-SI Unit)	
ENERGY			Unit, J (joule, kg $m^2 s^{-2}$)	
erg	1×10^{-7}	(*)	1×10^{7}	(*)
cal_{IT} (I.T. calorie)	$41\ 868 \times 10^{-4}$	(*)	$2\ 388\ 459 \times 10^{-7}$	
cal_{th} (thermochemical calorie)	$4\ 184 \times 10^{-3}$	(*)	$2\ 390\ 057 \times 10^{-7}$	
kW h (kilowatt hour)	36×10^{5}	(*)	$2\ 777\ 778 \times 10^{-13}$	
l atm	$101\ 325 \times 10^{-3}$	(*)	$9\ 869\ 233 \times 10^{-9}$	
ft lbf	$1\ 355\ 818 \times 10^{-6}$		$7\ 375\ 622 \times 10^{-7}$	
hp h (horse power hour)	$2\ 684\ 519$		$3\ 725\ 062 \times 10^{-13}$	
Btu (British thermal unit)	$1\ 055\ 056 \times 10^{-3}$		$9\ 478\ 172 \times 10^{-10}$	

An asterisk (*) denotes an exact relationship

COMPONENTS:	EVALUATOR:
1. Krypton; Kr; 7439-90-9 2. Water; H_2O; 7732-18-5	Rubin Battino, Department of Chemistry, Wright State University Dayton, Ohio, 45431 U.S.A. June 1978

CRITICAL EVALUATION:

The experimental solubility data produced by three workers was considered to be of sufficient reliability to use for the smoothing equation. In fitting the data to the equation those points which were different by about two standard deviations or more from the smoothed values were rejected. Most of the low temperature values of Morrison and Johnstone were thrown out, but since they had the only higher temperature data these were retained. The transition between their high temperature data and the more reliable low temperature data of the other two workers was smooth. We used 30 points for the final smoothing and they were obtained as follows (reference - number of data points used from that reference): 1-8; 2-6; 3-16. The fitting equation used was

$$\ln x_1 = A + B/(T/100K) + C \ln (T/100K) + DT/100K \qquad (1)$$

where x_1 is the mole fraction solubility of krypton at 101,325 Pa (1 atm) partial pressure of gas. The fit in $\ln x_1$ gave a standard deviation of 0.32% taken at the middle of the temperature range. Table 1 gives the smoothed values of the mole fraction solubility (at 101,325 Pa partial pressure of gas) and the Ostwald coefficient at 5K intervals.

Table 1 also gives the thermodynamic functions $\Delta \bar{G}_1^\circ$, $\Delta \bar{H}_1^\circ$, $\Delta \bar{S}_1^\circ$, $\Delta \bar{C}_{p_1}^\circ$ for the transfer of gas from the vapor phase at 101,325 Pa partial gas pressure to the (hypothetical) solution phase of unit mole fraction. These thermodynamic properties were calculated from the smoothing equation according to the following equations:

$$\Delta \bar{G}_1^\circ = - RAT - 100RB - RCT \ln (T/100) - RDT^2/100 \qquad (2)$$

$$\Delta \bar{S}_1^\circ = RA + RC \ln (T/100) + RC + 2RDT/100 \qquad (3)$$

$$\Delta \bar{H}_1^\circ = - 100RB + RCT + RDT^2/100 \qquad (4)$$

$$\Delta \bar{C}_{p_1}^\circ = RC + 2RDT/100 \qquad (5)$$

The best fit for the 30 data points gave the following equation

$$\ln x_1 = -66.9928 + 91.0166/(T/100K) + 24.2207 \ln (T/100K) \qquad (6)$$

The data from six other workers were rejected for the following reasons. Yeh and Peterson's values (4) were 3 to 4 per cent low. Antropoff's data (5) were very far off, being all much too low. König's results (6) were also low, being off by about 4 to 12 per cent. The two values by Van Liempt (7) were rather imprecise. Kitani (8) determined the solubility of krypton-85 in water, saline, lipids and blood at 37°C by a radiochemical technique. Seventeen determinations at the one temperature were reproducible to ± 4% and were 5% lower than the selected values in this section. Wood and Caputi (9) determined the solubilities from 1 to 48°C in water and sea water, but their results are 2 to 5% high.

COMPONENTS:	EVALUATOR:
1. Krypton; Kr; 7439-90-9 2. Water; H_2O; 7732-18-5	Rubin Battino Department of Chemistry Wright State University Dayton, Ohio 45431 U.S.A. June 1978

CRITICAL EVALUATION:

Table 1. Smoothed values of the solubility of krypton in water and thermodynamic functions[a] at 5K intervals using equation 6 at 101,325 Pa partial pressure of krypton.

T/K	Mol Fraction $X_1 \times 10^5$	Ostwald Coefficient $L \times 10^2$	$\Delta \bar{G}_1^{\circ}$[b]/KJ mol^{-1}	$\Delta \bar{H}_1^{\circ}$/KJ mol^{-1}	$\Delta \bar{S}_1^{\circ}$/JK^{-1}mol^{-1}
273.15	8.842	11.000	21.20	-20.67	-153
278.15	7.537	9.550	21.95	-19.66	-150
283.15	6.511	8.396	22.69	-18.65	-146
288.15	5.696	7.470	23.41	-17.65	-142
293.15	5.041	6.720	24.12	-16.64	-139
298.15	4.512	6.109	24.80	-15.63	-136
303.15	4.079	5.609	25.47	-14.63	-132
308.15	3.725	5.197	26.13	-13.62	-129
313.15	3.432	4.858	26.76	-12.61	-126
318.15	3.190	4.578	27.39	-11.61	-123
323.15	2.990	4.348	27.99	-10.60	-119
328.15	2.823	4.160	28.58	-9.59	-116
333.15	2.686	4.007	29.15	-8.58	-113
338.15	2.572	3.885	29.71	-7.58	-110
343.15	2.480	3.790	30.26	-6.57	-107
348.15	2.405	3.718	30.79	-5.56	-104
353.15	2.346	3.668	31.30	-4.56	-102

[a] $\Delta \bar{C}_{p_1}^{\circ}$ was independent of temperature and has the value 201 JK^{-1}mol^{-1}.

[b] cal_{th} = 4.184 joule.

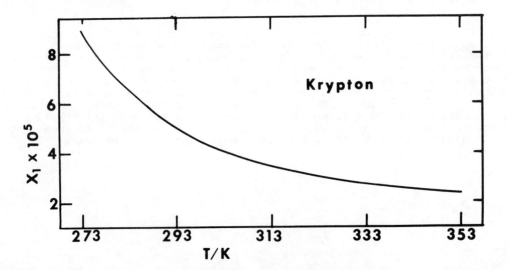

FIGURE 1. The mole fraction solubility of krypton in water at a krypton partial pressure of 101.325 kPa (1 atm).

COMPONENTS:	EVALUATOR:
1. Krypton; Kr; 7439-90-9 2. Water; H_2O; 7732-18-5	Rubin Battino Department of Chemistry Wright State University Dayton, Ohio 45431 U.S.A. June 1978

CRITICAL EVALUATION:

   Anderson, Keeler, and Klach (10) measured the solubility of krypton + krypton-85 between 373 and 573K at partial pressures of 1.5 to 60×10^{-4} psia in the presence of oxygen gas at pressures between 270 and 1930 psia. The mole fraction solubility calculated at 101.325 kPa krypton partial pressure appears to be a consistent extension of the krypton solubilities below 353K. The combined data indicates the minimum solubility is at 380 K. See the data sheets for the Anderson, Keeler and Klach paper for additional information.

   Alexander (11) made seven determinations of the heat of solution of krypton in water at 298.15 K. He obtained an enthalpy of solution of $-(15.8_2 \pm 0.6_8)$ kJ mol^{-1}. The value agrees well with the value of -15.63 kJ mol^{-1} derived from the solubility data.

   Popov and Drakin (12, 13) measured the apparent molar volume of krypton dissolved in water under a krypton partial pressure of 4.7 to 20.1 atm at 298.15 K by a density method.

p/atm	4.70	12.00	15.20	18.27	20.10
$\bar{V}$/cm^3 mol^{-1}	38 ± 5	32 ± 2	31.2 ± 1.4	31.4 ± 1.2	31.3 ± 1.1

They recommended a value of 31.3 cm^3 mol^{-1}.

References

1. Morrison, T. J.; Johnstone, N. B. J. Chem. Soc. 1954, 3441.

2. Benson, B. B.; Krause, D. J. Chem. Phys. 1976, 64, 689.

3. Weiss, R. F.; Kyser, T. K. J. Chem. Eng. Data 1978, 23, 69.

4. Yeh, S-Y.; Peterson, R. E. J. Pharm. Sci. 1964, 53, 822.

5. Antropoff, A. Proc. Roy. Soc. (London) 1910, 83, 474;
    Z. Elektrochem. 1919, 25, 269.

6. König, H. Z. Naturforsch. 1963, 18a, 363.

7. van Liempt, J. A. M.; van Wijk, W. Rec. Trav. Chim. 1937, 56, 632.

8. Kitani, K. Scand. J. Clinical Lab. Invest. 1972, 29, 167.

9. Wood, D.; Caputi, R. AD631557, Avail. CFSTI, 1966, 17 pp.

10. Anderson, C. J.; Keeler, R. A.; Klach, S. J.
    J. Chem. Eng. Data 1962, 7, 290.

11. Alexander, D. M. J. Phys. Chem. 1959, 63, 994.

12. Popov, G. A.; Drakin, S. I. Moskov. Khimiko-technol. Inst. Trudy
    1972, 71, 43.

13. Popov, G. A.; Drakin, S. I. Zh. Fiz. Khim. 1974, 48, 631.

The last three paragraphs of the evaluation were added by the Editor.

COMPONENTS:	ORIGINAL MEASUREMENTS:
1. Krypton; Kr; 7439-90-9 2. Water; H_2O; 7732-18-5	Morrison, T. J.; Johnstone, N. B. J. Chem. Soc. 1954, 3441 - 3446.

VARIABLES:	PREPARED BY:
T/K: 279.75 - 348.05	R. Battino

EXPERIMENTAL VALUES:

T/K	Mol Fraction $X_1 \times 10^5$	Kunenen Coefficient $S_0 \times 10^3$	T/K	Mol Fraction $X_1 \times 10^5$	Kunenen Coefficient $S_0 \times 10^3$
279.75	6.591	82.0	319.15	3.110	38.3
281.15	6.382	79.4	325.05	2.923*	35.9
285.75	5.589	69.5	331.05	2.719*	33.3
288.65	5.277	65.6	332.95	2.681*	32.8
291.35	4.966	61.7	337.85	2.590*	31.6
294.35	4.687	58.2	340.65	2.520*	30.7
297.85	4.312	53.5	344.65	2.444*	29.7
303.35	3.948	48.9	347.05	2.431*	29.5
308.55	3.582	44.3	348.05	2.416*	29.3
314.55	3.290	40.6			

*Solubility values which were used in the final smoothing equation for the recommended solubility values in the critical evaluation.

The authors reported the solubility at cm^3 at 273.15 K and 101.325 kPa (1 atm) per Kg water. We have labeled it Kunenen coefficient x 10^3.

The mole fraction solubility of gas at 101.325 kPa (1 atm) partial pressure of Krypton was calculated by the compiler.

AUXILIARY INFORMATION

METHOD:	SOURCE AND PURITY OF MATERIALS:
Flowing the previously degassed liquid in a thin film through the gas in an absorption spiral. Volume changes are measured in burets. See ref. 1.	1. Krypton. 99-100% pure; the residue being xenon. 2. Water. no comments made by authors.

APPARATUS/PROCEDURE:	ESTIMATED ERROR:
Apparatus of Morrison and Billett (1) used. The authors smoothing equation is $\log_{10}S_0 = -60.434 + (3410)/(T/K)$	Some data at lower temperatures are several per cent low.
	REFERENCES: 1. Morrison, T. J.; Billett, F. J. Chem. Soc. 1952, 3819.

COMPONENTS:	ORIGINAL MEASUREMENTS:
1. Krypton; Kr; 7439-90-9 2. Water; H_2O; 7732-18-5	Benson, B. B.; Krause, D. J. Chem. Phys. 1976, 64, 689 - 709.

VARIABLES:
 	
        T/K: 273.151 - 318.154
	PREPARED BY:
	
	        R. Battino

EXPERIMENTAL VALUES:

T/K	Mol Fraction X_1 x 10^5	Bunsen Coefficient α x 10^3
275.151	8.3389	103.224
278.141	7.5712	93.727
288.147	5.7156*	70.704
293.155	5.0492*	64.424
298.147	4.5339*	55.978
303.150	4.0881*	50.408
308.155	3.7407*	46.052
318.154	3.2119*	39.396

*Solubility values which were used in the final smoothing equation for
the recommended values given in the critical evaluation.

The mole fraction solubility of gas at 101.325 kPa (1 atm) partial pressure
of Krypton was calculated by the compiler.

AUXILIARY INFORMATION

METHOD /APPARATUS/PROCEDURE:

   Gas-free water and the pure gas
are equilibrated, and volumetric
samples of the liquid and gaseous
phases are isolated. The gas dissolved in the water is extracted and
the number of moles determined on a
special mercury anometer. After
removal of the water vapor, the
number of moles in the sample of the
gaseous phase is measured with the
same manometer; from which the pressure (and fugacity) above the solution may be calculated. Real gas
corrections are made. Predicted
maximum error is 0.02%. No
drawings of the apparatus are
given in the original paper.

SOURCE AND PURITY OF MATERIALS:

1. Krypton. No comment by authors.

2. Water. No comment by authors.

ESTIMATED ERROR:
Smoothed data fit to 0.11% rms in
the solubility. Calculated error
in measurements is 0.02%.

REFERENCES:

COMPONENTS:	ORIGINAL MEASUREMENTS:
1. Krypton; Kr; 7439-90-9 2. Water; H_2O; 7732-18-5	Weiss, R. F.; Kyser, T. K. J. Chem. Eng. Data 1978, 23, 69-72.

VARIABLES:	PREPARED BY:
T/K: 273.72 - 313.25	R. Battino

EXPERIMENTAL VALUES:

T/K	Mol Fraction $X_1 \times 10^5$	Bunsen Coefficient α	T/K	Mol Fraction $X_1 \times 10^5$	Bunsen Coefficient α
273.72	8.663*	0.10778	293.39	5.009*	0.06221
273.72	8.673*	0.10790	293.40	5.013*	0.06226
273.74	8.663*	0.10778	304.12	3.999*	0.04953
273.73	8.669	0.10786	304.13	4.003*	0.04957
283.48	6.416	0.07981	304.13	4.001*	0.04955
283.47	6.420	0.07985	304.12	4.006*	0.04961
283.47	6.415	0.07979	313.24	3.414*	0.04215
293.39	5.010*	0.06222	313.25	3.417*	0.04218
293.39	5.008*	0.06220	313.25	3.415*	0.04216
293.40	5.014*	0.06227			

*Solubility values which were used in the final smoothing equation for the recommended equation for the recommended values given in the critical evaluation.

The mole fraction solubility at 101.325 kPa (1 atm) partial pressure of krypton calculated by the compiler.

AUXILIARY INFORMATION

METHOD/APPARATUS/PROCEDURE:	SOURCE AND PURITY OF MATERIALS:
Gas-free water and the pure gas are equilibrated, and volumetric samples of the liquid and gaseous phases are isolated. The gas dissolved in the water is extracted and the number of moles determined on a special mercury manometer. After removal of the water vapor, the number of moles in the sample of the gaseous phase is measured with the same manometer; from which the pressure (and fugacity) above the solution may be calculated. Real gas corrections are made. Predicted maximum error is 0.02%. No drawings of the apparatus are given in the original paper.	1. Krypton. Matheson Gas Products. greater than 99.995 % pure. Gas chromatographic checks showed less than 0.01 % air. 2. Water. Distilled.
	ESTIMATED ERROR:
	REFERENCES:

COMPONENTS:	ORIGINAL MEASUREMENTS:

1. Krypton; Kr; 7439-90-9
 	
   Krypton-85; ^{85}Kr; 13983-27-2
 	
2. Water; H_2O; 7732-18-5
	Anderson, C. J.; Keeler, R. A.;
	   Klach, S. J.
	
	J. Chem. Eng. Data 1962, 7, 290-294.

| VARIABLES:
   T/K: 373.15 - 573.15
Kr P/kPa: 1.0135×10^{-3} - 41.8×10^{-3}
   (1.47×10^{-4}-60.6×10^{-4} psia) | PREPARED BY:
   A.L. Cramer
   H.L. Clever |

EXPERIMENTAL VALUES:

T/K	Pressure/psia Total $O_2 + H_2O + Kr$	Partial x 10^4 Kr	Henry's Constant $K = (P/psia)/X_1$	Mol Fraction $X_1 \times 10^5$ at 101.325 kPa Kr
373.15	270	1.99	6.63×10^5	2.22
	400	60.6	6.00×10^5	2.45
398.15	280	1.53	5.88×10^5	2.50
	400	60.4	5.86×10^5	2.51
423.15	295	1.68	5.42×10^5	2.71
	410	55.4	5.18×10^5	2.84
453.15	400	40.4	4.83×10^5	3.04
473.15	435	1.47	3.87×10^5	3.80
523.15	1000	32.0	2.67×10^5	5.51
	1340	3.48	2.54×10^5	5.79
	1410	6.6	2.64×10^5	5.57
548.15	1560	2.77	2.01×10^5	7.31
	1615	6.03	1.63×10^5	9.02
573.15	1860	2.67	1.21×10^5	12.2
	1930	5.27	1.22×10^5	12.1

The mole fraction solubility at a partial pressure of 101.325 kPa krypton was calculated by the compiler.

AUXILIARY INFORMATION

METHOD/APPARATUS/PROCEDURE:	SOURCE AND PURITY OF MATERIALS:
A Kr/^{85}Kr stock mixture and O_2 are added separately to a thermostated stainless steel vessel containing about 275 ml of water. Equilibration time varied from 1.5 to 22 hours without appreciable difference in results (one run went for 64 hours). At equilibrium both liquid and vapor were sampled and the krypton in each phase was determined by counting the ^{85}Kr tag.	
Henry's law constant is linear in temperature. The authors give $K = (p/psia)/X_1 = (0.126-0.0263(t/^{o}C)) \times 10^5$ with a standard deviation of 0.218×10^5.	ESTIMATED ERROR: At the 95% confidence level the uncertainity in K is $\pm 0.471 \times 10^5 (0.0677 - (T - 202)^2/75334)$ $\delta T/K = 0.1$

COMPONENTS:	ORIGINAL MEASUREMENTS:

1. Krypton; Kr; 7439-90-9
 	
   Krypton-85; ^{85}Kr; 13983-27-2
 	
2. Water; H_2O; 7732-18-5
	Anderson, C. J.; Keeler, R. A.;
	   Klach, S. J.
	
	$\underline{J}$. $\underline{Chem}$. $\underline{Eng}$. $\underline{Data}$ 1962, $\underline{7}$, 290-294.
	
	Continued from previous page.

Smoothed data: A graph of the krypton mole fraction solubility at a partial
krypton pressure of 101.325 kPa was constructed using the solubility data at
temperatures between 273 and 348 K used in the critical evaluation of
krypton + water, as well as the data from this paper. The data give a
single smooth curve with a minimum near 380 K.

The thirty data points used in the critical evaluation of krypton solubility
at temperature below 348 K and the 15 data points from the present paper
were combined in a linear regression to obtain the equation

$$\ln X_1 = -61.52677 + 82.72769/(T/100) + 21.7672 \ln (T/100)$$

The equation is a tentative equation for use between 373.15 and 573.15 K.
It is not recommended for use at lower temperatures. It gives a mole fraction solubility at 273.15 K that is 3.3% lower, at 293.15 K 0.02% lower,
and 348.15 K 2.5% higher than the recommended equation for that temperature
interval.

Smoothed values of the mole fraction solubility, Ostwald coefficients, Gibbs
energy, enthalpy and entropy of solution at 25 K intervals between 373.15
and 573.15 K are given in Table 1. A comparison of the two equations and
the data from this paper is given in Figure 1.

TABLE 1.   Tentative values of the solubility of krypton in water at a
           krypton partial pressure of 101.325 kPa and values of the
           thermodynamic properties on solution between 373.15 and 573.15 K.

T/K	Mol Fraction X_1 x 10^5	Ostwald Coefficient L	ΔG^O/kJ mol^{-1}	ΔH^O/kJ mol^{-1}	ΔS^O/J K^{-1}mol^{-1}
373.15	2.27	0.0370	33.18	−1.25	−92.3
398.15	2.31	0.0394	35.33	+3.27	−80.5
423.15	2.55	0.0451	37.21	+7.80	−69.5
448.15	2.99	0.0545	38.81	+12.3	−59.1
473.15	3.68	0.0684	40.17	+16.8	−49.3
498.15	4.69	0.0884	41.28	+21.4	−40.0
523.15	6.16	0.116	42.17	+25.9	−31.1
548.15	8.27	0.155	42.84	+30.4	−22.7
573.15	11.3	0.207	43.31	+34.9	−14.6

ΔC_p^O is constant with a value of 181. J K^{-1} mol^{-1}.

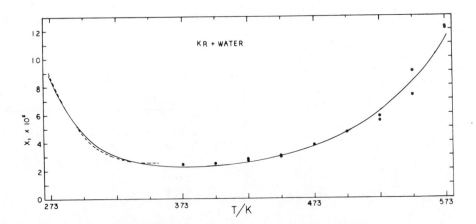

FIGURE 1. The mole fraction solubility of krypton in water at a krypton
          partial pressure of 101.325 kPa (1 atm). Recommended low tempera-
          ture equation ----,tentative high temperature equation ――― .

COMPONENTS:	EVALUATOR:
1. Krypton; Kr; 7439-90-9 2. Sea Water	H. L. Clever Chemistry Department Emory University Atlanta, GA 30322 U.S.A.

CRITICAL EVALUATION:

There are three reports of the solubility of krypton in sea water. König (1) reports krypton solubility values at eight temperatures between 273.15 and 297.15 K. Wood and Caputi (2) report krypton solubility values at three temperatures between 274.45 and 321.25 K. Weiss and Kyser (3) report three to five krypton solubility values at each of seven temperatures between 273.23 and 313.25 K.

Weiss and Kyser appear to have carried out the most reliable work and we recommend use of their values. Weiss and Kyser corrected these data for the real molar value of krypton and for the fugacity of the krypton. Their equation for the Bunsen solubility coefficient of krypton, corrected for nonideal behavior, is given by the equation

$$\ln[\text{Bunsen}] = -57.2596 + 87.4242(100/T) + 22.9332 \ln(T/100)$$
$$+ S\text{‰}[-0.008723 - 0.002793(T/100) + 0.0012398(T/100)^2]$$

where T is the absolute temperature and S‰ is the salinity in parts per thousand.

Weiss and Kyser give equations for the solubility of krypton from moist air at 101.325 kPa (1 atm) total pressure in units of cm^3 Kr(STP) dm^{-3} sea water and cm^3 Kr(STP) kg^{-1} sea water assuming that krypton mole fraction in dry air is 1.141×10^{-6} (4). The equations are

$$\ln[cm^3 \text{ Kr(STP) } dm^{-3}] = -109.9320 + 149.8152(100/T) + 72.8393 \ln(T/100)$$
$$- 9.9217(T/100) + S\text{‰}[-0.006953 - 0.004085(T/100) + 0.0014759(T/100)^2]$$

and

$$\ln[cm^3 \text{ Kr(STP) } kg^{-1}] = -112.6840 + 153.5817(100/T) + 74.4690 \ln(T/100)$$
$$- 10.0189(T/100) + S\text{‰}[-0.011213 - 0.001844(T/100) + 0.0011201(T/100)^2]$$

Extensive tables of krypton Bunsen coefficients and cm^3 Kr(STP) kg^{-1} as a function of temperature and salinity as calculated from the above equations are given in the original paper.

Weiss and Kyser compare the earlier work with their results. They show that teh Köenig data are as much as 14 per cent lower at low temperatures and that the Wood and Caputi data averages 1.9 percent lower than their own data at 274.45 and 299.55 K. Data sheets follow for both the Weiss and Kyser and the Wood and Caputi solubility values. The Wood and Caputi values extend the temperature range another seven degrees.

1. Köenig, H. Z. Naturforsch. 1963, 18a, 363.
2. Wood, D.; Caputi, R. USNRDL-TR-988, Feb. 1966.
3. Weiss, R. F.; Kyser, T. K. J. Chem. Eng. Data, 1978, 23, 69.
4. Glueckauf, E.; Kitt, G. A. Proc. Roy. Soc. London, 1956, 234A, 557.

COMPONENTS:	ORIGINAL MEASUREMENTS:
1. Krypton; Kr; 7439-90-9 2. Sea Water	Weiss, R. F.; Kyser, T. K. J. Chem. Eng. Data 1978, 23, 69 - 72.
VARIABLES: T/K: 273.22 - 313.25 Kr P/kPa: 101.325 (1 atm) Salinity/mil^{-1}: 0 - 36.595	**PREPARED BY:** A. L. Cramer

EXPERIMENTAL VALUES: Salinity/$^{0}/_{00}$

0.0		19.046		36.595	
T/K	Bunsen/α	T/K	Bunsen/α	T/K	Bunsen/α
273.74	0.10778			273.24	0.08451
273.72	0.10778			273.22	0.08438
273.72	0.10790			273.24	0.08442
273.73	0.10786			273.23	0.08444
		278.05	0.08200	278.47	0.07198
283.48	0.07981	278.05	0.08212	278.46	0.07204
283.47	0.07985	278.05	0.08220	278.47	0.07193
283.47	0.07979			283.57	0.06273
293.39	0.06222			283.57	0.06276
293.39	0.06220			283.56	0.06270
293.40	0.06227			288.25	0.05566
293.39	0.06221			288.25	0.05547
293.40	0.06226			288.25	0.05554
				293.49	0.04920
304.12	0.04953			293.49	0.04933
304.13	0.04957			293.49	0.04935
304.13	0.04955	303.23	0.04525	303.11	0.04084
304.12	0.04961	303.23	0.04510	303.12	0.04093
		303.23	0.04517	303.11	0.04085
313.24	0.04215	303.23	0.04532	313.25	0.03475
313.25	0.04218			313.25	0.03469
313.25	0.04216			313.25	0.03481
				313.24	0.03475

AUXILIARY INFORMATION

METHOD: Solubility determinations by the Scholander microgasometric technique as used by Douglas (1), with minor modification by Weiss (2). Real krypton molar volume at STP was used and a fugacity correction was applied.	SOURCE AND PURITY OF MATERIALS: 1. Krypton. Matheson Gas Products. > 99.995 % pure. Gas chromatographic checks showed $\leqq$ 0.01 % air. 2. Sea Water. Collected from surface, evaporated to increase salinity $\backsim$ 2$\%_0$, passed through 0.45 μ millipore filter and poisoned with 1 mg/l of HgCl$_2$.
APPARATUS/PROCEDURE: An equilibrium chamber, containing pure gas saturated with water vapor, is separated by mercury from a closed side chamber containing degassed water. The apparatus is tipped on its side, allowing degassed water to flow into the equilibrium chamber. Dissolution is aided by mechanical shaking.	ESTIMATED ERROR: $\delta T/K = 0.01$ δsalinity = 0.004
	REFERENCES: 1. Douglas, E. J. Phys. Chem. 1964, 68, 169. ibid. 1965, 69, 2608. 2. Weiss, R. F. J. Chem. Eng. Data 1971, 16, 235-241.

COMPONENTS:	ORIGINAL MEASUREMENTS:
1. Krypton; Kr; 7439-90-9 2. Sea Water	Wood, D.; Caputi, R. U.S.N.R.D.L.-TR-988, 27 Feb. 1966 Chem. Abstr. 1967, 66, 118693u.

VARIABLES:
 	
    T/K: 274.45 - 320.95
    P/kPa: 101.325 (1 atm)
	PREPARED BY:
	
	    A. L. Cramer

EXPERIMENTAL VALUES:

T/K	Henry's Constant $K = (P_1/mmHg)/X_1$	Percent Error*	Number of Determinations	Mol Fraction $X_1 \times 10^3$	Bunsen Coefficient α
		Water; H_2O; 7732-18-5			
274.45	0.866×10^7	0.3	3	0.0878	0.1098
298.15	1.653×10^7	0.4	2	0.0450	0.0571
320.95	2.340×10^7	2.1	4	0.0325	0.0400
		Artificial Sea Water(1); S‰ = 34.727			
274.45	1.181×10^7	0.3	2	0.0644	0.0803
299.55	2.163×10^7	0.1	2	0.0351	0.0435
320.35	2.792×10^7	0.3	2	0.0272	0.0335

* Percent error is the maximum spread in Henry's constant times 100 divided by average Henry's constant.

The mole fractions were calculated by the compiler from the average Henry's constant. The Bunsen coefficients were calculcated by the compiler from the mole fractions using a solvent mean molecular weight of 18.4823 and sea water densities from the International Critical Tables.

AUXILIARY INFORMATION

METHOD/APPARATUS/PROCEDURE:

Degassed water was introduced into an evacuated apparatus (< 50 µ Hg) and gas bled into burette. After the system was isolated, gas was admitted to equilibrium column and the water was circulated through the column, flowing over packing of 4 mm Berl saddles at 110 ml/min for 4-5 hr.

Dissolved gas was reclaimed and measured by evacuating the system → < 1 µ Hg and allowing water to distill and condense in a cold trap. Water was melted and refrozen until all the gas was recovered. Gas was then transferred to a gas burette and the pressure was measured with a Hg manometer. Purity was checked by gas chromatography.

SOURCE AND PURITY OF MATERIALS:

1. Krypton. AIRCO. Certified 0.015% Xe. Air contamination 0.0001 determined by gas chromatography.

2. Water. Distilled three times before degassing. Sea Water. Artificial, modified from (1).

ESTIMATED ERROR: δT/K = 0.1, 0.005, 0.03 (as T increases) δP/P = 0.001 δH/H = 0.005 (author's error analysis)

REFERENCES:

1. Lyman, J.; Fleming, R. H. J. Mar. Res. 1940, 3, 134.

COMPONENTS:	EVALUATOR:
1. Krypton; Kr; 7439-90-9	H. L. Clever
	Chemistry Department
2. Water; H_2O; 7732-18-5	Emory University
	Atlanta, GA 30322
3. Electrolyte	U. S. A.
	September 1978

CRITICAL EVALUATION:

The solubility of krypton in aqueous electrolyte solutions.

The results of studies of the solubility of krypton in aqueous salt solution can be classified as no better than tentative. Körösy (1) reports a
solubility of krypton in aqueous 20 weight per cent $CaCl_2$ solution which
is of dubious value. Morrison and Johnstone (2) report Setschenow salt
effect parameters for 13 aqueous electrolyte solutions. All of their
values are based on only one measurement in a one molal salt solution.
Although Morrison and Johnstone's Setschenow parameter values for other
gases usually accord well with the results of more extensive studies of
other workers, their values based on only one measurement in an electrolyte
solution must be classed as tentative.

Yeh and Peterson (3), Kitani (4), and Kirk, Parrish and Morken (5) have
measured the solubility of krypton in 0.9 weight percent NaCl solution.
Kirk et al. did not report a krypton solubility in water so it is not
possible to convert their solubility value to a Setschenow parameter.
The Setschenow salt effect parameters for krypton in aqueous NaCl solutions from the data of Morrison and Johnstone, Yeh and Peterson, and
Kitani do not agree well (Table 1). The values based on Yeh and Peterson's
data appear to be high and somewhat erratic as a function of temperature.
The values of Morrison and Johnstone and of Kitani are probably more
reliable.

Anderson, Keeler and Klach (6) measured the solubility of krypton in
water, and in aqueous solution containing UO_2SO_4, $CuSO_4$ and H_2SO_4 at
temperatures between 373 and 573 K. They found no statistical difference
in the Henry's constant of krypton in water and in a solution that was
0.02 mol dm^{-3} UO_2SO_4, 0.005 mol dm^{-3} $CuSO_4$ and 0.005 mol dm^{-3} H_2SO_4.
There were differences in the Henry's constant for krypton in water and in
a little more concentrated solution containing 0.04 mol dm^{-3} UO_2SO_4,
0.01 mol dm^{-3} $CuSO_4$ and 0.01 mol dm^{-3} H_2SO_4.

COMPONENTS:	EVALUATOR:
1. Krypton; Kr; 7439-90-9	H. L. Clever
	Chemistry Department
2. Water; H_2O; 7732-18-5	Emory University
	Atlanta, GA 30322
3. Electrolyte	U. S. A.
	September 1978

CRITICAL EVALUATION:

Electrolyte	T/K	mol salt Kg^{-1} H_2O	$k_s =$ (1/m)log(S^o/S)	$K_{sx} =$ (1/m)log(X^o/X)	Ref.
NH_4Cl	298.15	1.0	0.065	0.080	2
$(CH_3)_4NI$	298.15	1.0	-0.016	-0.001	2
$(C_2H_5)_4NBr$	298.15	1.0	-0.032	-0.017	2
HCl	298.15	1.0	0.028	0.043	2
HNO_3	298.15	1.0	-0.003	0.012	2
$BaCl_2$	298.15	1.0	0.151	0.174	2
LiCl	298.15	1.0	0.116	0.131	2
NaCl	298.15	1.0	0.146	0.161	2
	298.15	0.155	0.195	0.210	3
	303.15	0.155	0.216	0.231	3
	310.15	0.155	0.224	0.239	3
	310.15	0.155	0.137	0.152	4
	310.15	0.155	*	--	5
	318.15	0.155	0.197		3
Na_2SO_4	298.15	1.0	0.203	0.226	2
Na_3PO_4**	298.15	0.066	0.266	--	3
	303.15	0.066	0.287	--	3
	310.15	0.066	0.265	--	3
	318.15	0.066	0.368	--	3
KCl	298.15	1.0	0.124	0.139	2
KBr	298.15	1.0	0.120	0.135	2
KI	298.15	1.0	0.120	0.135	2
KNO_3	298.15	1.0	0.093	0.108	2
UO_2SO_4	373-573	0.02-0.04	--	--	6

*k_s ranges between 0.05 and 0.20 depending on the value of water
 solubility used.

** The Na_3PO_4 concentration in mol dm^{-3} solution.

REFERENCES

1. Korossy, F. Trans. Faraday Soc. 1937, 33, 416.

2. Morrison, T. J.; Johnstone, N. B. B. J. Chem. Soc. 1955, 3655.

3. Yeh, S. Y.; Peterson, R. E. J. Pharm. Sci. 1964, 53, 822.

4. Kitani, K. Scand. J. Clin. Lab. Invest. 1972, 29, 167.

5. Kirk, W. P.; Parish, P. W.; Morken, D. A. Health Physics 1975, 28, 249.

6. Anderson, C. J.; Keeler, R. A.; Klach, S. S. J. Chem. Eng. Data
   1962, 7, 290.

COMPONENTS:	ORIGINAL MEASUREMENTS:
1. Krypton; Kr; 7439-90-9 2. Water; H_2O; 7732-18-5 3. Ammonium Chloride; NH_4Cl; 12125-02-9	Morrison, T.J.; Johnstone, N.B.B. J. Chem. Soc. 1955, 3655-3659.

VARIABLES:	PREPARED BY:
T/K: 298.15 P/kPa: 101.325 (1 atm)	T.D. Kittredge H.L. Clever

EXPERIMENTAL VALUES:

T/K	$k_s = (1/m) \log (S^O/S)$	$k_{sX} = (1/m) \log (X^O/X)$
298.15	0.065	0.080

The value of the Setschenow salt effect parameter , k_s, was apparently
determined from only two solubility measurements. They were the solubility
of krypton in pure water, S^O, and the solubility of krypton in a near one
equivalent of salt per kg of water solution, S. No solubility values are
given in the paper. The S^O/S ratio was referenced to a solution containing
one kg of water. The compiler calculated the salt effect parameter k_{sX}
from the mole fraction solubility ratio X^O/X. The electrolyte was
assumed to be 100 per cent dissociated and both cation and anion were used
in the mole fraction calculation.

AUXILIARY INFORMATION

METHOD:	SOURCE AND PURITY OF MATERIALS:
Gas absorption in a flow system.	1. Krypton. British Oxygen Co. Ltd. 2. Water. No information given. 3. Electrolyte. No information given.

APPARATUS/PROCEDURE:	ESTIMATED ERROR:
The previously degassed solvent flows in a thin film down an absorp- tion spiral containing Kr gas plus solvent vapor at a total pressure of one atm. The volume of gas absorbed is measured in attached calibrated burets (1).	$\delta k_s = 0.010$ REFERENCES: 1. Morrison, T.J.; Billett, F. J. Chem. Soc. 1952, 3819.

COMPONENTS:	ORIGINAL MEASUREMENTS:
1. Krypton; Kr; 7439-90-9 2. Water; H_2O; 7732-18-5 3. Ammonium Type Salts	Morrison, T.J.; Johnstone, N.B.B. $\underline{J}$. $\underline{Chem}$. $\underline{Soc}$. 1955, 3655-3659.

VARIABLES:	PREPARED BY:
T/K: 298.15 P/kPa: 101.325 (1 atm)	T.D. Kittredge H.L. Clever

EXPERIMENTAL VALUES:

$$T/K \qquad k_s = (1/m) \log (S^O/S) \qquad k_{sX} = (1/m) \log (X^O/X)$$

N,N,N-Trimethylmethanaminium iodide (Tetramethyl-ammonium iodide); $C_4H_{12}NI$; 75-58-1

298.15	-0.016	-0.001

N,N,N-Triethylethanaminium bromide (Tetraethyl-ammonium bromide); $C_8H_{20}NBr$; 71-91-0

298.15	-0.032	-0.017

The values of the Setschenow salt effect parameter, k_s, were apparently determined from only two solubility measurements. They were the solubility of krypton in pure water, S^O, and the solubility of krypton in a near one equivalent of salt per kg of water solution, S. No solubility values are given in the paper. The S^O/S ratio was referenced to a solution containing one kg of water. The compiler calculated the salt effect parameter k_{sX} from the mole fraction solubility ratio X^O/X. The electrolytes were assumed to be 100 per cent dissociated and both cation and anion were used in the mole fraction calculation.

AUXILIARY INFORMATION

METHOD:	SOURCE AND PURITY OF MATERIALS:
Gas absorption in a flow system.	1. Krypton. British Oxygen Co. Ltd. 2. Water. No information given. 3. Electrolyte. No information given.

APPARATUS/PROCEDURE:	ESTIMATED ERROR:
The previously degassed solvent flows in a thin film down an absorption spiral containing the gas plus solvent vapor at a total pressure of one atm. The volume of gas absorbed is measured in attached calibrated burets (1).	$\delta k_s = 0.010$
	REFERENCES: 1. Morrison, T.J.; Billett, F. $\underline{J}$. $\underline{Chem}$. $\underline{Soc}$. 1952, 3819.

COMPONENTS:	ORIGINAL MEASUREMENTS:
1. Krypton; Kr; 7439-90-9 2. Water; H_2O; 7732-18-5 3. Acids	Morrison, T.J.; Johnstone, N.B.B. J. Chem. Soc. 1955, 3655-3659.

VARIABLES:
    T/K: 298.15
    P/kPa: 101.325 (1 atm)
	PREPARED BY:
	    T.D. Kittredge
	    H.L. Clever

EXPERIMENTAL VALUES:

T/K	k_s = (1/m) log (S^O/S)	k_{sX} = (1/m) log (X^O/X)
Hydrochloric Acid; HCl; 7647-01-0		
298.15	0.028	0.043
Nitric Acid; HNO_3; 7697-37-2		
298.15	-0.003	0.012

The values of the Setschenow salt effect parameter, k_s, were apparently determined from only two solubility measurements. They were the solubility of krypton in pure water, S^O, and the solubility of krypton in a near one equivalent of salt per kg of water solution, S. No solubility values are given in the paper. The S^O/S ratio was referenced to a solution containing one kg of water. The compiler calculated the salt effect parameter k_{sX} from the mole fraction solubility ratio X^O/X. The electrolytes were assumed to be 100 per cent dissociated and both cation and anion were used in the mole fraction calculation.

AUXILIARY INFORMATION

METHOD:	SOURCE AND PURITY OF MATERIALS:
Gas absorption in a flow system.	1. Krypton. British Oxygen Co. Ltd. 2. Water. No information given. 3. Electrolyte. No information given.

APPARATUS/PROCEDURE:
 	
   The previously degassed solvent flows in a thin film down an absorption spiral containing the gas plus solvent vapor at a total pressure of one atm. The volume of gas absorbed is measured in attached calibrated burets (1).
	ESTIMATED ERROR:
	
	        δk_s = 0.010

	REFERENCES: 1. Morrison, T.J.; Billett, F. J. Chem. Soc. 1952, 3819.

COMPONENTS:	ORIGINAL MEASUREMENTS:

1. Krypton; Kr; 7439-90-9
   Krypton-85; ^{85}Kr; 13983-27-3
2. Water; H_2O; 7732-18-5
3. Uranium dioxosulfato-(Uranyl sulfate); UO_2SO_4;1314-64-3
4. Copper sulfate; $CuSO_4$; 7758-98-7
5. Sulfuric Acid; H_2SO_4; 7664-93-9
	Anderson, C. J.; Keeler, R. A.;
	   Klach, S.J.
	
	J. Chem. Eng. Data 1962, 7, 290-294.

VARIABLES:	PREPARED BY:
T/K: 373.15 - 578.15 Kr P/pa: 0.407 - 36.680 (0.59 - 53.2 x 10^{-4} psia)	A.L. Cramer H.L. Clever

EXPERIMENTAL VALUES:

T/K	Pressure/psia Total $O_2 + H_2O + Kr$	Pressure/psia Partial x 10^4 Kr	Henry's Constant $K = (p/psia)/X_1$ K x 10^{-5}	Mol Fraction X_1 x 10^5 at 101.325 kPa Kr
0.02 mol dm^{-3} UO_2SO_4, 0.005 mol dm^{-3} $CuSO_4$ and 0.005 mol dm^{-3} H_2SO_4				
373.15	115	0.93	6.37	2.31
	365	5.03	6.66	2.21
398.15	110	0.59	5.78	2.54
	350	33.9	5.93	2.48
423.15	125	0.40	5.1	2.88
	315	28.2	5.20	2.83
523.15	1120	15.0	2.60	5.65
	1165	15.0	2.70	5.44
	1310	6.20	2.80	5.25
	1330	7.20	2.60	5.65
	1495	39.0	2.38	6.18
524.15	1095	53.2	2.08	7.07
543.15	1235	44.1	1.75	8.40
548.15	1340	11.0	1.9	7.74
	1485	4.20	1.8	8.17
	1565	6.00	2.0	7.35
	1715	42.9	1.90	7.74
573.15	1577	7.0	1.0	14.7
	1665	9.1	1.1	13.4
	1736	3.0	1.20	12.3
	1870	4.5	1.3	11.3
	2015	3.77	1.49	9.87
578.15	1625	2.5	1.01	14.55

*See note below

AUXILIARY INFORMATION

METHOD /APPARATUS/PROCEDURE:	SOURCE AND PURITY OF MATERIALS:
A Kr/^{85}Kr stock mixture and O_2 are added separately to a thermostated stainless steel vessel containing about 275 ml of solution. Equilibration time varied from 1.5 to 22 hours without appreciable difference in results (one run went for 64 hours). At equilibrium both liquid and vapor were sampled and the krypton in each phase was determined by counting the ^{85}Kr tag.	No information

METHOD (cont.)	ESTIMATED ERROR:
Henry's law constant is linear in temperature. The authors give $K = (9.188 - 0.0267t/^oC)x10^5$ psia x_1^{-1} with a standard deviation about the line of 0.170 x 10^5.	The 95% confidence limits are given by $K \pm 0.357 \times 10^5 \left[0.05 + \frac{(t-230)^2}{130,000}\right]^{\frac{1}{2}}$

The solubility of Kr in water and in the uranyl sulfate solution above appears to be the same. The Henry's constant for the two sets of data pooled for one equation is
$K = (9.162-0.0265t/^oC)x10^5$ psia X_1^{-1}

REFERENCES:

*The mole fraction solubility at a krypton partial pressure of 101.325 kPa was calculated by the compilers.

COMPONENTS:	ORIGINAL MEASUREMENTS:

1. Krypton; Kr; 7439-90-9
   Krypton-85; ^{85}Kr; 13983-27-3
2. Water; H_2O; 7732-18-5
3. Uranium dioxosulfato-(Uranyl sulfate); UO_2SO_4; 1314-64-3
4. Copper sulfate; $CuSO_4$; 7758-98-7
5. Sulfuric Acid; H_2SO_4; 7664-93-9
	Anderson, C. J.; Keeler, R. A.;
	   Klach, S. J.
	
	J. Chem. Eng. Data 1962, 7, 290-294.

VARIABLES: T/K: 373.15 - 573.15
   Kr P/pa: 2.261 - 50-332
   (3.28 - 73.0 x 10^{-4} psia)
	PREPARED BY:
	   A.L. Cramer
	   H.L. Clever

EXPERIMENTAL VALUES:

T/K	Pressure/psia		Henry's Constant $K = (p/psia)/X_1$ $K \times 10^{-5}$	Mol Fraction $X_1 \times 10^5$ at 101.325 kPa Kr
	Total $O_2 + H_2O + Kr$	Partial x 10^4 Kr		
	over: 0.04 mol dm^{-3} UO_2SO_4, 0.01 mol dm^{-3} $CuSO_4$, 0.01 mol dm^{-3} H_2SO_4			
373.15	355	29.8	6.31	2.33
	350	7.25	6.20	2.37
	380	68.0	5.96	2.47
423.15	400	27.6	5.19	2.83
	410	6.93	5.73	2.57
	445	73.0	5.18	2.84
523.15	825	43.7	1.90	7.74
	860	5.02	2.03	7.24
	860	25.1	2.20	6.68
	905	55.7	2.11	6.97
573.15	1440	26.0	0.97	15.15
	1480	3.28	1.17	12.6
	1485	13.3	1.34	11.0
	1505	36.0	0.889	16.5

The mole fraction solubility at a krypton partial pressure of 101.325 kPa was calculated by the compilers.

AUXILIARY INFORMATION

METHOD/APPARATUS/PROCEDURE:

A Kr/^{85}Kr stock mixture and O_2 are added separately to a thermostated stainless steel vessel containing about 275 ml of solution. Equilibration time varied from 1.5 to 22 hours without appreciable difference in results (one run went for 64 hours). At equilibrium both liquid and vapor were sampled and the krypton in each phase was determined by counting the ^{85}Kr tag.

Henry's law constant is not linear in temperature as it appears to be for water and the more dilute uranyl sulfate solution.

The mean values of K at the four temperatures are

T/K	K/psia X_1^{-1} x 10^{-5}
373.15	(6.16 $\pm$ 0.26)
423.15	(5.36 $\pm$ 0.26)
473.15	(2.06 $\pm$ 0.23)
523.15	(1.09 $\pm$ 0.23)

SOURCE AND PURITY OF MATERIALS:

No information

ESTIMATED ERROR:

REFERENCES:

COMPONENTS:	ORIGINAL MEASUREMENTS:
1. Krypton; Kr; 7439-90-9 2. Water; H_2O; 7732-18-5 3. Barium Chloride; $BaCl_2$; 10361-37-2	Morrison, T.J.; Johnstone, N.B.B. J. Chem. Soc. 1955, 3655-3659.

VARIABLES:
   T/K: 298.15
   P/kPa: 101.325 (1 atm)
	PREPARED BY:
	
	   T.D. Kittredge
	   H.L. Clever

EXPERIMENTAL VALUES:

T/K	k_s = (1/m) log (S^O/S)	k_{sX} = (1/m) log (X^O/X)
298.15	0.151	0.166

The value of the Setschenow salt effect parameter, k_s, was apparently determined from only two solubility measurements. They were the solubility of krypton in pure water, S^O, and the solubility of krypton in a near one equivalent of salt per kg of water solution, S. No solubility values are given in the paper. The S^O/S ratio was referenced to a solution containing one kg of water. The compiler calculated the salt effect parameter k_{sX} from the mole fraction solubility ratio X^O/X. The electrolyte was assumed to be 100 per cent dissociated and both cation and anion were used in the mole fraction calculation.

AUXILIARY INFORMATION

METHOD:	SOURCE AND PURITY OF MATERIALS:
Gas absorption in a flow system.	1. Krypton. British Oxygen Co. Ltd. 2. Water. No information given. 3. Electrolyte. No information given.
APPARATUS/PROCEDURE:	ESTIMATED ERROR: δk_s = 0.010
The previously degassed solvent flows in a thin film down an absorption spiral containing the gas plus solvent vapor at a total pressure of one atm. The volume of gas absorbed is measured in attached calibrated burets (1).	REFERENCES: 1. Morrison, T.J.; Billett, F. J. Chem. Soc. 1952, 3819.

COMPONENTS:	ORIGINAL MEASUREMENTS:
1. Krypton; Kr; 7439-90-9 2. Water; H_2O; 7732-18-5 3. Sodium Chloride; NaCl; 7647-14-5	Yeh, S.Y.; Peterson, R.E. J. Pharm. Sci. 1964, 53, 822 - 824.

VARIABLES:
  T/K: 298.15 - 318.15
  P/kPa: 101.325 (1 atm)
	PREPARED BY:
	
	 H.L. Clever

EXPERIMENTAL VALUES:

T/K	Ostwald Coefficient L	ΔH^{O}/cal mol^{-1}	ΔS^{O}/cal K^{-1} mol^{-1}
298.15	0.0542	-5017	-28.9
303.15	0.0499	-4252	-26.4
310.15	0.0444	-3181	-22.9
318.15	0.0411	-1957	-19.0

For comparison, the authors' Ostwald coefficients in water were 0.0581, 0.0539, 0.0481 and 0.0441 at the four temperatures.

The sodium chloride solution is 0.9 weight percent which is about 0.155 mole NaCl kg^{-1} water.

Each solubility value is the average of three or four measurements. The standard deviation of each measurement closely approximates 1.0 per cent.

The thermodynamic changes are for the transfer of one mole of krypton from the gas phase at a concentration of one mole dm^{-3} to the solution at a concentration of one mole dm^{-3}.

AUXILIARY INFORMATION

METHOD/APPARATUS/PROCEDURE:	SOURCE AND PURITY OF MATERIALS:
Freshly boiled solution was introduced into 125 ml. absorption flask of solubility apparatus (1), then frozen and boiled under vacuum three times. Water-saturated gas was introduced and equilibrated (2) and weight of solution was determined. Thermodynamic constants were calculated from equations (3): $\log S = A/T + B \log T - C$ $\Delta H^{O} = R(-2.3A + BT - T)$ $\Delta S^{O} = R(-B-B\ln T + 2.3C +$ $\ln(0.082T) + 1$	1. Krypton. Matheson Co. 2. Water. Distilled from glass apparatus. 3. Sodium chloride. Analytical grade.

	ESTIMATED ERROR:
	  $\delta T/K = 0.05$
	  $\delta P/mmHg = 0.2$
	  $\delta L/L = 0.01$

	REFERENCES: 1. Geffken, G. Z. Physik Chem. 1904, 49, 257. 2. Yeh, S.Y.; Peterson, R.E. J. Pharm. Sci. 1963, 52, 453-8. 3. Eley, E.E. Trans. Faraday Soc. 1939, 35, 1281.

COMPONENTS:	ORIGINAL MEASUREMENTS:
1. Krypton; Kr; 7439-90-9 2. Water; H_2O; 7732-18-5 3. Alkali Halides	Morrison, T.J.; Johnstone, N.B.B. $\underline{J}$. $\underline{Chem}$. $\underline{Soc}$. 1955, 3655-3659.
VARIABLES: T/K: 298.15 P/kPa: 101.325 (1 atm)	PREPARED BY: T.D. Kittredge H.L. Clever

EXPERIMENTAL VALUES:

T/K	$k_s = (1/m) \log (S^O/S)$	$k_{sX} = (1/m) \log (X^O/X)$
Lithium Chloride; LiCl; 7447-41-8		
298.15	0.116	0.0131
Sodium Chloride; NaCl; 7647-14-5		
298.15	0.146	0.161
Potassium Chloride; KCl; 7447-40-7		
298.15	0.124	0.139
Potassium Bromide; KBr; 7758-02-3		
298.15	0.120	0.135
Potassium Iodide; KI; 7681-11-0		
298.15	0.120	0.135

The values of the Setschenow salt effect parameter, k_s, were apparently determined from only two solubility measurements. They were the solubility of krypton in pure water, S^O, and the solubility of krypton in a near one equivalent of salt per kg of water solution, S. No solubility values are given in the paper. The S^O/S ratio was referenced to a solution containing one kg of water. The compiler calculated the salt effect parameter k_{sX} from the mole fraction solubility ratio X^O/X. The electrolytes were assumed to be 100 per cent dissociated and both cation and anion were used in the mole fraction calculation.

AUXILIARY INFORMATION

METHOD: Gas absorption in a flow system.	SOURCE AND PURITY OF MATERIALS: 1. Krypton. British Oxygen Co. Ltd. 2. Water. No information given. 3. Electrolyte. No information given.
APPARATUS/PROCEDURE: The previously degassed solvent flows in a thin film down an absorption spiral containing the gas plus solvent vapor at a total pressure of one atm. The volume of gas absorbed is measured in attached calibrated burets (1).	ESTIMATED ERROR: $\delta k_s = 0.010$
	REFERENCES: 1. Morrison, T.J.; Billett, F. $\underline{J}$. $\underline{Chem}$. $\underline{Soc}$. 1952, 3819.

COMPONENTS:	ORIGINAL MEASUREMENTS:
1. Krypton; Kr; 7439-90-9 2. Water; H_2O; 7732-18-5 3. Sodium Sulfate; Na_2SO_4; 7757-82-6	Morrison, T.J.; Johnstone, N.B.B. $\underline{J}$. $\underline{Chem}$. $\underline{Soc}$. 1955, 3655-3659.

VARIABLES:
     T/K: 298.15
     P/kPa: 101.325 (1 atm)
	PREPARED BY:
	
	     T.D. Kittredge
	     H.L. Clever

EXPERIMENTAL VALUES:

T/K	$k_s = (1/m) \log (S^o/S)$	$k_{sX} = (1/m) \log (X^o/X)$
298.15	0.203	0.226

The value of the Setschenow salt effect parameter, k_s, was apparently determined from only two solubility measurements. They were the solubility of krypton in pure water, S^o, and the solubility of krypton in a near one equivalent of salt per kg of water solution, S. No solubility values are given in the paper. The S^o/S ratio was referenced to a solution containing one kg of water. The compiler calculated the salt effect parameter k_{sX} from the mole fraction solubility ratio X^o/X. The electrolyte was assumed to be 100 per cent dissociated and both cation and anion were used in the mole fraction calculation.

AUXILIARY INFORMATION

METHOD:	SOURCE AND PURITY OF MATERIALS:
Gas absorption in a flow system.	1. Krypton. British Oxygen Co. Ltd. 2. Water. No information given. 3. Electrolyte. No information given.

APPARATUS/PROCEDURE:	ESTIMATED ERROR: $\delta k_s = 0.010$
The previously degassed solvent flows in a thin film down an absorption spiral containing the gas plus solvent vapor at a total pressure of one atm. The volume of gas absorbed is measured in attached calibrated burets (1).	REFERENCES: 1. Morrison, T.J.; Billett, F. $\underline{J}$. $\underline{Chem}$. $\underline{Soc}$. 1952, 3819.

COMPONENTS:	ORIGINAL MEASUREMENTS:
1. Krypton; Kr; 7439-90-9 2. Water; H_2O; 7732-18-5 3. Sodium Phosphate (phosphate buffer); Na_3PO_4; 7601-54-9	Yeh, S-Y.; Peterson, R.E. J. Pharm. Sci. 1964, 53, 822 - 824.

VARIABLES:	PREPARED BY:
T/K: 298.15 - 318.15 P/kPa: 101.325 (1 atm)	H.L. Clever

EXPERIMENTAL VALUES:

T/K	Ostwald Coefficient L	ΔH^o/cal mol^{-1}	ΔS^o/cal K^{-1} mol^{-1}
298.15	0.0558	−4012	−25.5
303.15	0.0516	−3744	−24.6
310.15	0.0462	−3369	−23.4
318.15	0.0417	−2941	−22.0

For comparison, the authors Ostwald coefficients in water were 0.0581, 0.0539, 0.0481 and 0.0441 at the four temperatures.

The sodium phosphate solution is an 0.066 molar buffer adjusted to pH 7.0. The solution might be better described as $Na_xH_yPO_4$; 7632-05-5.

Each solubility value is the average of three to four measurements. The standard deviation of each measurement closely approximates 1.0 per cent.

The thermodynamic changes are for the standard state transfer of one mole of krypton from the gas at a concentration of one mole dm^{-3} to the solution at a concentration of one mole dm^{-3}.

AUXILIARY INFORMATION

METHOD /APPARATUS/PROCEDURE:	SOURCE AND PURITY OF MATERIALS:
Freshly boiled solution was introduced into 125 ml. absorption flask of solubility apparatus (1), then frozen and boiled under vacuum three times. Water-saturated gas was introduced and equilibrated (2) and weight of solution was determined. Thermodynamic constants were calculated from equations (3): $\log S = A/T + B \log T - C$ $\Delta H^o = R(-2.3A + BT - T)$ $\Delta S^o = R(-B-B\ln T + 2.3C +$ $\ln(0.082T) + 1)$	1. Krypton. Matheson Co. 2. Water. Distilled from glass apparatus. 3. Sodium phosphate. Analytical grade.

ESTIMATED ERROR:
$\delta T/K = 0.05$
$\delta P/mmHg = 0.2$
$\delta L/L = 0.01$

REFERENCES:

1. Geffken, G. Z. Physik Chem. 1904, 49, 257.

2. Yeh, S.Y.; Peterson, R.E. J. Pharm. Sci. 1963, 52, 453-8.

3. Eley, E.E. Trans. Faraday Soc. 1939, 35, 1281.

24

COMPONENTS:	ORIGINAL MEASUREMENTS:
1. Krypton; Kr; 7439-90-9 2. Water; H_2O; 7732-18-5 3. Potassium Nitrate; KNO_3; 7757-79-1	Morrison, T.J.; Johnstone, N.B.B. J. Chem. Soc. 1955, 3655-3659.
VARIABLES: T/K: 298.15 P/kPa: 101.325 (1 atm)	PREPARED BY: T.D. Kittredge H.L. Clever

EXPERIMENTAL VALUES:

T/K	$k_s = (1/m) \log (S^O/S)$	$k_{sX} = (1/m) \log (X^O/X)$
298.15	0.093	0.108

The value of the Setschenow salt effect parameter, k_s, was apparently determined from only two solubility measurements. They were the solubility of krypton in pure water, S^O, and the solubility of krypton in a near one equivalent of salt per kg of water solution, S. No solubility values are given in the paper. The S^O/S ratio was referenced to a solution containing one kg of water. The compiler calculated the salt effect parameter k_{sX} from the mole fraction solubility ratio X^O/X. The electrolyte was assumed to be 100 per cent dissociated and both cation and anion were used in the mole fraction calculation.

AUXILIARY INFORMATION

METHOD:	SOURCE AND PURITY OF MATERIALS:
Gas absorption in a flow system.	1. Krypton. British Oxygen Co. Ltd. 2. Water. No information given. 3. Electrolyte. No information given.

APPARATUS/PROCEDURE:	ESTIMATED ERROR: $\delta k_s = 0.010$
The previously degassed solvent flows in a thin film down an absorption spiral containing the gas plus solvent vapor at a total pressure of one atm. The volume of gas absorbed is measured in attached calibrated burets (1).	REFERENCES: 1. Morrison, T.J.; Billett, F. J. Chem. Soc. 1952, 3819.

COMPONENTS:	ORIGINAL MEASUREMENTS:
1. Krypton; Kr; 7439-90-9 2. Pentane; C_5H_{12}; 109-66-0	Makranczy, J.; Megyery-Balog, K.; Rusz, L.; Patyi, L. Hung. J. Ind. Chem. 1976, 4, 269-280.

VARIABLES:
     T/K: 298.15
     P/kPa: 101.325 (1 atm)
	PREPARED BY:
	
	        S.A. Johnson

EXPERIMENTAL VALUES:

T/K	Mol Fraction $x_1 \times 10^3$	Bunsen Coefficient α	Ostwald Coefficient L
298.15	7.85	1.528	1.668

The mole fraction and Bunsen coefficient were calculated by the compiler.

AUXILIARY INFORMATION

METHOD:	SOURCE AND PURITY OF MATERIALS:
Volumetric method. The apparatus of Bodor, Bor, Mohai, and Sipos (1) was used.	Both the gas and liquid were analytical grade reagents of Hungarian or foreign origin. No further information.
APPARATUS/PROCEDURE:	ESTIMATED ERROR: $\delta X_1/X_1 = 0.03$

	REFERENCES:
	 1. Bodor, E.; Bor, Gy.; Mohai, B.; Sipos, G.
	    Veszpremi Vegyip. Egy. Kozl. 1957, 1, 55.
	    Chem. Abstr. 1961, 55, 3175h.

COMPONENTS:	EVALUATOR:
1. Krypton; Kr; 7439-90-9 2. Hexane; C_6H_{14}; 110-54-3	H. L. Clever Chemistry Department Emory University Atlanta, GA 30322 U.S.A.

CRITICAL EVALUATION:

The solubility of krypton in hexane at 101.325 kPa was measured in three laboratories. Clever, Battino, Saylor and Gross (1) report three solubilities between 289.30 and 313.75 K. Steinberg and Manowitz (2) report solubilities at 183.15 and 298.15 K. Makranczy, Megyery-Balog, Rusz, and Patyi (3) report a solubility at 298.15. The solubility values near 298.15 K from the three laboratories differ by about 30 percent.

Clever et al. probably used materials of better purity than the materials used by Steinberg and Manowitz. The single value of Makranczy et al. is difficult to judge, but their results tend to scatter for other gases dissolved in hydrocarbons. The values of Clever et al. are to be preferred for the 288.15 - 313.15 K temperature range. The smoothed data is given on the Clever et al. data sheet which follows this page.

The 183.15 K solubility value of Steinberg and Manowitz and the three solubility values of Clever et al. were combined in a linear least squares fit to an equation for the Gibbs energy of solution as a function of temperature. The result is a highly tentative equation for the 183.15 - 313.15 K temperature interval.

The tentative values for the transfer of one mole of krypton from the gas at a pressure of 101.325 kPa to the hypothetical unit mole fraction solution are

$$\Delta G^\circ / J \; mol^{-1} = - RT \ln X_1 = -3,340.9 + 52.564 \; T$$

$$Std. \; Dev. \; \Delta G^\circ = 63, \qquad Coef. \; Corr. = 0.9998$$

$$\Delta H^\circ / J \; mol^{-1} = -3,340.9, \quad \Delta S^\circ / J \; K^{-1} mol^{-1} = -52.564$$

The tentative solubility values and the Gibbs energy of solution at 101.325 kPa as a function of temperature are in Table 1.

TABLE 1. Solubility of krypton in hexane. Tentative values of the mole fraction solubility and Gibbs energy at 101.325 kPa (1 atm) as a function of temperature.

T/K	Mol Fraction X_1 x 10^3	$\Delta G^\circ / J \; mol^{-1}$
183.15	16.1	6,286.3
203.15	13.0	7,337.6
223.15	10.9	8,388.9
243.15	9.38	9,440.2
263.15	8.27	10,491
273.15	7.82	11,017
283.15	7.42	11,543
293.15	7.07	12,068
298.15	6.91	12,331
303.15	6.76	12,594
313.15	6.48	13,120

1. Clever, H. L.; Battino, R.; Saylor, J. H.; Gross, P. M. J. Phys. Chem. 1957, 61, 1078.
2. Steinberg, M.; Manowitz, B. Ind. Eng. Chem. 1959, 51, 47.
3. Makranczy, J.; Megyery-Balog, K.; Rusz, L.; Patyi, L. Hung. J. Ind. Chem. 1976, 4, 269.

COMPONENTS:	ORIGINAL MEASUREMENTS:
1. Krypton; Kr; 7439-90-9 2. Hexane; C_6H_{14}; 110-54-3	Clever, H.L.; Battino, R.; Saylor, J.H.; Gross, P.M. _J. Phys. Chem._ 1957, _61_, 1078-1083.

VARIABLES:	PREPARED BY:
T/K: 289.30 - 313.75 P/kPa: 101.325 (1 atm)	P.L. Long

EXPERIMENTAL VALUES:

T/K	Mol Fraction X_1 x 10^3	Bunsen Coefficient α	Ostwald Coefficient L
289.30	7.48	1.30	1.38
298.25	6.76	1.16	1.27
313.75	6.40	1.07	1.23

Smoothed data: $\Delta G^O = -RT\ln x_1 = -4,560.3 + 56.612\ T$

Std. Dev. $\Delta G^O = 58.2$, Coef. Corr. $= 0.9966$

$\Delta H^O/J\ mol^{-1} = -4,560.3$, $\Delta S^O/JK^{-1}\ mol^{-1} = -56.612$

T/K	Mol Fraction X_1 x 10^3	$\Delta G^O/J\ mol^{-1}$
288.15	7.41	11,752
293.15	7.17	12,035
298.15	6.95	12,319
303.15	6.74	12,602
308.15	6.54	12,885
313.15	6.36	13,168
318.15	6.19	13,451

See the evaluation of the krypton + hexane system on page 26 for a tentative Gibbs energy equation and solubility values covering the 183.15 to 313.15 K range.

AUXILIARY INFORMATION

METHOD:	SOURCE AND PURITY OF MATERIALS:
Volumetric. The solvent is saturated with gas as it flows through an 8 mm x 180 cm glass helix attached to a gas buret. The total pressure of solute gas plus solvent vapor is maintained at 1 atm as the gas is absorbed.	1. Krypton. Linde Air Products Co. Pure grade. 2. Hexane. Humphrey-Wilkinson, Inc. Shaken with H_2SO_4, washed, dried over Na, distilled.

APPARATUS/PROCEDURE:	ESTIMATED ERROR: $\delta T/K = 0.05$ $\delta P/mmHg = 3$ $\delta X_1/X_1 = 0.03$
The apparatus is a modification of the apparatus of Morrison and Billett (1). The modifications include the addition of a helical storage for the solvent, a manometer for a reference pressure, and an extra buret for highly soluble gases. The solvent is degassed by a modification of the method of Baldwin and Daniel (2).	REFERENCES: 1. Morrison, T.J.; Billett, F. _J. Chem. Soc._ 1948, 2033; _ibid._ 1952, 3819. 2. Baldwin, R.R.; Daniel, S.G. _J. Appl. Chem._ 1952, _2_, 161.

COMPONENTS:	ORIGINAL MEASUREMENTS:
1. Krypton; Kr; 7439-90-9 2. Hexane; C_6H_{14}; 110-54-3	Steinberg, M.; Manowitz, B. Ind. Eng. Chem. 1959, 51, 47-51.

VARIABLES:
    T/K: 183.15 - 298.15
    P/kPa: 101.325 (1 atm)
	PREPARED BY:
	    H.L. Clever
	    A.L. Cramer

EXPERIMENTAL VALUES:

T/K	Mol Fraction $X_1 \times 10^3$	Bunsen Coefficient α	Absorption Coefficient β
183.15	16.0	2.80	2.95
298.15	5.73	0.995	1.05

The authors define the Absorption coefficient as the volume of gas, corrected to 288.15 K and 101.325 kPa, absorbed under a total system pressure of 101.325 kPa per unit volume of solvent at 288.15 K.

The mole fraction solubilities and Bunsen coefficients were calculated by the compiler.

For an evaluation of the krypton + hexane system see page 26.

AUXILIARY INFORMATION

METHOD/APPARATUS/PROCEDURE:	SOURCE AND PURITY OF MATERIALS:
Absorption coefficient determined by a modified McDaniel method (1).	1. Krypton. Matheson Co. Technical grade. 2. Hexane. Technically or chemically pure.

ESTIMATED ERROR:

$$\delta\beta/\beta = 0.05 - 0.10$$

REFERENCES:

1. Furman, N.H. "Scott's Standard Methods of Chemical Analysis" Van Nostrand Co., NY, 1939, 5th ed., Vol. II, p. 2587.

COMPONENTS:	ORIGINAL MEASUREMENTS:

COMPONENTS:

1. Krypton; Kr; 7439-90-9

2. Hexane; C_6H_{14}; 110-54-3

ORIGINAL MEASUREMENTS:

Makranczy, J.; Megyery-Balog, K.; Rusz, L.; Patyi, L.

Hung. J. Ind. Chem. 1976, 4, 269-280.

VARIABLES:

       T/K: 298.15
       P/kPa: 101.325 (1 atm)

PREPARED BY:

       S.A. Johnson

EXPERIMENTAL VALUES:

T/K	Mol Fraction X_1 x 10^3	Bunsen Coefficient α	Ostwald Coefficient L
298.15	7.53	1.293	1.411

The mole fraction and Bunsen coefficient were calculated by the compiler.

See the krypton + hexane evaluation on page 26 for the recommended values of solubility and thermodynamic change.

AUXILIARY INFORMATION

METHOD/APPARATUS/PROCEDURE:

   Volumetric method. The apparatus of Bodor, Bor, Mohai, and Sipos (1) was used.

SOURCE AND PURITY OF MATERIALS:

Both the gas and liquid were analytical grade reagents of Hungarian or foreign origin. No further information.

ESTIMATED ERROR:

$$\delta X_1/X_1 = 0.03$$

REFERENCES:

1. Bodor, E.; Bor, Gy.; Mohai, B.; Sipos, G.

    Veszpremi Vegyip. Egy. Kozl. 1957, 1, 55.
    Chem. Abstr. 1961, 55, 3175h.

COMPONENTS:	ORIGINAL MEASUREMENTS:
1. Krypton; Kr; 7439-90-9 2. Heptane; C_7H_{16}; 142-82-5	Clever, H.L.; Battino, R.; Saylor, J.H.; Gross, P.M. J. Phys. Chem. 1957, 61, 1078-1083

VARIABLES:
     T/K: 289.25 - 313.45
 Total P/kPa: 101.325 (1 atm)
	PREPARED BY:
	
	         P.L. Long

EXPERIMENTAL VALUES:

T/K	Mol Fraction $X_1 \times 10^3$	Bunsen Coefficient α	Ostwald Coefficient L
289.25	7.59	1.175	1.244
298.35	7.16	1.096	1.197
313.45	6.38	0.956	1.097

Smoothed Data: $\Delta G^O/J\ mol^{-1} = -\ RT\ \ln X_1 = -5469.3 + 59.454\ T$

                Std. Dev. $\Delta G^O = 14.6$, Coef. Corr. = .9998

                $\Delta H^O/J\ mol^{-1} = -5469.3$, $\Delta S^O/J\ K^{-1}\ mol^{-1} = -59.454$

T/K	Mol Fraction $X_1 \times 10^3$	$\Delta G^O/J\ mol^{-1}$
288.15	7.69	11662
293.15	7.40	11960
298.15	7.12	12257
303.15	6.87	12554
308.15	6.63	12851
313.15	6.41	13149
318.15	6.20	13446

The solubility values were adjusted to a partial pressure of 101.325 kPa (1 atm) by Henry's law.

The Bunsen coefficients were calculated by the compiler.

AUXILIARY INFORMATION

METHOD:	SOURCE AND PURITY OF MATERIALS:
Volumetric. The solvent is satu- rated with gas as it flows through an 8 mm x 180 cm glass spiral attached to a gas buret. The total pressure of solute gas plus solvent vapor is main- tained at 1 atm as the gas is absorbed. ADDED NOTE: Makranczy, J.; Megyery- Balog, K.; Rusz,L.; Patyi,L. Hung. J. Ind. Chem. 1976, 4, 269 report an Ostwald coefficient of 1.229 at 298.15 K for this system. The value was not used in the smoothed data fit above.	1. Krypton. Linde Air Products Co. Pure grade. 2. Heptane. Phillips Petroleum Co. Used as received.

APPARATUS/PROCEDURE:	

The apparatus is a modification of that of Morrison and Billett (1). The modifications include the addition of a spiral storage for the solvent, a manometer for a constant reference pressure, and an extra buret for highly soluble gases. The solvent is degassed by a modification of the method of Baldwin and Daniel (2).	ESTIMATED ERROR:
	         $\delta T/K = 0.05$
	       $\delta P/mmHg = 3$
	       $\delta X_1/X_1 = 0.03$
	
	 REFERENCES:
	
	 1. Morrison, T.J.; Billett, F. J. Chem. Soc. 1948, 2033; ibid. 1952, 3819.
	
	 2. Baldwin, R.R.; Daniel, S.G. J. Appl. Chem. 1952, 2, 161.

COMPONENTS:	ORIGINAL MEASUREMENTS:
1. Krypton; Kr; 7439-90-9 2. 3-Methylheptane; C_8H_{18}; 589-81-1	Clever, H.L.; Battino, R.; Saylor, J.H.; Gross, P.M. J. Phys. Chem. 1957, 61, 1078-1083
VARIABLES: T/K: 289.15 - 313.75 Total P/kPa: 101.325 (1 atm)	PREPARED BY: P.L. Long

EXPERIMENTAL VALUES:

T/K	Mol Fraction X_1 x 10^3	Bunsen Coefficient α	Ostwald Coefficient L
289.15	7.40	1.038	1.099
298.40	7.18	0.995	1.087
313.75	6.55	0.891	1.024

Smoothed Data: ΔG^O/J mol^{-1} = $- RT \ln X_1$ = -3844.8 + 54.026 T

Std. Dev. ΔG^O = 25.6, Coef. Corr. = 0.9993

ΔH^O/J mol^{-1} = -3844.8, ΔS^O/J K^{-1} mol^{-1} = -54.026

T/K	Mol Fraction X_1 x 10^3	ΔG^O/J mol^{-1}
288.15	7.50	11723
293.15	7.30	11993
298.15	7.11	12263
303.15	6.93	12533
308.15	6.76	12803
313.15	6.60	13073
318.15	6.44	13344

The solubility values were adjusted to a partial pressure of 101.325 kPa (1 atm) by Henry's law.

The Bunsen coefficients were calculated by the compiler.

AUXILIARY INFORMATION

METHOD: Volumetric. The solvent is saturated with gas as it flows through an 8 mm x 180 cm glass spiral attached to a gas buret. The total pressure of solute gas plus solvent vapor is maintained at 1 atm as the gas is absorbed.	SOURCE AND PURITY OF MATERIALS: 1. Krypton. Linde Air Products Co. Pure grade. 2. 3-Methylheptane. Humphrey-Wilkinson, Inc. Shaken with H_2SO_4, washed, dried over Na, distilled through a vacuum column.
APPARATUS/PROCEDURE: The apparatus is a modification of that of Morrison and Billett (1). The modifications include the addition of a spiral storage for the solvent, a manometer for a constant reference pressure, and an extra buret for highly soluble gases. The solvent is degassed by a modification of the method of Baldwin and Daniel (2).	ESTIMATED ERROR: $\delta T/K$ = 0.05 δP/mmHg = 3 $\delta X_1/X_1$ = 0.03 REFERENCES: 1. Morrison, T.J.; Billett, F. J. Chem. Soc. 1948, 2033; ibid. 1952, 3819. 2. Baldwin, R.R.; Daniel, S.G. J. Appl. Chem. 1952, 2, 161.

COMPONENTS:	ORIGINAL MEASUREMENTS:

1. Krypton; Kr; 7439-90-9
 	
 2. 2, 3-Dimethylhexane; C_8H_{18};
     584-94-1
	Clever, H.L.; Battino, R.;
	     Saylor, J.H.; Gross, P.M.
	
	 J. Phys. Chem. 1957, 61, 1078-1083

| VARIABLES:
       T/K: 289.15 - 313.65
 Total P/kPa: 101.325 (1 atm) | PREPARED BY:

         P.L. Long |

EXPERIMENTAL VALUES:

T/K	Mol Fraction X_1 x 10^3	Bunsen Coefficient α	Ostwald Coefficient L
289.15	7.46	1.056	1.118
297.95	7.04	0.987	1.077
313.65	6.54	0.899	1.032

Smoothed Data: $\Delta G^O/J\ mol^{-1} = - RT \ln X_1 = -3998.6 + 54.583\ T$

                Std. Dev. ΔG^O = 11.5, Coef. Corr. = 0.9999

                $\Delta H^O/J\ mol^{-1} = -3998.6$, $\Delta S^O/J\ K^{-1}\ mol^{-1} = -54.583$

T/K	Mol Fraction X_1 x 10^3	$\Delta G^O/J\ mol^{-1}$
288.15	7.48	11730
293.15	7.27	12002
298.15	7.07	12275
303.15	6.88	12548
308.15	6.71	12821
313.15	6.54	13094
318.15	6.39	13367

The solubility values were adjusted to a partial pressure of 101.325 kPa (1 atm) by Henry's law.

The Bunsen coefficients were calculated by the compiler.

AUXILIARY INFORMATION

METHOD:	SOURCE AND PURITY OF MATERIALS:
Volumetric. The solvent is satu-rated with gas as it flows through an 8 mm x 180 cm glass spiral attached to a gas buret. The total pressure of solute gas plus solvent vapor is main-tained at 1 atm as the gas is absorbed.	1. Krypton. Linde Air Products Co. Pure grade. 2. 2,3 - Dimethylhexane. Humphrey-Wilkinson, Inc. Shaken w/H_2SO_4, washed, dried over Na, distilled thru a vacuum column.

APPARATUS/PROCEDURE:	ESTIMATED ERROR:
	         $\delta T/K = 0.05$
	       $\delta P/mmHg = 3$
	       $\delta X_1/X_1 = 0.03$

The apparatus is a modification of that of Morrison and Billett (1). The modifications include the addition of a spiral storage for the solvent, a manometer for a constant reference pressure, and an extra buret for highly soluble gases. The solvent is degassed by a modification of the method of Baldwin and Daniel (2).	REFERENCES:
	 1. Morrison, T.J.; Billett, F.
	     J. Chem. Soc. 1948, 2033;
	     ibid. 1952, 3819.
	
	 2. Baldwin, R.R.; Daniel, S.G.
	     J. Appl. Chem. 1952, 2, 161.

COMPONENTS:	ORIGINAL MEASUREMENTS:
1. Krypton; Kr; 7439-90-9 2. 2, 4-Dimethylhexane; C_8H_{18}; 589-43-5	Clever, H.L.; Battino, R.; Saylor, J.H.; Gross, P.M. J. Phys. Chem. 1957, 61, 1078-1083.
VARIABLES: T/K: 289.25 - 313.75 Total P/kPa: 101.325 (1 atm)	PREPARED BY: P.L. Long

EXPERIMENTAL VALUES:

T/K	Mol Fraction X_1 x 10^3	Bunsen Coefficient α	Ostwald Coefficient L
289.25	7.72	1.074	1.137
298.15	7.33	1.010	1.102
313.75	6.79	0.916	1.052

Smoothed Data: ΔG^O/J mol^{-1} = - RT ln X_1 = -3936.0 + 54.058 T

Std. Dev. ΔG^O = 3.9, Coef. Corr. = .9999

ΔH^O/J mol^{-1} = -3936.0, ΔS^O/J K^{-1} mol^{-1} = -54.058

T/K	Mol Fraction X_1 x 10^3	ΔG^O/J mol^{-1}
288.15	7.76	11641
293.15	7.54	11911
298.15	7.34	12181
303.15	7.15	12452
308.15	6.97	12722
313.15	6.81	12992
318.15	6.65	13262

The solubility values were adjusted to a partial pressure of 101.325 kPa (1 atm) by Henry's law.

The Bunsen coefficients were calculated by the compiler.

AUXILIARY INFORMATION

METHOD:
   Volumetric. The solvent is saturated with gas as it flows through an 8 mm x 180 cm glass spiral attached to a gas buret. The total pressure of solute gas plus solvent vapor is maintained at 1 atm as the gas is absorbed.

SOURCE AND PURITY OF MATERIALS:
1. Krypton. Linde Air Products Co. Pure grade.

2. 2, 4-Dimethylhexane. Humphrey-Wilkinson, Inc. Shaken w/H_2SO_4, washed, dried over Na, distilled thru a vacuum column.

APPARATUS/PROCEDURE:
   The apparatus is a modification of that of Morrison and Billett (1). The modifications include the addition of a spiral storage for the solvent, a manometer for a constant reference pressure, and an extra buret for highly soluble gases. The solvent is degassed by a modification of the method of Baldwin and Daniel (2).

ESTIMATED ERROR:
       $\delta T/K$ = 0.05
     δP/mmHg = 3
       $\delta X_1/X_1$ = 0.03

REFERENCES:
1. Morrison, T.J.; Billett, F. J. Chem. Soc. 1948, 2033; Ibid. 1952, 3819.

2. Baldwin, R.R.; Daniel, S.G. J. Appl. Chem. 1952, 2, 161.

COMPONENTS:	ORIGINAL MEASUREMENTS:

1.   Krypton; Kr; 7439-90-9
 	
2.   2, 2, 4-Trimethylpentane; C_8H_{18};
      540-84-1
	Clever, H.L.; Battino, R.;
	      Saylor, J.H.; Gross, P.M.
	
	$\underline{J}$. $\underline{Phys}$. $\underline{Chem}$. 1957, $\underline{61}$, 1078-1083.

VARIABLES:	PREPARED BY:
T/K: 289.15 - 313.60 Total P/kPa: 101.325 (1 atm)	P.L. Long

EXPERIMENTAL VALUES:

T/K	Mol Fraction X_1 x 10^3	Bunsen Coefficient α	Ostwald Coefficient L
289.15	8.49	1.171	1.24
298.25	7.88	1.072	1.171
313.60	7.12	0.950	1.091

Smoothed Data: ΔG^O/J mol^{-1} = - RT ln X_1 = -5392.8 + 58.320 T

Std. Dev. ΔG^O = 8.1, Coef. Corr. = .9999

ΔH^O/J mol^{-1} = -5392.8, ΔS^O/J K^{-1} mol^{-1} = -58.320

T/K	Mol Fraction X_1 x 10^3	ΔG^O/J mol^{-1}
288.15	8.54	11412
293.15	8.21	11704
298.15	7.92	11995
303.15	7.64	12287
308.15	7.38	12579
313.15	7.13	12870
318.15	6.90	13162

The solubility values were adjusted to a partial pressure of 101.325 kPa (1 atm) by Henry's law.

The Bunsen coefficients were calculated by the compiler.

AUXILIARY INFORMATION

METHOD:

   Volumetric. The solvent is saturated with gas as it flows through an 8 mm x 180 cm glass spiral attached to a gas buret. The total pressure of solute gas plus solvent vapor is maintained at 1 atm as the gas is absorbed.

SOURCE AND PURITY OF MATERIALS:

1.   Krypton. Linde Air Products Co. Pure grade.

2.   2, 2, 4-Trimethylpentane. Enjay Co. Used as received.

APPARATUS/PROCEDURE:

   The apparatus is a modification of that of Morrison and Billett (1). The modifications include the addition of a spiral storage for the solvent, a manometer for a constant reference pressure, and an extra buret for highly soluble gases. The solvent is degassed by a modification of the method of Baldwin and Daniel (2).

ESTIMATED ERROR:
δT/K = 0.05
δP/mmHg = 3
$\delta X_1/X_1$ = 0.03

REFERENCES:
1.   Morrison, T.J.; Billett, F. J. Chem. Soc. 1948, 2033; ibid. 1952, 3819.

2.   Baldwin, R.R.; Daniel, S.G. J. Appl. Chem. 1952, 2, 161.

COMPONENTS:	EVALUATOR:
1. Krypton; Kr; 7439-90-9	H. L. Clever
2. Octane; C_8H_{18}; 111-65-9	Chemistry Department
Decane; $C_{10}H_{22}$; 124-18-5	Emory University
	Atlanta, GA 30322
	U. S. A.
	October 1978

CRITICAL EVALUATION:

Solubilities in the octane + krypton and decane + krypton systems were reported from three laboratories. Clever, Battino, Saylor and Gross (1) reported solubility values at three temperatures between 289 and 313 K. Both Makranczy, Megyery-Balog, Rusz and Patyi (2) and Wilcock, Battino, Danforth and Wilhelm (3) report one value at or very near 298.15 K.

Octane + Krypton

The solubility values of krypton in octane from the three laboratories at 298.15 K shows a range of values of 3.8 per cent. The value of Wilcock, et al. (3), determined with improved apparatus for degassing and solubility measurement, should be the most reliable value. However, since they report a solubility at only one temperature, we recommend as the tentative values the data of Clever, et al. (2) (Page 36) with the reservation that the smoothed value at 298.15 K may be about one percent high.

Decane + Krypton

The solubility values for krypton in decane at 298.15 K from the three laboratories shows a 4.3 per cent range. The average value from the three laboratories agrees within 0.2 per cent of the value of Clever, et al. (2). We recommend as the tentative solubility values the smoothed data of Clever, et al. (2) (Page 40). The single value of Wilcock, et al. (3) at 298.16 K is 2 per cent higher and may be the most reliable experimental value.

REFERENCES

1. Clever, H. L.; Battino, R.; Saylor, J. H.; Gross, P. M. J. Phys. Chem. 1957, 61, 1078.
2. Makranczy, J.; Megyery-Balog, K.; Rusz, L.; Patyi, L. Hung. J. Ind. Chem. 1976, 4, 269.
3. Wilcock, R. J.; Battino, R.; Danforth, W. F; Wilhelm, E. J. Chem. Thermodyn. 1978, 10, 817.

COMPONENTS:	ORIGINAL MEASUREMENTS:
1. Krypton; Kr; 7439-90-9 2. Octane; C_8H_{18}; 111-65-9	Clever, H.L.; Battino, R.; Saylor, J.H.; Gross, P.M. J. Phys. Chem. 1957, 61, 1078-1083.
VARIABLES: T/K: 289.35 - 313.45 P/kPa: 101.325 (1 atm)	PREPARED BY: P.L. Long

EXPERIMENTAL VALUES:

T/K	Mol Fraction $X_1 \times 10^3$	Bunsen Coefficient α	Ostwald Coefficient L
289.35	7.52	1.05	1.11
298.35	7.06	0.974	1.064
313.45	6.40	0.868	0.996

Smoothed Data: $\Delta G^o / J\ mol^{-1} = - RT \ln X_1 = -5046.9 + 58.100\ T$

Std. Dev. ΔG^o = 0.2, Coef. Corr. = .9999

$\Delta H^o / J\ mol^{-1} = -5046.9$, $\Delta S^o / J\ K^{-1}\ mol^{-1} = -58.100$

T/K	Mol Fraction $X_1 \times 10^3$	$\Delta G^o / J\ mol^{-1}$
288.15	7.59	11695
293.15	7.32	11985
298.15	7.07	12276
303.15	6.84	12566
308.15	6.62	12857
313.15	6.41	13147
318.15	6.22	13438

Solubility values were adjusted to a partial pressure of krypton of 101.325 kPa (1 atm).

The Bunsen coefficients were calculated by the compiler.

AUXILIARY INFORMATION

METHOD/APPARATUS/PROCEDURE:	SOURCE AND PURITY OF MATERIALS:
Solvent was degassed by a modification of the method of Baldwin and Daniel (1). Saturation apparatus was that of Morrison and Billett (2), modified to include spiral storage for the solvent, a manometer for constant reference pressure, and an extra buret. Solvent was saturated with gas as it flowed through an 8 mm x 180 cm glass spiral attached to a buret.	1. Krypton. Linde Air Products Co. Pure. 2. Octane. Humphrey-Wilkinson, Inc. Shaken w/H_2SO_4, washed, dried over Na, distilled.

ESTIMATED ERROR:
$$\delta T/K = 0.05$$
$$\delta P/mmHg = 3$$
$$\delta X_1/X_1 = 0.03$$

REFERENCES:

1. Baldwin, R.R.; Daniel, S.G. J. Appl. Chem. 1952, 2, 161.

2. Morrison, T.J.; Billett, F. J. Chem. Soc. 1948, 2033; Ibid. 1952, 3819.

COMPONENTS:	ORIGINAL MEASUREMENTS:
1. Krypton; Kr; 7439-90-9 2. Octane; C_8H_{18}; 111-65-9	Makranczy, J.; Megyery-Balog, K.; Rusz, L.; Patyi, L. Hung. J. Ind. Chem. 1976, 4, 269-280.

VARIABLES:	PREPARED BY:
T/K: 298.15 P/kPa: 101.325 (1 atm)	S.A. Johnson

EXPERIMENTAL VALUES:

T/K	Mol Fraction X_1 x 10^3	Bunsen Coefficient α	Ostwald Coefficient L
298.15	7.27	1.003	1.095

The mole fraction and Bunsen coefficient were calculated by the compiler.

See the evaluation of the krypton + octane system on page 35 for the recommended Gibbs energy equation and the smoothed solubility values.

AUXILIARY INFORMATION

METHOD /APPARATUS/PROCEDURE:	SOURCE AND PURITY OF MATERIALS:
Volumetric method. The apparatus of Bodor, Bor, Mohai, and Sipos (1) was used.	Both the gas and liquid were analytical grade reagents of Hungarian or foreign origin. No further information.

APPARATUS/PROCEDURE:	ESTIMATED ERROR: $\delta X_1/X_1 = 0.03$
	REFERENCES: 1. Bodor, E.; Bor, Gy.; Mohai, B.; Sipos, G. Veszpremi Vegyip. Egy. Kozl. 1957, 1, 55. Chem. Abstr. 1961, 55, 3175h.

COMPONENTS:	ORIGINAL MEASUREMENTS:
1. Krypton; Kr; 7439-90-9 2. Octane; C_8H_{18}; 111-65-9	Wilcock, R.J.;Battino, R.; Danforth, W.F. J.Chem.Thermodyn. 1978, <u>10</u>, 817 - 822.

VARIABLES:
     T/K: 298.18
     P/kPa: 101.325 (1 atm)
	PREPARED BY:
	
	     A.L. Cramer

EXPERIMENTAL VALUES:

T/K	Mol Fraction X_1 x 10^3	Bunsen Coefficient α	Ostwald Coefficient L
298.18	7.001	0.9646	1.053

A preliminary account of this work appeared in <u>Conf</u>. <u>Int</u>. <u>Thermodyn</u>. <u>Chim</u>. (C.R.) 4th 1975, <u>6</u>, 122-128.

The solubility value was adjusted to a partial pressure of krypton of 101.325 kPa (1 atm) by Henry's law.

The Bunsen coefficient was calculated by the compiler.

See the evaluation of krypton + octane for the recommended Gibbs energy equation and smoothed solubility values.

AUXILIARY INFORMATION

METHOD /APPARATUS/PROCEDURE:

The solubility apparatus is based on the design of Morrison and Billett (1) and the version used is described by Battino, Evans, and Danforth (2). The degassing apparatus is that described by Battino, Banzhof, Bogan, and Wilhelm (3).

Degassing. Up to 500 cm^3 of solvent is placed in a flask of such size that the liquid is about 4 cm deep. The liquid is rapidly stirred, and vacuum is applied intermittently through a liquid N_2 trap until the permanent gas residual pressure drops to 5 microns.

Solubility Determination. The degassed solvent is passed in a thin film down a glass spiral tube containing solute gas plus the solvent vapor at a total pressure of one atm. The volume of gas absorbed is found by difference between the initial and final volumes in the buret system. The solvent is collected in a tared flask and weighed.

SOURCE AND PURITY OF MATERIALS:

1. Krypton. The Matheson Co., Inc., or Air Products and Chemicals. Purest commercially available.

2. Octane. Phillips Petroleum Co. 99 mol per cent minimum.

ESTIMATED ERROR:
$$\delta T/K = 0.03$$
$$\delta P/mmHg = 0.5$$
$$\delta X_1/X_1 = 0.005$$

REFERENCES:
1. Morrison,T.J.;Billett,F. J. Chem. Soc. 1948, 2033.
2. Battino,R.;Evans,F.D.;Danforth,W.F. J.Am.Oil Chem.Soc. 1968, <u>45</u>, 830.
3. Battino,R.;Banzhof,M.;Bogan, M.; Wilhelm, E. Anal. Chem. 1971, <u>43</u>, 806.

COMPONENTS:	ORIGINAL MEASUREMENTS:
1. Krypton; Kr; 7439-90-9 2. Nonane; C_9H_{20}; 111-84-2	Clever, H.L.; Battino, R.; Saylor, J.H.; Gross, P.M. J. Phys. Chem. 1957, 61, 1078-1083.

| VARIABLES:
      T/K: 289.50 - 313.50
 Total P/kPa: 101.325 (1 atm) | PREPARED BY:

        P.L. Long |

EXPERIMENTAL VALUES:

T/K	Mol Fraction $X_1 \times 10^3$	Bunsen Coefficient α	Ostwald Coefficient L
289.50	7.55	0.960	1.017
297.95	7.08	0.890	0.971
313.50	6.52	0.804	0.923

Smoothed Data: $\Delta G^{o}/J\ mol^{-1} = - RT \ln X_1 = -4544.4 + 56.358\ T$

         Std. Dev. $\Delta G^{o} = 14.0$, Coef. Corr. $= 0.9998$

      $\Delta H^{o}/J\ mol^{-1} = -4544.4$, $\Delta S^{o}/J\ K^{-1}\ mol^{-1} = -56.358$

T/K	Mol Fraction $X_1 \times 10^3$	$\Delta G^{o}/J\ mol^{-1}$
288.15	7.58	11695
293.15	7.34	11977
298.15	7.12	12259
303.15	6.90	12541
308.15	6.71	12822
313.15	6.52	13104
318.15	6.34	13386

The solubility values were adjusted to a partial pressure of 101.325 kPa (1 atm) by Henry's law.

The Bunsen coefficients were calculated by the compiler.

AUXILIARY INFORMATION

METHOD:
   Volumetric. The solvent is saturated with gas as it flows through an 8 mm x 180 cm glass spiral attached to a gas buret. The total pressure of solute gas plus solvent vapor is maintained at 1 atm as the gas is absorbed.

ADDED NOTE: Makranczy, J.; Megyery-Balog, K.; Rusz,L.; Patyi, L. Hung. J. Ind. Chem. 1976, 4, 269 report an Ostwald coefficient of 0.988 at 298.15 K for this system. The value was not used in the smoothed data fit above.

SOURCE AND PURITY OF MATERIALS:
1. Krypton. Linde Air Products Co. Pure grade.

2. Nonane. Phillips Petroleum Co. Used as received.

APPARATUS/PROCEDURE:
   The apparatus is a modification of that of Morrison and Billett (1). The modifications include the addition of a spiral storage for the solvent, a manometer for a constant reference pressure, and an extra buret for highly soluble gases. The solvent is degassed by a modification of the method of Baldwin and Daniel (2).

ESTIMATED ERROR:
       $\delta T/K = 0.05$
     $\delta P/mmHg = 3$
     $\delta X_1/X_1 = 0.03$

REFERENCES:
1. Morrison, T.J.; Billett, F. J. Chem. Soc. 1948, 2033; Ibid. 1952, 3819.

2. Baldwin, R.R.; Daniel, S.G. J. Appl. Chem. 1952, 2, 161.

COMPONENTS:	ORIGINAL MEASUREMENTS:
1. Krypton; Kr; 7439-90-9 2. Decane. $C_{10}H_{22}$; 124-18-5	Clever, H.L.; Battino, R.; Saylor, J.H.; Gross, P.M. J. Phys. Chem. 1957, 61, 1078-1083.

VARIABLES:
      T/K: 289.75 - 313.35
      P/kPa: 101.325 (1 atm)
	PREPARED BY:
	
	      P.L. Long

EXPERIMENTAL VALUES:

T/K	Mol Fraction $X_1 \times 10^3$	Bunsen Coefficient α	Ostwald Coefficient L
289.75	7.59	0.883	0.937
298.15	7.22	0.832	0.908
313.35	6.53	0.740	0.849

Smoothed Data:

$$\Delta G^O/J\ mol^{-1} = -RT \ln X_1 = -4853.9 + 57.311\ T$$

Std. Dev. ΔG^O = 8.9, Coef. Corr. = 0.9999

$$\Delta H^O/J\ mol^{-1} = -4853.9, \quad \Delta S^O/J\ K^{-1}\ mol^{-1} = -57.311$$

T/K	Mol Fraction $X_1 \times 10^3$	$\Delta G^O/J\ mol^{-1}$
288.15	7.70	11660
293.15	7.43	11947
298.15	7.19	12233
303.15	6.96	12520
308.15	6.75	12806
313.15	6.55	13093
318.15	6.36	13380

The solubility values were adjusted to a partial pressure of krypton of 101.325 kPa (1 atm) by Henry's law.

The Bunsen coefficients were calculated by the compiler.

AUXILIARY INFORMATION

METHOD /APPARATUS/PROCEDURE:

     Solvent was degassed by a modification of the method of Baldwin and Daniel (1). Saturation apparatus was that of Morrison and Billett (2), modified to include spiral storage for the solvent, a manometer for constant reference pressure, and an extra buret. Solvent was saturated with gas as it flowed through an 8 mm x 180 cm glass spiral attached to a buret.

For an evaluation of the krypton + decane system see page 35.

SOURCE AND PURITY OF MATERIALS:

1. Krypton. Linde Air Products Co. Pure.

2. Decane. Humphrey-Wilkinson, Inc. Shaken with H_2SO_4, washed, dried over Na.

ESTIMATED ERROR:
$$\delta T/K = 0.05$$
$$\delta P/mmHg = 3$$
$$\delta X_1/X_1 = 0.03$$

REFERENCES:

1. Baldwin, R.R.; Daniel, S.G. J. Appl. Chem. 1952, 2, 161.

2. Morrison, T.J.; Billett, F. J. Chem. Soc. 1948, 2033; ibid. 1952, 3819.

COMPONENTS:	ORIGINAL MEASUREMENTS:
1. Krypton; Kr; 7439-90-9 2. Decane; $C_{10}H_{22}$; 124-18-5	Makranczy, J.; Megyery-Balog, K.; Rusz, L.; Patyi, L. Hung. J. Ind. Chem. 1976, 4, 269-280.

VARIABLES:
    T/K: 298.15
    P/kPa: 101.325 (1 atm)
	PREPARED BY:
	
	    S.A. Johnson

EXPERIMENTAL VALUES:

T/K	Mol Fraction X_1 x 10^3	Bunsen Coefficient α	Ostwald Coefficient L
298.15	7.05	0.812	0.886

The mole fraction and Bunsen coefficient were calculated by the compiler.

See the evaluation of the krypton + decane system on page 35 for the recommended Gibbs energy equation and smoothed solubility values.

AUXILIARY INFORMATION

METHOD /APPARATUS/PROCEDURE:

Volumetric method. The apparatus of Bodor, Bor, Mohai, and Sipos (1) was used.

SOURCE AND PURITY OF MATERIALS:

Both the gas and liquid were analytical grade reagents of Hungarian or foreign origin. No further information.

ESTIMATED ERROR:

$$\delta X_1/X_1 = 0.03$$

REFERENCES:

1. Bodor, E.; Bor, Gy.; Mohai, B.; Sipos, G.
Veszpremi Vegyip. Egy. Kozl. 1957, 1, 55.
Chem. Abstr. 1961, 55, 3175h.

COMPONENTS:	ORIGINAL MEASUREMENTS:
1. Krypton; Kr; 7439-90-9 2. Decane; $C_{10}H_{22}$; 124-18-5	Wilcock, R. J.; Battino, R.; Danforth, W. F. J. Chem. Thermodyn. 1978, 10, 817 - 822.

VARIABLES:	PREPARED BY:
T/K: 298.16 P/kPa: 101.325 (1 atm)	A. L. Cramer

EXPERIMENTAL VALUES:

T/K	Mol Fraction $X_1 \times 10^3$	Bunsen Coefficient α	Ostwald Coefficient L
298.16	7.361	0.8467	0.9242

A preliminary account of this work appeared in Conf. Int. Thermodyn. Chim. (C.R.) 4th 1975, 6, 122-128.

The solubility value was adjusted to a partial pressure of 101.325 kPa (1 atm) of krypton by Henry's law.

The Bunsen coefficient was calculated by the compiler.

See the evaluation of krypton + decane for the recommended Gibbs energy equation and smoothed solubility values.

AUXILIARY INFORMATION

METHOD/APPARATUS/PROCEDURE:	SOURCE AND PURITY OF MATERIALS:
The solubility apparatus is based on the design of Morrison and Billett (1) and the version used is described by Battino, Evans, and Danforth (2). The degassing apparatus is that described by Battino, Banzhof, Bogan, and Wilhelm (3). See Krypton + n-Octane data sheet for more detail.	1. Krypton. The Matheson Co., Inc., or Air Products and Chemicals. Purest commercially available. 2. Decane. Phillips Petroleum Co. 99 mol per cent minimum.

ESTIMATED ERROR:
$$\delta T/K = 0.03$$
$$\delta P/mmHg = 0.5$$
$$\delta X_1/X_1 = 0.005$$

REFERENCES:
1. Morrison,T.J.; Billett, F.
   J. Chem. Soc. 1948, 2033.
2. Battino,R.;Evans,F.D.;Danforth,W.F.
   J.Am.Oil Chem.Soc. 1968, 45, 830.
3. Battino, R.; Banzhof, M.; Bogan, M.;
   Wilhelm, E.
   Anal. Chem. 1971, 43, 806.

COMPONENTS:	ORIGINAL MEASUREMENTS:
1. Krypton; Kr; 7439-90-9 2. Undecane; $C_{11}H_{24}$; 1120-21-4	Makranczy, J.; Megyery-Balog, K.; Rusz, L.; Patyi, L. Hung. J. Ind. Chem. 1976, 4, 269-280.
VARIABLES: T/K: 298.15 P/kPa: 101.325 (1 atm)	PREPARED BY: S.A. Johnson

EXPERIMENTAL VALUES:

T/K	Mol Fraction x_1 x 10^3	Bunsen Coefficient α	Ostwald Coefficient L
298.15	7.10	0.756	0.825

The mole fraction and Bunsen coefficient were calculated by the compiler.

AUXILIARY INFORMATION

METHOD:

Volumetric method. The apparatus of Bodor, Bor, Mohai, and Sipos (1) was used.

SOURCE AND PURITY OF MATERIALS:

Both the gas and liquid were analytical grade reagents of Hungarian or foreign origin. No further information.

APPARATUS/PROCEDURE:

ESTIMATED ERROR:

$$\delta X_1/X_1 = 0.03$$

REFERENCES:
1. Bodor, E.; Bor, Gy.; Mohai, B.; Sipos, G.

Veszpremi Vegyip. Egy. Kozl. 1957, 1, 55.
Chem. Abstr. 1961, 55, 3175h.

COMPONENTS:	ORIGINAL MEASUREMENTS:
1. Krypton; Kr; 7439-90-9 2. Dodecane; $C_{12}H_{26}$; 112-40-3	Clever, H.L.: Battino, R.; Saylor, J.H.; Gross, P.M. J. Phys. Chem. 1957, 61, 1082-1083.

VARIABLES:	PREPARED BY:
T/K: 289.25 - 313.65 P/kPa: 101.325 (1 atm)	P.L. Long

EXPERIMENTAL VALUES:

T/K	Mol Fraction X_1 x 10^3	Bunsen Coefficient α	Ostwald Coefficient L
289.25	7.92	0.788	0.834
298.35	7.65	0.754	0.824
313.65	6.91	0.672	0.772

Smoothed data: $\Delta G^O = -RT \ln x_1 = -4328.10 + 55.125\ T$

Std. Dev. $\Delta G^O = 26.6$, Coef. Corr. = 0.9992

ΔH^O/J mol^{-1} = -4328.1, ΔS^O/J K^{-1} mol^{-1} = -55.125

T/K	Mole Fraction X_1 x 10^3	ΔG^O/J mol^{-1}
288.15	8.04	11,556
293.15	7.79	11,832
298.15	7.57	12,107
303.15	7.35	12,383
313.15	6.96	12,934
318.15	6.78	13,210

Steinberg and Manowitz, and Makranczy,Megyery-Balog, Rusz, and Patyi each report a krypton in dodecane solubility value at 298.15 K (see next two pages). The tentative recommendation is to use the data above.

The Bunsen coefficients were calculated by the compiler. Solubility values were corrected to a krypton partial pressure of 101.325 kPa by Henry's law.

AUXILIARY INFORMATION

METHOD:	SOURCE AND PURITY OF MATERIALS:
Volumetric. The solvent is saturated with gas as it flows through an 8 mm x 180 cm glass helix attached to a gas buret. The total pressure of solute gas plus solvent vapor is maintained at 1 atm as the gas is absorbed.	1. Krypton. Linde Air Products Co. 2. Dodecane. Humphrey-Wilkinson, Inc. Shaken with conc. H_2SO_4, water washed, dried over Na.

APPARATUS/PROCEDURE:	ESTIMATED ERROR:
The apparatus is a modification of the apparatus of Morrison and Billett (1). The modifications include the addition of a helical storage for the solvent, a manometer for a reference pressure, and an extra buret for highly soluble gases. The solvent is degassed by a modification of the method of Baldwin and Daniel (2).	$\delta T/K = 0.05$ δP/mmHg = 3 $\delta X_1/X_1 = 0.03$
	REFERENCES: 1. Morrison, T.J.; Billett, F. J. Chem. Soc. 1948, 2033; Ibid. 1952, 3819. 2. Baldwin, R.R.; Daniel, S.G. J. Appl. Chem. 1952, 2, 161.

COMPONENTS:	ORIGINAL MEASUREMENTS:
1. Krypton; Kr; 7439-90-9 2. Dodecane; $C_{12}H_{26}$; 112-40-3	Steinberg, M.; Manowitz, B. Ind. Eng. Chem. 1959, 51, 47-51.

VARIABLES: T/K: 298.15 P/kPa: 101.325 (1 atm)	PREPARED BY: H.L. Clever A.L. Cramer

EXPERIMENTAL VALUES:

T/K	Mol Fraction $X_1 \times 10^3$	Bunsen Coefficient α	Absorption Coefficient β
298.15	7.8	0.78	0.82

The authors define the Absorption coefficient as the volume of gas, corrected to 288.15 K and 101.325 kPa, absorbed under a total system pressure of 101.325 kPa per unit volume of solvent at 288.15 K.

The mole fraction solubilities and Bunsen coefficients were calculated by the compiler.

For an evaluation note on the krypton + dodecane system see page 44.

AUXILIARY INFORMATION

METHOD /APPARATUS/PROCEDURE:	SOURCE AND PURITY OF MATERIALS:
Absorption coefficient determined by a modified McDaniel method (1).	1. Krypton. Technical grade. The Matheson Co. 2. Dodecane. Technically or chemically pure.

	ESTIMATED ERROR: $\delta\beta/\beta = 0.05 - 0.10$

	REFERENCES: 1. Furman, N.H. "Scott's Standard Methods of Chemical Analysis" Van Nostrand Co., NY, 1939, 5th ed., Vol. II, p. 2587.

COMPONENTS:	ORIGINAL MEASUREMENTS:
1. Krypton; Kr; 7439-90-9 2. Dodecane; $C_{12}H_{26}$; 112-40-3	Makranczy, J.; Megyery-Balog, K.; Rusz, L.; Patyi, L. Hung. J. Ind. Chem. 1976, 4, 269-280.

VARIABLES:
    T/K: 298.15
    P/kPa: 101.325 (1 atm)
	PREPARED BY:
	    S.A. Johnson

EXPERIMENTAL VALUES:

T/K	Mol Fraction X_1 x 10^3	Bunsen Coefficient α	Ostwald Coefficient L
298.15	7.01	0.693	0.756

The mole fraction and Bunsen coefficient were calculated by the compiler.

For an evaluation note on the krypton + dodecane system see page 44.

AUXILIARY INFORMATION

METHOD/APPARATUS/PROCEDURE:	SOURCE AND PURITY OF MATERIALS:
Volumetric method. The apparatus of Bodor, Bor, Mohai, and Sipos (1) was used.	Both the gas and liquid were analytical grade reagents of Hungarian or foreign origin. No further information.
	ESTIMATED ERROR: $\delta X_1/X_1 = 0.03$
	REFERENCES: 1. Bodor, E.; Bor, Gy.; Mohai, B.; Sipos, G. Veszpremi Vegyip. Egy. Kozl. 1957, 1, 55. Chem. Abstr. 1961, 55, 3175h.

COMPONENTS:	ORIGINAL MEASUREMENTS:
1. Krypton; Kr; 7439-90-9 2. Tridecane; $C_{13}H_{28}$; 629-50-5	Makranczy, J.; Megyery-Balog, K.; Rusz, L.; Patyi, L. Hung. J. Ind. Chem. 1976, 4, 269-280.

VARIABLES:
    T/K: 298.15
    P/kPa: 101.325 (1 atm)
	PREPARED BY:
	
	  S.A. Johnson

EXPERIMENTAL VALUES:

T/K	Mol Fraction $x_1 \times 10^3$	Bunsen Coefficient α	Ostwald Coefficient L
298.15	6.96	0.641	0.700

The mole fraction and Bunsen coefficient were calculated by the compiler.

AUXILIARY INFORMATION

METHOD:	SOURCE AND PURITY OF MATERIALS:
Volumetric method. The apparatus of Bodor, Bor, Mohai, and Sipos (1) was used.	Both the gas and liquid were analytical grade reagents of Hungarian or foreign origin. No further information.
APPARATUS/PROCEDURE:	ESTIMATED ERROR: $\delta X_1/X_1 = 0.03$

	REFERENCES:
	
	1. Bodor, E.; Bor, Gy.; Mohai, B.;
	    Sipos, G.
	  Veszpremi Vegyip. Egy. Kozl.
	  1957, 1, 55.
	  Chem. Abstr. 1961, 55, 3175

COMPONENTS:	ORIGINAL MEASUREMENTS:
1. Krypton; Kr; 7439-90-9 2. Tetradecane; $C_{14}H_{30}$; 629-59-4	Clever, H.L.; Battino, R.; Saylor, J.H.; Gross, P.M. J. Phys. Chem. 1957, 61, 1078-1083.

VARIABLES:
     T/K: 289.15 - 313.45
 Total P/kPa: 101.325 (1 atm)
	PREPARED BY:
	
	          P.L. Long

EXPERIMENTAL VALUES:

T/K	Mol Fraction X_1 x 10^3	Bunsen Coefficient α	Ostwald Coefficient L
289.15	8.34	0.747	0.791
298.15	7.81	0.693	0.756
313.45	7.03	0.614	0.705

Smoothed Data: ΔG^O/J mol^{-1} = - RT ln X_1 = -5304.3 + 58.139 T

            Std. Dev. ΔG^O = 1.2, Coef. Corr. = .9999

            ΔH^O/J mol^{-1} =-5304.3 ΔS^O/J K^{-1} mol^{-1} = -58.139

T/K	Mol Fraction X_1 x 10^3	ΔG^O/J mol^{-1}
288.15	8.41	11449
293.15	8.10	11739
298.15	7.81	12030
303.15	7.53	12321
308.15	7.28	12611
313.15	7.05	12902
318.15	6.82	13193

The solubility values were adjusted to a partial pressure of 101.325 kPa (1 atm) by Henry's law.

The Bunsen Coefficients were calculated by the compiler.

AUXILIARY INFORMATION

METHOD:

    Volumetric. The solvent is saturated with gas as it flows through an 8 mm x 180 cm glass spiral attached to a gas buret. The total pressure of solute gas plus solvent vapor is maintained at 1 atm as the gas is absorbed.
ADDED NOTE: Makranczy, J.; Megyery-Balog, K.; Rusz,L.; Patyi,L. Hung. J. Ind. Chem. 1976, 4, 269 report an Ostwald coefficient of 0.657 at 298.15 K for this system. The value was not used in the smoothed data fit above.

APPARATUS/PROCEDURE:

    The apparatus is a modification of that of Morrison and Billett (1). The modifications include the addition of a spiral storage for the solvent, a manometer for a constant reference pressure, and an extra buret for highly soluble gases. The solvent is degassed by a modification of the method of Baldwin and Daniel (2).

SOURCE AND PURITY OF MATERIALS:

1. Krypton. Linde Air Products Co. Pure grade.

2. Tetradecane. Humphrey-Wilkinson, Inc. Shaken w/H_2SO_4, washed, dried over Na.

ESTIMATED ERROR:
          $\delta T/K$ = 0.05
          δP/mmHg = 3
          $\delta X_1/X_1$ = 0.03

REFERENCES:
1. Morrison, T.J.; Billett, F. J. Chem. Soc. 1948, 2033; ibid. 1952, 3819.

2. Baldwin, R.R.; Daniel, S.G. J. Appl. Chem. 1952, 2, 161.

COMPONENTS:	ORIGINAL MEASUREMENTS:
1. Krypton; Kr; 7439-90-9 2. Pentadecane; $C_{15}H_{32}$; 629-62-9 or Hexadecane; $C_{16}H_{34}$; 544-76-3	Makranczy, J.; Megyery-Balog, K.; Rusz, L.; Patyi, L. Hung. J. Ind. Chem. 1976, 4, 269-280.

VARIABLES:	PREPARED BY:
T/K: 298.15 P/kPa: 101.325 (1 atm)	S.A.Johnson

EXPERIMENTAL VALUES:

T/K	Mol Fraction x_1 x 10^3	Bunsen Coefficient α	Ostwald Coefficient L
Pentadecane; $C_{15}H_{32}$; 629-62-9			
298.15	7.04	0.573	0.625
Hexadecane; $C_{16}H_{34}$; 544-76-3			
298.15	6.96	0.534	0.583

The mole fraction and Bunsen coefficient were calculated by the compiler.

AUXILIARY INFORMATION

METHOD:	SOURCE AND PURITY OF MATERIALS:
Volumetric method. The apparatus of Bodor, Bor, Mohai, and Sipos (1) was used.	Both the gas and liquid were analytical grade reagents of Hungarian or foreign origin. No further information.

APPARATUS/PROCEDURE:	ESTIMATED ERROR: $\delta X_1 / X_1 = 0.03$
	REFERENCES: 1. Bodor, E.; Bor, Gy.; Mohai, B.; Sipos, G. Veszpremi Vegyip. Egy. Kozl. 1957, 1, 55. Chem. Abstr. 1961, 55, 3175h.

50

COMPONENTS:	ORIGINAL MEASUREMENTS:
1. Krypton; Kr; 7439-90-9 2. Propene; C$_3$H$_6$; 74-98-6	Orobinsky, N. A., Blagoy, Yu. P. and Semyannikova, E. L., *Ukr. Fiz. Zhur.*, 1968, *13*, 372.
VARIABLES: Temperature, pressure	PREPARED BY: C. L. Young

EXPERIMENTAL VALUES:

T/K	P/bar	Mole fraction of krypton in liquid, x_{Kr}	T/K	P/bar	Mole fraction of krypton in liquid, x_{Kr}
130	1.10	0.291	150	4.95	0.669
	1.30	0.417		5.34	0.763
	1.47	0.530		5.83	0.854
	1.64	0.663		5.98	0.873
	1.81	0.784		6.28	0.958
	1.91	0.865	160	2.40	0.133
	1.96	0.925		2.57	0.147
140	1.11	0.121		4.30	0.302
	1.32	0.177		4.69	0.345
	1.89	0.355		5.42	0.411
	2.37	0.426		6.23	0.503
	2.55	0.506		6.96	0.578
	2.82	0.569		7.45	0.635
	3.06	0.679		8.92	0.820
	3.29	0.785		9.32	0.889
	3.33	0.785		9.66	0.935
	3.51	0.887	170	2.70	0.103
	3.73	0.948		5.52	0.278
150	1.91	0.172		7.13	0.387
	2.89	0.310		8.48	0.455
	3.46	0.378		9.12	0.474
	3.66	0.425		9.84	0.550
	4.10	0.515		10.59	0.588
	4.63	0.605		13.53	0.828
	4.83	0.667		14.22	0.904

AUXILIARY INFORMATION

METHOD/APPARATUS/PROCEDURE:	SOURCE AND PURITY OF MATERIALS:
Recirculating vapor flow apparatus fitted with magnetic pump. Temperature measured with platinum resistance thermometer; pressure measured with Bourdon gauge. Samples of liquid phase analysed by gas chromatography.	1. Sample purity 99.97 mole per cent. 2. Purified sample contained about 0.02 mole per cent nitrogen, oxygen and carbon dioxide and less than 0.5 mole per cent hydrocarbon impurity.
	ESTIMATED ERROR: $\delta T/K = \pm 0.03$; $\delta P/bar = \pm 0.4\%$; $\delta x_{Kr} = \pm 0.5\%$.
	REFERENCES:

COMPONENTS:

(1)   Krypton; Kr; 7439-90-9

(2)   Propene; C_3H_6; 74-98-6

ORIGINAL MEASUREMENTS:

Orobinsky, N. A., Blagoy, Yu. P.
and Semyannikova, E. L., *Ukr. Fiz.
Zhur.*, 1968, *13*, 372.

EXPERIMENTAL VALUES:

T/K	P/bar	Mole fraction of krypton in liquid, x_{Kr}	T/K	P/bar	Mole fraction of krypton in liquid, x_{Kr}
180	2.78	0.086	190	11.13	0.307
	3.66	0.110		14.42	0.403
	6.62	0.234		16.5	0.477
	9.12	0.355		17.3	0.510
	10.84	0.404		21.5	0.656
	12.61	0.495		22.8	0.724
	14.42	0.579		23.5	0.760
	14.57	0.590	200	4.90	0.097
	17.46	0.744		7.97	0.160
	17.7	0.769		12.32	0.255
	18.3	0.799		16.5	0.333
	19.3	0.865		17.8	0.351
190	2.78	0.063		23.3	0.503
	7.33	0.189		29.4	0.693

COMPONENTS:	EVALUATOR:
1. Krypton; Kr; 7439-90-9 2. Cyclohexane; C_6H_{12}; 110-82-7	H.L. Clever Chemistry Department Emory University Atlanta, GA 30322 U.S.A.

CRITICAL EVALUATION:
The solubility of krypton in cyclohexane at 101.325 kPa was measured in two laboratories. Clever, Battino, Saylor and Gross (1) report three values between 289.15 and 313.65 K. Dymond (2) reports four values between 293.50 and 309.35 K. Each data set was fitted to a Gibbs energy equation linear in temperature by the method of least squares. The smoothed data agreed within 1.7 per cent at 293.15 K and 0.5 per cent at 313.15 K, with the Dymond solubility values being the higher values over the temperature range. The two data sets agree well within experimental error.

The seven solubility values from the two laboratories were combined and fitted to a Gibbs energy equation linear in temperature by the method of least squares to obtain the recommended equation.

The recommended thermodynamic values for the transfer of krypton from the gas at 101.325 kPa (1 atm) to the hypothetical unit mole fraction solution are

$$\Delta G^O/J\ mol^{-1} = -RT\ ln\ X_1 = -3,576.0 + 56.557\ T$$
$$Std.\ Dev.\ \Delta G^O = 15,\ Coef.\ Corr. = 0.9995$$
$$\Delta H^O/J\ mol^{-1} = -3,576.0,\ \Delta S^O/J\ K^{-1}\ mol^{-1} = -56.557$$

The recommended values of the mole fraction solubility at 101.325 kPa and the Gibbs energy of solution as a function of temperature are given in Table 1.

TABLE 1. Solubility of krypton in cyclohexane. Recommended mole fraction solubility and Gibbs energy of solution as a function of temperature.

T/K	Mol Fraction $X_1 \times 10^3$	$\Delta G^O/J\ mol^{-1}$
288.15	4.94	12,721
293.15	4.82	13,004
298.15	4.70	13,286
303.15	4.59	13,569
308.15	4.49	13,852
313.15	4.39	14,135
318.15	4.29	14,418

1. Clever, H.L.; Battino, R.; Saylor, J.H.; Gross, P.M. J. Phys. Chem. 1957, 61, 1078.

2. Dymond, J.H. J. Phys. Chem. 1967, 71, 1829.

COMPONENTS:	ORIGINAL MEASUREMENTS:
1. Krypton; Kr; 7439-90-9 2. Cyclohexane; C_6H_{12}; 110-82-7	Clever, H.L.; Battino, R.; Saylor, J.H.; Gross, P.M. J. Phys. Chem. 1957, 61, 1078-1083.

VARIABLES:
     T/K: 289.15 - 313.65
     P/kPa: 101.325 (1 atm)
	PREPARED BY:
	
	     P.L. Long

EXPERIMENTAL VALUES:

T/K	Mol Fraction X_1 x 10^3	Bunsen Coefficient α	Ostwald Coefficient L
289.15	4.88	1.021	1.081
298.15	4.67	0.967	1.055
313.65	4.36	0.885	1.016

Smoothed Data: $\Delta G^O = -RT \ln x_1 = -3465.1 + 56.239\ T$

              Std. Dev. ΔG^O = 0.6, Coef. Corr. = .9999

For an evaluation of the krypton + cyclohexane system see page 53 with a recommendation of a Gibbs energy equation and solubility values.

The solubility values were adjusted to a partial pressure of krypton of 101.325 kPa (1 atm) by Henry's law. The Bunsen coefficients were calculated by the compiler.

AUXILIARY INFORMATION

METHOD /APPARATUS/PROCEDURE:	SOURCE AND PURITY OF MATERIALS:

Solvent was degassed by a modifi-cation of the method of Baldwin and Daniel (1). Saturation apparatus was that of Morrison and Billett (2), modified to include spiral storage for the solvent, a manometer for con-stant reference pressure, and an extra buret. Solvent was saturated with gas as it flowed through an 8mm x 180 cm glass spiral attached to a buret.	1. Krypton. Linde Air Products Co.
	    Pure.
	
	 2. Cyclohexane. Phillips Petroleum
	    Co. Used as received.

	ESTIMATED ERROR:
	           $\delta T/K$ = 0.05
	         $\delta P/mmHg$ = 3
	         $\delta X_1/X_1$ = 0.03

	REFERENCES:
	 1. Baldwin, R.R.; Daniel, S.G.
	    J. Appl. Chem. 1952, 2, 161.
	
	 2. Morrison, T.J.; Billett, F.
	    J. Chem. Soc. 1948, 2033;
	    ibid. 1952, 3819.

COMPONENTS:	ORIGINAL MEASUREMENTS:

COMPONENTS:

1.   Krypton; Kr; 7439-90-9

2.   Cyclohexane; C_6H_{12}; 110-82-7

ORIGINAL MEASUREMENTS:

Dymond, J. H.

J. Phys. Chem. 1967, 71, 1829 - 1831.

VARIABLES:

   T/K: 293.50 - 309.35
   P/kPa: 101.325 (1 atm)

PREPARED BY:

   M. E. Derrick

EXPERIMENTAL VALUES:

T/K	Mol Fraction $X_1 \times 10^3$	Bunsen Coefficient α	Ostwald Coefficient L
293.50	4.85	1.01	1.08
298.15	4.73	0.979	1.07
304.75	4.57	0.939	1.05
309.35	4.47	0.909	1.03

Smoothed Data: $\Delta G°/J\ mol^{-1} = - RT \ln X_1 = -3893.4 + 57.572\ T$

   Std. Dev. $\Delta G°$ = 0.8,   Coef. Corr. = 0.9999

See the evaluation of krypton + cyclohexane for the recommended Gibbs energy equation and smoothed solubility values.

The Bunsen and Ostwald coefficients were calculated by the compiler.

AUXILIARY INFORMATION

METHOD:

Saturation of liquid with gas at partial pressure of gas equal to 1 atm.

SOURCE AND PURITY OF MATERIALS:

1.   Krypton. Matheson Co., dried.

2.   Cyclohexane. Matheson, Coleman, and Bell, chromatoquality reagent. Dried and fractionally frozen. m.p. 6.45° C.

APPARATUS/PROCEDURE:

Dymond-Hildebrand apparatus (1) using an all-glass pumping system to spray slugs of degassed solvent into the gas. Amount of gas dissolved calculated from initial and final gas pressures.

ESTIMATED ERROR:

   $\delta X_1/X_1 = 0.01$

REFERENCES:

1.   Dymond, J.; Hildebrand, J. H. Ind. Eng. Chem. Fundam. 1967, 6, 130.

COMPONENTS:	ORIGINAL MEASUREMENTS:
1. Krypton; Kr; 7439-90-9 2. Methylcyclohexane; C_7H_{14}; 108-87-2	Clever, H.L.; Saylor, J.H.; Gross, P.M. J. Phys. Chem. 1958, 62, 89-91.

VARIABLES:	PREPARED BY:
T/K: 289.15 - 316.25 Total P/kPa: 101.325 (1 atm)	P.L. Long

EXPERIMENTAL VALUES:

T/K	Mol Fraction $X_1 \times 10^3$	Bunsen Coefficient α	Ostwald Coefficient L
289.15	6.14	1.086	1.15
303.15	5.56	0.971	1.078
316.25	5.02	0.863	0.999

Smoothed Data: $\Delta G^O/J \ mol^{-1} = - RT \ln X_1 = -5652.1 + 61.865 \ T$

Std. Dev. ΔG^O = 13.5, Coef. Corr. = 0.9999

$\Delta H^O/J \ mol^{-1} = -5652.1$, $\Delta S^O/J \ K^{-1} \ mol^{-1} = -61.865$

T/K	Mol Fraction $X_1 \times 10^3$	$\Delta G^O/J \ mol^{-1}$
288.15	6.21	12,174
293.15	5.96	12,484
298.15	5.74	12,793
303.15	5.53	13,102
308.15	5.33	13,412
313.15	5.14	13,721
318.15	4.97	14,030

The solubility values were adjusted to a partial pressure of krypton of 101.325 kPa (1 atm) by Henry's law.

The Bunsen coefficients were calculated by the compiler.

AUXILIARY INFORMATION

METHOD:

Volumetric. The apparatus (1) is a modification of that used by Morrison and Billett (2). Modifications include the addition of a spiral solvent storage tubing, a manometer for constant reference pressure, and an extra gas buret for highly soluble gases.

SOURCE AND PURITY OF MATERIALS:

1. Krypton. Matheson Co., Inc. Both standard and research grades were used.

2. Methylcyclohexane. Eastman Kodak Co., white label. Dried over Na and distilled; corrected b.p. 100.95 to 100.97°, lit. b.p. 100.93°C.

APPARATUS/PROCEDURE:

(a) Degassing. 700 ml of solvent is shaken and evacuated while attached to a cold trap, until no bubbles are seen; solvent is then transferred through a 1 mm capillary tubing, released as a fine mist into a continuously evacuated flask. (b) Solvent is saturated with gas as it flows through 8 mm x 180 cm of tubing attached to a gas buret. Pressure is maintained at 1 atm as the gas is absorbed.

ESTIMATED ERROR:
$\delta T/K = 0.05$
$\delta P/mmHg = 3$
$\delta X_1/X_1 = 0.03$

REFERENCES:
1. Clever, H.L.; Battino, R.; Saylor, J.H.; Gross, P.M. J. Phys. Chem. 1957, 61, 1078.

2. Morrison, T.J.; Billett, F. J. Chem. Soc. 1948, 2033; Ibid. 1952, 3819.

COMPONENTS:	ORIGINAL MEASUREMENTS:
1. Krypton; Kr; 7439-90-9 2. Cyclooctane; C_8H_{16}; 292-64-8	Wilcock, R.J.; Battino, R.; Wilhelm, E. J. Chem. Thermodyn. 1977, 9, 111-115.

VARIABLES:	PREPARED BY:
T/K: 298.26 P/kPa: 101.325	H.L. Clever

EXPERIMENTAL VALUES:

T/K	Mol Fraction X_1 x 10^3	Bunsen Coefficient α	Ostwald Coefficient L
298.26	3.442	0.5727	0.6254

The solubility values were adjusted to a partial pressure of krypton of 101.325 kPa (1 atm) by Henry's law. The Bunsen coefficients were calculated by the compiler.

AUXILIARY INFORMATION

METHOD/APPARATUS/PROCEDURE:

The apparatus is based on the design by Morrison and Billett (1) and the version used is described by Battino, Evans, and Danforth (2).

Degassing. Up to 500 cm³ of solvent is placed in a flask of such size that the liquid is about 4 cm deep. The liquid is rapidly stirred and vacuum is applied intermittently through a liquid N_2 trap until the permanent gas residual pressure drops to 5 microns.

Solubility Determination. The degassed solvent passes in a thin film down a glass spiral containing the solute gas and solvent vapor at a total pressure of one atm. The volume of gas absorbed is measured in the attached gas buret, and the solvent is collected in a tared flask and weighed.

SOURCE AND PURITY OF MATERIALS:

1. Krypton. Matheson Co., Inc. Minimum purity 99.995 mol per cent.

2. Cyclooctane. Chemical Samples Co. 99 mol per cent, fractionally distilled, n(Na D, 298.15 K) = 1.4562.

ESTIMATED ERROR:
$$\delta T/K = 0.03$$
$$\delta P/mmHg = 0.5$$
$$\delta X_1/X_1 = 0.03$$

REFERENCES:
1. Morrison, T.J.; Billett, F. J. Chem. Soc. 1948, 2033.

2. Battino, R.; Evans, F.D.; Danforth, W.F. J. Am.Oil Chem. Soc. 1968, 45, 830.

COMPONENTS:	ORIGINAL MEASUREMENTS:

COMPONENTS:
1. Krypton; Kr; 7439-90-9
2. cis-1,2-Dimethylcyclohexane; C_8H_{16}; 2207-01-4

ORIGINAL MEASUREMENTS:
Geller, E.B.; Battino, R.; Wilhelm, E.

J. Chem. Thermodyn. 1976, 8, 197-202.

VARIABLES:
T/K: 297.92
P/kPa: 101.325 (1 atm)

PREPARED BY:
H.L. Clever

EXPERIMENTAL VALUES:

T/K	Mol Fraction X_1 x 10^3	Bunsen Coefficient α	Ostwald Coefficient L
297.92	5.447	0.8684	0.9472

The solubility values were adjusted to a partial pressure of krypton of 101.325 kPa (1 atm) by Henry's law.

The Bunsen coefficients were calculated by the compiler.

AUXILIARY INFORMATION

METHOD/APPARATUS/PROCEDURE:

The apparatus is based on the design by Morrison and Billett (1) and the version used is described by Battino, Evans, and Danforth (2).
Degassing. Up to 500 cm³ of solvent is placed in a flask of such size that the liquid is about 4 cm deep. The liquid is rapidly stirred, and vacuum is applied intermittently through a liquid N_2 trap until the permanent gas residual pressure drops to 5 microns.
Solubility Determination. The degassed solvent is passed in a thin film down a glass spiral tube containing the solute gas plus the solvent vapor at a total pressure of one atm. The volume of gas absorbed is found by difference between the initial and final gas volume in the buret system. The solvent is collected in a tared flask and weighed.

SOURCE AND PURITY OF MATERIALS:
1. Krypton. Either Air Products & Chemicals, Inc., or Matheson Co., Inc., 99 mol % or better.
2. cis-1,2-Dimethylcyclohexane. Chemical Samples Co., fractionally distilled and stored in dark. n_D(298.15 K) 1.4337.

ESTIMATED ERROR:
$\delta T/K = 0.03$
$\delta P/mmHg = 0.5$
$\delta X_1/X_1 = 0.2$

REFERENCES:
1. Morrison, T.J.; Billett, F. J. Chem. Soc. 1948, 2033.
2. Battino, R.; Evans, F.D.; Danforth, W.F. J. Am. Oil Chem. Soc. 1968, 45, 830.

K.X.R.—F

COMPONENTS:	ORIGINAL MEASUREMENTS:

1. Krypton; Kr; 7439-90-9
 	
2. trans-1,2-Dimethylcyclohexane;
   C_8H_{16}; 6876-23-9
	Geller, E.B.; Battino, R.;
	   Wilhelm, E.
	
	J. Chem. Thermodyn. 1976, 8, 197-202.

VARIABLES:
   T/K: 297.91
   P/kPa: 101.325 (1 atm)
	PREPARED BY:
	
	   H.L. Clever

EXPERIMENTAL VALUES:

T/K	Mol Fraction $X_1 \times 10^3$	Bunsen Coefficient α	Ostwald Coefficient L
297.91	5.972	0.9270	1.011

The solubility values were adjusted to a partial pressure of krypton of 101.325 kPa (1 atm) by Henry's law.

The Bunsen coefficients were calculated by the compiler.

AUXILIARY INFORMATION

METHOD/APPARATUS/PROCEDURE:

The apparatus is based on the design by Morrison and Billett (1) and the version used is described by Battino, Evans, and Danforth (2).

See krypton + 1,2 dimethylcyclohexane data sheet for more detail.

SOURCE AND PURITY OF MATERIALS:

1. Krypton. Either Air Products & Chemicals Inc., or Matheson Co., Inc., 99 mol % or better.

2. trans-1,2-Dimethylcyclohexane. Chemical Samples Co., fractionally distilled and stored in dark. n_D(298.15) 1.4248.

ESTIMATED ERROR:
$$\delta T/K = 0.03$$
$$\delta P/mmHg = 0.5$$
$$\delta X_1/X_1 = 0.02$$

REFERENCES:

1. Morrison, T.J.; Billett, F. J. Chem. Soc. 1948, 2033.

2. Battino, R.; Evans, F.D.; Danforth, W.F. J. Am. Oil Chem. Soc. 1968, 45, 830.

COMPONENTS:	ORIGINAL MEASUREMENTS:
1. Krypton; Kr; 7439-90-9 2. <u>cis</u>-1,3-Dimethylcyclohexane, 59 <u>mol</u> %; C_8H_{16}; 638-04-0 3. <u>trans</u>-1,3-Dimethylcyclohexane,41 <u>mol</u> %; C_8H_{16}; 2207-03-6	Geller, E.B.; Battino, R.; Wilhelm, E. J. Chem. Thermodyn. 1976, 8, 197-202.

VARIABLES:	PREPARED BY:
T/K: 298.21 P/kPa: 101.325 (1 atm)	H.L. Clever

EXPERIMENTAL VALUES:

T/K	Mol Fraction $X_1 \times 10^3$	Bunsen Coefficient α	Ostwald Coefficient L
298.21	5.883	0.9094	0.9928

The solubility values were adjusted to a partial pressure of krypton of 101.325 kPa (1 atm) by Henry's law.

The Bunsen coefficients were calculated by the compiler.

AUXILIARY INFORMATION

METHOD /APPARATUS/PROCEDURE:	SOURCE AND PURITY OF MATERIALS:
The apparatus is based on the design by Morrison and Billett (1) and the version used is described by Battino, Evans, and Danforth (2). See krypton + 1,2-Dimethylcyclohexane data sheet for more detail.	1. Krypton. Either Air Products & Chemicals, Inc., or Matheson Co., Inc. 99 mol % or better. 2. <u>cis</u>-1,3-Dimethylcyclohexane. Chemical Samples Co., binary mixture, analysed by R. I. by authors, used as received. 3. <u>trans</u>-1,3-Dimethylcyclohexane. Chemical Samples Co., binary mixture, analysed by R. I. by authors, used as received.

ESTIMATED ERROR:
$$\delta T/K = 0.03$$
$$\delta P/mmHg = 0.5$$
$$\delta X_1/X_1 = 0.02$$

REFERENCES:

1. Morrison, T.J.; Billett, F. J. Chem. Soc. 1948, 2033.

2. Battino, R.; Evans, F.D.; Danforth, W.F. J. Am. Oil Chem. Soc. 1968, 45, 830.

COMPONENTS:	ORIGINAL MEASUREMENTS:
1. Krypton; Kr; 7439-90-9	Geller, E.B.; Battino, R.; Wilhelm, E.
2. cis-1,4-Dimethylcyclohexane, 70 mol %; C_8H_{16}; 624-29-3	
3. trans-1,4-Dimethylcyclohexane, 30 mol %; C_8H_{16}; 2207-04-7	J. Chem. Thermodyn. 1976, 8, 197-202.

VARIABLES:	PREPARED BY:
T/K: 298.35 P/kPa: 101.325 (1 atm)	H.L. Clever

EXPERIMENTAL VALUES:

T/K	Mol Fraction $X_1 \times 10^3$	Bunsen Coefficient α	Ostwald Coefficient L
298.35	6.005	0.9320	1.018

The solubility value was adjusted to a partial pressure of krypton of 101.325 kPa (1 atm) by Henry's law.

The Bunsen coefficient was calculated by the compiler.

AUXILIARY INFORMATION

METHOD/APPARATUS/PROCEDURE:

The apparatus is based on the design by Morrison and Billett (1) and the version used is described by Battino, Evans, and Danforth (2).
See krypton + 1,2 dimethylcyclohexane data sheet for more detail.

SOURCE AND PURITY OF MATERIALS:

1. Krypton. Either Air Products & Chemicals, Inc., or Matheson Co., Inc. 99 mol % or better.

2. cis-1,4-Dimethylcyclohexane. Chemical Samples Co., binary mixture, analysed by R. I. by authors, used as received.

3. trans-1,4-Dimethylcyclohexane. Chemical Samples Co., binary mixture, analysed by R. I. by authors, used as received.

ESTIMATED ERROR:
$$\delta T/K = 0.03$$
$$\delta P/mmHg = 0.5$$
$$\delta X_1/X_1 = 0.02$$

REFERENCES:
1. Morrison, T.J.; Billett, F. J. Chem. Soc. 1948, 2033.

2. Battino, R.; Evans, F.D.; Danforth, W.F. J. Am. Oil Chem. Soc. 1968, 45, 830.

COMPONENTS:	ORIGINAL MEASUREMENTS:
1. Krypton; Kr; 7439-90-9 2. 1,2,3,4-Tetrahydronaphthalene (Tetralin); $C_{10}H_{12}$; 119-64-2	Körösy, F. Trans. Faraday Soc. 1937, 33, 416-425.

VARIABLES:	PREPARED BY:
T/K: 297.15 P/kPa: 101.325 (1 atm)	H.L. Clever

EXPERIMENTAL VALUES:

T/K	Mol Fraction $X_1 \times 10^3$	Bunsen Coefficient α	Ostwald Coefficient L
297.15	2.7	0.45	0.49

The mole fraction solubility and the Bunsen coefficient were calculated by
the compiler. It was assumed that gas behavior is ideal, the Ostwald
coefficient is independent of pressure, and that Henry's law is obeyed.

AUXILIARY INFORMATION

METHOD:	SOURCE AND PURITY OF MATERIALS:
The apparatus and method of Winkler (1) were used. However, the apparatus was usually not thermostated, and degassing was by evacuating and shaking the solvent, not by evacuating and boiling the solvent as was done by Winkler.	1. Krypton. Source not given. The gas contained 5% xenon and 1% non-inert gases. 2. 1,2,3,4-Tetrahydronaphthalene. No information.

APPARATUS/PROCEDURE:	ESTIMATED ERROR: $\delta X_1/X_1 = 0.05$ REFERENCES: 1. Winkler, L.W. Ber. 1891, 24, 89.

COMPONENTS:	ORIGINAL MEASUREMENTS:
1. Krypton; Kr; 7439-90-9 2. Benzene; C_6H_6; 71-43-2	Clever, H.L.; Battino, R.; Saylor, J.H.; Gross, P.M. J. Phys. Chem. 1957, 61, 1078-1083.

VARIABLES:
     T/K: 289.35 - 313.45
     P/kPa: 101.325 (1 atm)
	PREPARED BY:
	
	      P.L. Long

EXPERIMENTAL VALUES:

T/K	Mol Fraction $X_1 \times 10^3$	Bunsen Coefficient α	Ostwald Coefficient L
289.35	2.81	0.714	0.756
298.35	2.73	0.685	0.748
313.45	2.64	0.654	0.750

Smoothed Data:

$$\Delta G^O/\text{J mol}^{-1} = -RT \ln X_1 = -1926.5 + 55.517\,T$$

Std. Dev. $\Delta G^O = 6.2$, Coef. Corr. $= .9999$

$$\Delta H^O/\text{J mol}^{-1} = -1926.5, \quad \Delta S^O/\text{J K}^{-1}\,\text{mol}^{-1} = -55.517$$

T/K	Mol Fraction $X_1 \times 10^3$	$\Delta G^O/\text{J mol}^{-1}$
288.15	2.81	14071
293.15	2.78	14348
298.15	2.74	14626
303.15	2.70	14903
308.15	2.67	15181
313.15	2.64	15458
318.15	2.61	15736

The solubility values were adjusted to a partial pressure of krypton of 101.325 kPa (1 atm) by Henry's law.

The Bunsen coefficients were calculated by the compiler.

AUXILIARY INFORMATION

METHOD /APPARATUS/PROCEDURE:

    Solvent was degassed by a modification of the method of Baldwin and Daniel (1). Saturation apparatus was that of Morrison and Billett (2), modified to include spiral storage for the solvent, a manometer for constant reference pressure, and an extra buret. Solvent was saturated with gas as it flowed through an 8 mm x 180 cm glass spiral attached to a buret.

NOTE: F. Körösy Trans. Faraday Soc. 1937, 33, 416 reports an Ostwald coefficient of 0.97 (mole fraction 3.55×10^{-3}) at 295.15 K. Van Liempt and van Wijk Recl. trav. Chim. Pays-Bas 1937, 56, 632 report a Bunsen coefficient of 0.67 (mole fraction 2.65×10^{-3}) at 292.15 K. Neither of the solubility of krypton in benzene values were considered in the smoothed data fit given above.

SOURCE AND PURITY OF MATERIALS:

1. Krypton. Linde Air Products Co. Pure.

2. Benzene. Jones & Laughlin Steel Co. Shaken with H_2SO_4, water washed, dried over Na, distilled.

ESTIMATED ERROR:
$$\delta T/K = 0.05$$
$$\delta P/\text{mmHg} = 3$$
$$\delta X_1/X_1 = 0.03$$

REFERENCES:

1. Baldwin, R.R.; Daniel, S.G. J. Appl. Chem. 1952, 2, 161.

2. Morrison, T.J.; Billett, F. J. Chem. Soc. 1948, 2033; ibid. 1952, 3819.

63

COMPONENTS:	ORIGINAL MEASUREMENTS:
1. Krypton; Kr; 7439-90-9 2. Methylbenzene (Toluene); C_7H_8; 108-88-3	Saylor, J. H.; Battino, R. J. Phys. Chem. 1958, 62, 1334-1337.

VARIABLES:	PREPARED BY:
T/K: 288.15 - 328.15 P/kPa: 101.325 (1 atm)	H. L. Clever

EXPERIMENTAL VALUES:

T/K	Mol Fraction $X_1 \times 10^3$	Bunsen Coefficient α	Ostwald Coefficient L
288.15	3.50	0.744	0.785
298.15	3.37	0.710	0.775
313.15	3.13	0.646	0.741
328.15	2.91	0.592	0.711

Smoothed Data: $\Delta G°/J\ mol^{-1} = - RT\ \ln X_1 = 3,686.3 + 59.752\ T$

$\Delta H°/J\ mol^{-1} = 3,686.3$, $\Delta S°/J\ K^{-1}\ mol^{-1} = -59.752$

T/K	Mol Fraction $X_1 \times 10^3$	$\Delta G°/J\ mol^{-1}$
288.15	3.52	13,531
293.15	3.43	13,830
298.15	3.35	14,129
303.15	3.27	14,428
308.15	3.19	14,726
313.15	3.12	15,025
318.15	3.05	15,324
323.15	2.98	15,623
328.15	2.92	15,922

The solubility values were adjusted to a partial pressure of krypton of 101.325 kPa (1 atm) by Henry's law.
The Bunsen coefficients were calculated by the compiler.

AUXILIARY INFORMATION

METHOD /APPARATUS/PROCEDURE:

The solvent was degassed by evacuating the space above it, shaking, and then passing it as a fine mist into another evacuated container. The degassed liquid was saturated as it passed as a thin film inside a glass helix which contained the solute gas plus solvent vapor at a total pressure of 1 atm (1,2). The volume of liquid and the volume of gas absorbed are determined in a system of burets.

Evaluator's Note: Van Liempt and van Wijk Recl. Trav. Chim. Pays-Bas, 1937, 56, 632 report a Bunsen coefficient of 0.84 (mole fraction 4.0 x 10^{-3}) at 291.15 K for this system. The value is not recommended for use.

SOURCE AND PURITY OF MATERIALS:

1. Krypton. Matheson Co., Inc. Research grade.

2. Toluene. Mallinckrodt. Reagent grade. Shaken over conc. H_2SO_4, water washed, dried over Drierite, distilled b.p. 110.40 - 110.60° C.

ESTIMATED ERROR:

$\delta T/K = 0.03$
$\delta P/torr = 3$
$\delta X_1/X_1 = 0.005$

REFERENCES:

1. Morrison, T. J.; Billett, F. J. Chem. Soc. 1948, 2033; ibid, 1952, 3819.

2. Baldwin, R. R.; Daniel, S. G. J. Appl. Chem. 1952, 2, 161.

COMPONENTS:	ORIGINAL MEASUREMENTS:
1. Krypton; Kr; 7439-90-9 2. 1,2-Dimethylbenzene (o-Xylene); C_8H_{10}; 95-47-6	Byrne, J.E.; Battino, R.; Wilhelm, E. J. Chem. Thermodyn. 1975, 7, 515-522.

VARIABLES:	PREPARED BY:
T/K: 298.12 - 298.15 P/kPa: 101.325 (1 atm)	H.L. Clever

EXPERIMENTAL VALUES:

T/K	Mol Fraction X_1 x 10^3	Bunsen Coefficient α	Ostwald Coefficient L
298.12	3.372	0.6257	0.6829
298.12	3.379	0.6270	0.6843
298.13	3.368	0.6248	0.6819
298.15	3.378	0.6266	0.6840

The solubility values were adjusted to a partial pressure of krypton of 101.325 kPa (1 atm) by Henry's law. The Bunsen coefficients were calculated by the compiler.

AUXILIARY INFORMATION

METHOD/APPARATUS/PROCEDURE:	SOURCE AND PURITY OF MATERIALS:
The apparatus is based on the design by Morrison and Billett (1) and the version used is described by Battino, Evans, and Danforth (2). Degassing. Up to 500 cm^3 of solvent is placed in a flask of such size that the liquid is about 4 cm deep. The liquid is rapidly stirred, and vacuum is applied intermittently through a liquid N_2 trap until the permanent gas residual pressure drops to 5 microns. Solubility Determination. The degassed solvent passes in a thin film down a glass spiral tube containing the solute gas plus the solvent vapor at a total pressure of one atm. The volume of gas absorbed is found by difference between the initial and final gas volume in the buret system. The solvent is collected in a tared flask and weighed.	1. Krypton. Either Air Products & Chemicals, Inc., or Matheson Co., Inc. 99 mol % or better. 2. 1,2-Dimethylbenzene. Phillips Petroleum Co. Pure grade.

ESTIMATED ERROR:
$\delta T/K$ = 0.03 $\delta P/mmHg$ = 0.5 $\delta X_1/X_1$ = 0.02

REFERENCES:
1. Morrison, T.J.; Billett, F. J. Chem. Soc. 1948, 2033.   2. Battino, R.; Evans, F.D.; Danforth, W.F. J. Am. Oil Chem. Soc. 1968, 45, 830.

COMPONENTS:	ORIGINAL MEASUREMENTS:
1. Krypton; Kr; 7439-90-9 2. 1,3-Dimethylbenzene (m-Xylene); C_8H_{10}; 108-38-3	Byrne, J.E.; Battino, R.; Wilhelm, E. J. Chem. Thermodyn. 1975, 7, 515-522.

VARIABLES:
  T/K: 298.18 - 298.21
  P/kPa: 101.325 (1 atm)
	PREPARED BY:
	
	   H.L. Clever

EXPERIMENTAL VALUES:

T/K	Mol Fraction X_1 x 10^3	Bunsen Coefficient α	Ostwald Coefficient L
298.18	3.624	0.6598	0.7203
298.20	3.629	0.6609	0.7215
298.21	3.629	0.6608	0.7214
298.21	3.632	0.6614	0.7221

The solubility values were adjusted to a partial pressure of krypton of 101.325 kPa (1 atm) by Henry's law. The Bunsen coefficients were calculated by the compiler.

AUXILIARY INFORMATION

METHOD

   The apparatus is based on the design by Morrison and Billett (1) and the version used is described by Battino, Evans, and Danforth (2).

APPARATUS/PROCEDURE:

   Degassing. Up to 500 cm^3 of solvent is placed in a flask of such size that the liquid is about 4 cm deep. The liquid is rapidly stirred, and vacuum is applied intermittently through a liquid N_2 trap until the permanent gas residual pressure drops to 5 microns.
   Solubility Determination. The degassed solvent passes in a thin film down a glass spiral tube containing the solute gas plus the solvent vapor at a total pressure of one atm. The volume of gas absorbed is found by difference between the initial and final gas volume in the buret system. The solvent is collected in a tared flask and weighed.

SOURCE AND PURITY OF MATERIALS:

1. Krypton. Either Air Products & Chemicals, Inc., or Matheson Co., Inc. 99 mol % or better.

2. 1,3-Dimethylbenzene. Phillips Petroleum Co., pure grade.

ESTIMATED ERROR:

$$\delta T/K = 0.03$$
$$\delta P/mmHg = 0.5$$
$$\delta X_1/X_1 = 0.02$$

REFERENCES:

1. Morrison, T.J.; Billett, F. J. Chem. Soc. 1948, 2033.

2. Battino, R.; Evans, F.D.; Danforth, W.F. J. Am. Oil Chem. Soc. 1968, 45, 830.

COMPONENTS:	ORIGINAL MEASUREMENTS:

COMPONENTS:

1. Krypton; Kr; 7439-90-9

2. 1,4-Dimethylbenzene (p-Xylene); C_8H_{10}; 106-42-3

ORIGINAL MEASUREMENTS:

Clever, H.L.

J. Phys. Chem. 1957, 61, 1082-1083.

VARIABLES:

T/K: 303.15
P/kPa: 101.325 (1 atm)

PREPARED BY:

C.E. Edelman
A.L. Cramer

EXPERIMENTAL VALUES:

T/K	Mol Fraction X_1 x 10^3	Bunsen Coefficient α	Ostwald Coefficient L
303.15	3.81	0.687	0.762

The solubility value was adjusted to a partial pressure of krypton of 101.325 kPa (1 atm) by Henry's law.

The Bunsen coefficient was calculated by the compiler.

AUXILIARY INFORMATION

METHOD:

 Volumetric. The solvent is saturated with gas as it flows through an 8 mm x 180 cm glass helix attached to a gas buret. The total pressure of solute gas plus solvent vapor is maintained at 1 atm as the gas is absorbed.

SOURCE AND PURITY OF MATERIALS:

1. Krypton. Linde Air Products Co.

2. 1,4-Dimethylbenzene. Eastman Kodak white label. Fractionally crystalized twice, dried over Na, distilled, b.p. 138.0-138.2°C.

APPARATUS/PROCEDURE:

 The apparatus is a modification of the apparatus of Morrison and Billett (1). The modifications include the addition of a helical storage for the solvent, a manometer for a reference pressure, and an extra buret for highly soluble gases. The solvent is degassed by a modification of the method of Baldwin and Daniel (2).

ESTIMATED ERROR:

$\delta T/K = 0.05$
$\delta P/mmHg = 3$
$\delta X_1/X_1 = 0.03$

REFERENCES:

1. Morrison, T.J.; Billett, F. J. Chem. Soc. 1948, 2033; ibid. 1952, 3819.

2. Baldwin, R.R.; Daniel, S.G. J. Appl. Chem. 1952, 2, 161.

COMPONENTS:	ORIGINAL MEASUREMENTS:
1. Krypton; Kr; 7439-90-9 2. 1,4-Dimethylbenzene (p-Xylene); C_8H_{10}; 106-42-3	Byrne, J.E.; Battino, R.; Wilhelm, E. $\underline{J}$. $\underline{Chem}$. $\underline{Thermodyn}$. 1975, $\underline{7}$, 515-522.

VARIABLES:	PREPARED BY:
T/K: 298.14 - 298.27 P/kPa: 101.325 (1 atm)	H.L. Clever

EXPERIMENTAL VALUES:

T/K	Mol Fraction X_1 x 10^3	Bunsen Coefficient α	Ostwald Coefficient L
298.14	3.779	0.6860	0.7488
298.17	3.816	0.6927	0.7562
298.19	3.810	0.6916	0.7550
298.19	3.844	0.6978	0.7618
298.21	3.806	0.6847	0.7574
298.27	3.796	0.6889	0.7523

The solubility values were adjusted to a partial pressure of krypton of 101.325 kPa (1 atm) by Henry's law. The Bunsen coefficients were calculated by the compiler.

AUXILIARY INFORMATION

METHOD:

   The apparatus is based on the design by Morrison and Billett (1) and the version used is described by Battino, Evans, and Danforth (2).

APPARATUS/PROCEDURE:

   Degassing. Up to 500 cm^3 of solvent is placed in a flask of such size that the liquid is about 4 cm deep. The liquid is rapidly stirred, and vacuum is applied intermittently through a liquid N_2 trap until the permanent gas residual pressure drops to 5 microns.
   Solubility Determination. The degassed solvent passes in a thin film down a glass spiral tube containing the solute gas plus the solvent vapor at a total pressure of one atm. The volume of gas absorbed is found by difference between the initial and final gas volume in the buret system. The solvent is collected in a tared flask and weighed.

SOURCE AND PURITY OF MATERIALS:

1. Krypton. Either Air Products & Chemicals, Inc., or Matheson Co., Inc. 99 mol % or better.

2. 1,4-Dimethylbenzene. Phillips Petroleum Co., pure grade.

ESTIMATED ERROR:
$$\delta T/K = 0.03$$
$$\delta P/mmHg = 0.5$$
$$\delta X_1/X_1 = 0.02$$

REFERENCES:
1. Morrison, T.J.; Billett, F. $\underline{J}$. $\underline{Chem}$. $\underline{Soc}$. 1948, 2033.

2. Battino, R.; Evans, F.D.; Danforth, W.F. $\underline{J}$. $\underline{Am}$. $\underline{Oil}$ $\underline{Chem}$. $\underline{Soc}$. 1968, $\underline{45}$, 830.

COMPONENTS:	ORIGINAL MEASUREMENTS:
1. Krypton; Kr; 7439-90-9 2. Industrial Hydrocarbons	van Liempt, J.A.M.; van Wijk, W. Rec.Trav.Chim.Pays-Bas 1937, 56, 632 - 634.

VARIABLES:	PREPARED BY:
T/K: 291.65 - 293.15 P/kPa: 101.325 (1 atm)	H. L. Clever

EXPERIMENTAL VALUES:

T/K	Bunsen Coefficient	Ostwald Coefficient
Gasoline		
292.15	0.89	0.95
Paraffin oil		
291.65	0.60	0.64
Petroleum		
293.15	1.00	1.07

The Ostwald coefficients were calculated by the compiler.

AUXILIARY INFORMATION

METHOD:	SOURCE AND PURITY OF MATERIALS:
The apparatus appears to be similar to the Winkler type apparatus used by Körösy (1).	1. Krypton. Source not given. Contained 5 % xenon. 2. Solvents. No information given.

APPARATUS/PROCEDURE:	ESTIMATED ERROR: $\delta X_1/X_1 = 0.10$ (Compiler)
	REFERENCES: 1. Körösy, F. Trans. Faraday Soc.1937, 33, 416.

COMPONENTS:	ORIGINAL MEASUREMENTS:
1. Krypton; Kr; 7439-90-9	Steinberg, M.; Manowitz, B.
2. Hydrocarbons	Ind. and Eng. Chem., 1959, 51, 47-50.

VARIABLES:
    T/K: 296.15 - 298.65
    P/kPa: 101.325 (1 atm)
	PREPARED BY:
	    H.L. Clever, P.L. Long, A.L. Cramer

EXPERIMENTAL VALUES:

T/K	Absorption Coefficient* β
296.15	"Ultrasene" (paraffins 80 wt %, napthenes 20 wt %) 0.58
298.65	1,3,5-Trimethylbenzene (Mesitylene); C_9H_{12}; 108-67-8 0.63
298.15	Diphenylbenzene (Terphenyl); $C_{18}H_{14}$; 26140-60-3 0.30

*Volume of gas (corrected to 288.15 K and 1 atm) absorbed under a total system pressure of 1 atm, per unit volume of solvent (corrected to 288.15 K).

AUXILIARY INFORMATION

METHOD/APPARATUS/PROCEDURE: Modified McDaniel (1) method.	SOURCE AND PURITY OF MATERIALS: 1. Krypton. Matheson Co. Technically pure. 2. Ultrasene. Atlantic Refining Co. Mesitylene. No source. Technical grade. Terphenyl. No source. Technical grade.
	ESTIMATED ERROR: $\delta\beta/\beta = 0.05 - 0.10$.
	REFERENCES: 1. Furman, N.H., 1939 "Scott's Standard Methods of Chemical Analysis" II, 5th Ed., Van Nostrand Co., N.Y., p. 2587.

COMPONENTS:	ORIGINAL MEASUREMENTS:
1. Krypton; Kr; 7439-90-9 2. Amsco 123-15	Steinberg, M.; Manowitz, B. Ind. Eng. Chem. 1959, 51, 47-50.

VARIABLES:
  T/K: 218.15 - 423.15
  P/kPa: 101.325 (1 atm)
	PREPARED BY:
	
	H.L. Clever

EXPERIMENTAL VALUES:

	Krypton at one atm		Krypton at low conc ln N_2	
T/K	Absorption Coefficient β	Henry's Constant K/atm	Krypton Initial ppm in N_2	Henry's Constant K/atm
218.15	1.50	70	41.6	71
297.15	0.56	190	48	173
333.15	0.44	238	92	251
373.15	0.30	-	-	-
383.15	0.34	313	92	290
423.15	0.19	544	92	497

The authors define the Absorption Coefficient as the volume of gas, corrected to 288.15 K and 101.325 kPa, absorbed under a total system pressure of 101.325 kPa per unit volume of solvent at 288.15 K.

AUXILIARY INFORMATION

METHOD/APPARATUS/PROCEDURE:

The absorption coefficient at one atm krypton was measured by modified McDaniel method (1).
The Henry's constant $(K = (P/atm)/X_1)$ at low concentration of krypton was measured by static and dynamic tracer techniques.
The authors state that log (Absorption Coefficient) vs 1/T is linear and gives an enthalpy of solution of -1700 cal mol^{-1}.

SOURCE AND PURITY OF MATERIALS:

1. Krypton. Matheson Co., Inc. Technical grade.

2. Amsco 123-15. American Mineral Spirits Co. No. 140. Paraffin 59.6 wt %, naphthene 27.3 wt %, aromatics 13.2 wt %.

ESTIMATED ERROR:
  $\delta K/K$ = 0.05 - 0.10 (McDaniel method)
  $\delta K/K$ = 0.18 (Tracer methods)

REFERENCES:

1. Furman, N.H. "Scott's Standard Methods of Chemical Analysis" Van Nostrand Co., NY 1939, 5th ed., Vol. II, p. 2587.

COMPONENTS:	EVALUATOR:
1. Krtypton; Kr; 7439-90-9 2. Methanol; CH_4O; 64-56-1	H. L. Clever Chemistry Department Emory University Atlanta, GA 30322 USA May 1978

CRITICAL EVALUATION:

The solubility of krypton in methanol was measured in several laboratories. Only one laboratory reports krypton in methanol solubility values at more than one temperature. Thus, it is not possible to intercompare data from the different laboratories and recommend a set of solubility values and thermodynamic parameters for the system.

Van Liempt and van Wijk (1) report a Bunsen coefficient of 0.52 (mole fraction 0.94×10^{-3}) at 291.15 K. Steinberg and Manowitz (2) report an absorption coefficient (volume of gas corrected to 288.15 K and 101.325 kPa absorbed under a total system pressure of 101.325 kPa per unit volume of solvent calculated at 288.15 K) of 1.06 at 144.15 to 153.15 K in the methanol + water eutectic (83 weight percent methanol). Komarenko and Manzhelii (3) report five solubility values between 176.15 and 213.15 K measured at a krypton pressure of 26.664 kPa (200 mmHg). The compiler has recalculated these at 101.325 kPa assuming Henry's law is obeyed.

It is not possible to judge the reliability of the single solubility values of van Liempt and van Wijk and of Steinberg and Manowitz. Data sheets on them are not included. Both the methanol and the krypton used by Komarenko and Manzhelii are of exceptional purity. Their degassing technique is thorough. The solubility method appears reliable, and their data should be of good quality. The Komarenko and Manzhelii solubility values were fitted to a three constant equation by the method of least squares. The smoothed solubility values at 101.325 kPa and the values of the thermodynamic changes for the transfer of krypton from the gas at 101.325 kPa to the hypothetical unit mole fraction solution for the temperature interval of 173.15 to 213.15 K are given on the data sheet which follows.

Although recommended solubility of krypton in methanol values are not given, a tentative set of values based on the five low temperature values of Komarenko and Manzhelii and the single room temperature value of von Liempt and van Wijk is given. This set of values should be used with extreme caution until more work is done on the system. The solubility equation is

$$\ln X_1 = -12.625 + 11.868/(T/100) + 1.4776 \ln (T/100)$$

Solubility values and values of thermodynamic parameters are given in Tables 1 and 2.

TABLE 1. Tentative values of the solubility of krypton in methanol at 101.325 kPa.

T/K	Mol Fraction $X_1 \times 10^3$	T/K	Mol Fraction $X_1 \times 10^3$
173.15	7.01	243.15	1.61
183.15	5.24	253.15	1.41
193.15	4.05	263.15	1.25
203.15	3.23	273.15	1.12
213.15	2.63	283.15	1.01
223.15	2.20	293.15	0.92
233.15	1.87		

TABLE 2. Tentative values for the transfer of one mole of krypton from the gas at 101.325 kPa to the hypothetical unit mole fraction solution.

T/K	$\Delta G^o/J\ mol^{-1}$	$\Delta H^o/J\ mol^{-1}$	$\Delta S^o/JK^{-1}mol^{-1}$	$\Delta Cp/JK^{-1}mol^{-1}$
193.15	8,845	−7,494	−84.60	12.29
293.15	17,031	−6,266	−79.47	12.29

1. van Liempt, J.A.M.; von Wijk, W.
   Rec. Trav. Chim. Pays-Bas 1937, 56, 632.
2. Steinberg, M.; Manowitz, B. Ind. Eng. Chem. 1959, 51, 47.
3. Komarenko, V.G.; Manzhelii, V.G. Ukr. Fiz. Zh. (Ukr. Ed.)
   1968, 13, 387.

COMPONENTS:	ORIGINAL MEASUREMENTS:
1. Krypton; Kr; 7439-90-9 2. Methanol; CH_4O; 64-56-1	Komarenko, V.G.; Manzhelii, V.G. Ukr.Fiz.Zh.(Ukr.Ed.) 1968,13,387-391. Ukr. Phys. J. 1968, 13, 273-276.

VARIABLES:
    T/K: 176.15 - 213.15
    P/kPa: 26.664 (200 mmHg)
	PREPARED BY:
	
	    T.D. Kittredge

EXPERIMENTAL VALUES:

T/K	Mol Fraction P/mmHg 200 X_1 x 10^3	Mol Fraction P/mmHg 760 X_1 x 10^3
176.15	1.768	6.718
183.15	1.307	4.967
193.15	1.045	3.971
203.15	0.846	3.21
213.15	0.718	2.73

Smoothed Data: $\ln X_1 = -40.0423 + 43.7246/(T/100) + 18.0139 \ln (T/100)$

T/K	Mol Fraction X_1 x 10^3	ΔG^o/kJmol^{-1}	ΔH^o/kJmol^{-1}	ΔS^o/JK^{-1}mol^{-1}	ΔC_p^o /JK^{-1}mol^{-1}
173.15	7.44	7.055	-10.42	-100.9	149.8
183.15	5.15	8.022	-8.923	-92.5	"
193.15	3.90	8.907	-7.425	-84.5	"
203.15	3.18	9.714	-5.928	-77.00	"
213.15	2.75	10.45	-4.430	-69.80	"

The mole fraction solubility at 101.325 kPa (760 mmHg) was calculated by Henry's law by the compiler.

AUXILIARY INFORMATION

METHOD /APPARATUS/PROCEDURE:

The solvent was degassed by vacuum. A thin layer of alcohol, cooled to 125-175 K, was kept for 20 hours in a vacuum maintained at 10^{-3} mmHg.

The degassed liquid was sealed under vacuum in an ampoule which was placed in the apparatus. The apparatus consisted of a manostat, a mercury compensator, and a solubility cell divided by a mercury seal. A gas pressure of 200 mmHg and the temperature were established. The foil ends of the ampoule were pierced. The gas dissolved as the liquid flowed through a series of small cups. The amount of gas dissolved was determined by the rise in mercury level in the compensator.

Some measurements were made at 400 mmHg gas pressure. The results confirmed that Henry's law was obeyed.

SOURCE AND PURITY OF MATERIALS:

1. Krypton. Source not given. Purity by chromatographic method was 99.89 per cent.

2. Methanol. Purified and analyzed in the All-Union Sci. Res. Inst. for Single Crystals & High-Purity Substances. Purity 99.97 weight per cent.

ESTIMATED ERROR:
$$\delta T/K = 0.05$$
$$\delta P/mmHg = 0.01$$
$$\delta X_1/X_1 = 0.005$$

REFERENCES:

73

COMPONENTS:	EVALUATOR:
1. Krypton; Kr; 7439-90-9	H. L. Clever
	Chemistry Department
2. Ethanol; C_2H_6O; 64-17-5	Emory University
	Atlanta, GA 30322
	USA
	May 1978

CRITICAL EVALUATION:

Komarenko and Manzhelii (1) report eight values of the solubility of krypton in ethanol at 26.664 kPa (200 mmHg) between 159.15 and 223.15 K. We are not aware of any other measurements of the solubility of krypton in pure ethanol. However, van Liempt and van Wijk (2) report a Bunsen coefficient of 0.51 at 292.15 K in 96 per cent ethanol and 0.67 in 97½ per cent ethanol. Krestov and Nedelko (4) have calculated the solubility and the themodynamic properties of krypton in ethanol + water mixtures over the 303 to 343 K temperature range by a comparative method using argon data as the reference. We have not been able to obtain a copy of their paper.

Komarenko and Manzhelii used krypton and ethanol of exceptional purity. Their degassing technique is thorough, the solubility method appears reliable, and their data should be of good quality. The Komarenko and Manzhelii values were fitted to a three constant equation by the method of least squares. The equation, the smoothed solubility values at 101.325 kPa, and the values of the Thermodynamic parameters for the transfer of one mole of krypton from the gas at 101.325 kPa to the hypothetical unit mole fraction solution are given on the Komarenko and Manzhelii data sheet which follows.

1. Komarenko, V.G.; Manzhelii, V.G. Ukr. Fiz. Zh. (Ukr. Ed.)
    1968, 13, 387.

2. van Liempt, J.A.M.; van Wijk, W.
    Rec. Trav. Chim. Pays-Bas 1937, 56, 632.

3. Körösy, F. Trans. Faraday Soc. 1937, 33, 416.

4. Krestov, G.A.; Nedelko, B.E.
    Tr. Ivanov. Khim. Tekhnol. Inst. 1970, No. 12, 38.
    Chem. Abstr. 1972, 77, 93469n; 1974, 80, 7706v.

K.X.R.—G

COMPONENTS:	ORIGINAL MEASUREMENTS:
1. Krypton; Kr; 7439-90-9 2. Ethanol; C_2H_6O; 64-17-5	Komarenko, V.G.; Manzhelii, V.G. Ukr.Fiz.Zh.(Ukr.Ed.) 1968,13,387-391. Ukr. Phys. J. 1968, 13, 273-276.

VARIABLES:
    T/K: 159.15 - 223.15
    P/kPa: 26.664 (200 mmHg)
	PREPARED BY:
	
	    T.D. Kittredge

EXPERIMENTAL VALUES:

T/K	Mol Fraction P/mmHg 200 $X_1 \times 10^3$	Mol Fraction P/mmHg 760 $X_1 \times 10^3$
159.15	4.23	16.1
163.15	3.72	14.1
173.15	2.80	10.6
183.15	2.155	8.189
193.15	1.687	6.411
203.15	1.357	5.157
213.15	1.130	4.294
223.15	0.960	3.648

Smoothed Data: $\ln X_1 = -9.80594 + 8.82054/(T/100) + 0.291181 \ln(T/100)$

T/K	Mol Fraction $X_1 \times 10^3$	$\Delta G^o/kJmol^{-1}$	$\Delta H^o/kJmol^{-1}$	$\Delta S^o/JK^{-1}mol^{-1}$	$\Delta Cp^o/JK^{-1}mol^{-1}$
163.15	14.2	5.775	-6.939	-77.92	2.42
173.15	10.6	6.553	-6.914	-77.78	"
183.15	8.12	7.330	-6.890	-77.64	"
193.15	6.43	8.105	-6.866	-77.52	"
203.15	5.21	8.880	-6.842	-77.39	"
213.15	4.31	9.654	-6.818	-77.28	"
223.15	3.63	10.431	-6.793	-77.17	"

The mole fraction solubility at 101.325 kPa (760 mmHg) was calculated by Henry's law by the compiler.

AUXILIARY INFORMATION

METHOD/APPARATUS/PROCEDURE:	SOURCE AND PURITY OF MATERIALS:

The solvent was degassed by vacuum. A thin layer of alcohol, cooled to 125-175 K, was kept for 20 hours in a vacuum maintained at 10^{-3} mmHg.
    The degassed liquid was sealed under vacuum in an ampoule which was placed in the apparatus. The apparatus consisted of a manostat, a mercury compensator, and a solubility cell divided by a mercury seal. A gas pressure of 200 mmHg and the temperature were established. The foil ends of the ampoule were pierced. The gas dissolved as the liquid flowed through a series of small cups. The amount of gas dissolved was determined by the rise in mercury level in the compensator.
    Some measurements were made at 400 mmHg gas pressure. The results confirmed that Henry's law was obeyed.
	1. Krypton. Source not given. Purity by chromatographic method was 99.89 per cent.
	
	2. Ethanol. Purified and analyzed in the All-Union Sci. Res. Inst. for Single Crystals & High-Purity Substances. Purity 99.97 weight per cent.

	ESTIMATED ERROR:
	    $\delta T/K = 0.05$
	    $\delta P/mmHg = 0.01$
	    $\delta X_1/X_1 = 0.005$

	REFERENCES:

COMPONENTS:	ORIGINAL MEASUREMENTS:
1. Krypton; Kr; 7439-90-9 2. 1-Propanol; C_3H_8O; 71-23-8	Komarenko, V.G.; Manzhelii, V.G. Ukr.Fiz.Zh.(Ukr.Ed.) 1968,13,387-391. Ukr. Phys. J. 1968, 13, 273-276.

VARIABLES:	PREPARED BY:
T/K: 163.15 - 243.15 P/kPa: 26.664 (200 mmHg)	T.D. Kittredge

EXPERIMENTAL VALUES:

T/K	Mol Fraction P/mmHg 200 X_1 x 10^3	Mol Fraction P/mmHg 760 X_1 x 10^3
163.15	4.84	18.4
173.15	3.64	13.8
183.15	2.79	10.6
193.15	2.18	8.28
203.15	1.775	6.745
213.15	1.473	5.597
223.15	1.240	4.712
233.15	1.064	4.043
243.15	0.983	3.73

Smoothed Data: $\ln X_1 = 17.3988 - 30.7777/(T/100) - 41.3191 \ln (T/100) + 10.8454 (T/100)$

T/K	Mol Fraction X_1 x 10^3	ΔG^o/kJmol^{-1}	ΔH^o/kJmol^{-1}	ΔS^o/JK^{-1}mol^{-1}	ΔCp^o/JK^{-1}mol^{-1}
163.15	18.4	5.422	-6.457	-72.81	-49.31
173.15	13.8	6.163	-6.859	-75.21	-31.27
183.15	10.6	6.923	-7.083	-76.47	-13.24
193.15	8.33	7.689	-7.125	-76.70	+4.79
203.15	6.705	8.453	-6.987	-76.00	22.83
213.15	5.544	9.206	-6.668	-74.48	40.86
223.15	4.711	9.940	-6.169	-72.19	58.90
233.15	4.115	10.65	-5.490	-69.22	76.93
243.15	3.695	11.32	-4.631	-65.61	94.97

AUXILIARY INFORMATION

METHOD/APPARATUS/PROCEDURE:

The solvent was degassed by vacuum. A thin layer of alcohol, cooled to 125-175 K, was kept for 20 hours in a vacuum maintained at 10^{-3} mmHg.

The degassed liquid was sealed under vacuum in an ampuole which was placed in the apparatus. The apparatus consisted of a manostat, a mercury compensator, and a solubility cell divided by a mercury seal. A gas pressure of 200 mmHg and the temperature were established. The foil ends of the ampuole were pierced. The gas dissolved as the liquid flowed through a series of small cups. The amount of gas dissolved was determined by the rise in mercury level in the compensator.

Some measurements were made at 400 mmHg gas pressure. The results confirmed that Henry's law was obeyed.

The mole fraction solubility at 101.325 kPa (760 mmHg) was calculated by Henry's law by the compiler.

SOURCE AND PURITY OF MATERIALS:

1. Krypton. Source not given. Purity by chromatographic method was 99.89 per cent.

2. 1-Propanol. Purified and analyzed in the All-Union Sci. Res. Inst. for Single Crystals & High-Purity Substances. Purity 99.97 weight per cent.

ESTIMATED ERROR:
$$\delta T/K = 0.05$$
$$\delta P/mmHg = 0.01$$
$$\delta X_1/X_1 = 0.005$$

REFERENCES:

COMPONENTS:	ORIGINAL MEASUREMENTS:
1. Krypton; Kr; 7439-90-9 2. 1-Butanol; $C_4H_{10}O$; 71-36-3	Komarenko, V.G.; Manzhelii, V.G. Ukr.Fiz.Zh.(Ukr.Ed.) 1968,13,387-391. Ukr. Phys. J. 1968, 13, 273-276.

VARIABLES:
      T/K: 184.15 - 243.15
      P/kPa: 26.664 (200 mmHg)
	PREPARED BY:
	
	      T.D. Kittredge
	      H.L.Clever

EXPERIMENTAL VALUES:

T/K	Mol Fraction P/mmHg 200 $X_1 \times 10^3$	Mol Fraction P/mmHg 760 $X_1 \times 10^3$
184.15	3.29	12.5
193.15	2.68	10.2
203.15	2.18	8.28
213.15	1.80	6.84
223.15	1.522	5.784
233.15	1.295	4.921
243.15	1.129	4.290

Smoothed Data: $\ln X_1 = -8.98016 + 8.35664/(T/100) + 0.100910 \ln (T/100)$

T/K	Mol Fraction $X_1 \times 10^3$	ΔG^O/kJmol^{-1}	ΔH^O/kJmol^{-1}	ΔS^O/JK^{-1}mol^{-1}	ΔCp^O/JK^{-1}mol^{-1}
183.15	12.8	6.633	-6.794	-73.32	0.839
193.15	10.2	7.367	-6.786	-73.27	"
203.15	8.27	8.099	-6.778	-73.23	"
213.15	6.85	8.831	-6.769	-73.19	"
223.15	5.77	9.563	-6.761	-73.15	"
233.15	4.939	10.29	-6.752	-73.12	"
243.15	4.281	11.03	-6.744	-73.08	"

The mole fraction solubility at 101.325 kPa (760 mmHg) was calculated by Henry's law by the compiler.

AUXILIARY INFORMATION

METHOD/APPARATUS/PROCEDURE:

The solvent was degassed by vacuum. A thin layer of alcohol, cooled to 125-175 K, was kept for 20 hours in a vacuum maintained at 10^{-3} mmHg.

The degassed liquid was sealed under vacuum in an ampoule which was placed in the apparatus. The apparatus consisted of a manostat, a mercury compensator, and a solubility cell divided by a mercury seal. A gas pressure of 200 mmHg and the temperature were established. The foil ends of the ampoule were pierced. The gas dissolved as the liquid flowed through a series of small cups. The amount of gas dissolved was determined by the rise in mercury level in the compensator.

Some measurements were made at 400 mmHg gas pressure. The results confirmed that Henry's law was obeyed.

SOURCE AND PURITY OF MATERIALS:

1. Krypton. Source not given. Purity by chromatographic method was 99.89 per cent.

2. 1-Butanol. Purified and analyzed in the All-Union Sci. Res. Inst. for Single Crystals & High-Purity Substances. Purity 99.97 weight per cent.

ESTIMATED ERROR:
$$\delta T/K = 0.05$$
$$\delta P/mmHg = 0.01$$
$$\delta X_1/X_1 = 0.005$$

REFERENCES:

COMPONENTS:	ORIGINAL MEASUREMENTS:
1. Krypton; Kr; 7439-90-9 2. 2-Methyl-1-propanol; $C_4H_{10}O$; 78-83-1	Battino, R.; Evans, F. D.; Danforth, W. F.; Wilhelm, E. $\underline{J}$. $\underline{Chem}$. $\underline{Thermodyn}$. 1971, $\underline{3}$, 743-751.
VARIABLES: T/K: 282.93 - 308.44 P/kPa: 101.325 (1 atm)	PREPARED BY: H. L. Clever

EXPERIMENTAL VALUES:

T/K	Mol Fraction X_1 x 10^3	Bunsen Coefficient α	Ostwald Coefficient L
282.93	2.90	0.711	0.737
297.89	2.57	0.622	0.678
308.44	2.34	0.559	0.632

Smoothed Data: $\Delta G° = - RT \ln X_1 = - 6,078.0 + 70.039\ T$

Std. Dev. $\Delta G° = 11.0$, Coef. Corr. = 0.9999

$\Delta H°/J\ mol^{-1} = -6078.0$, $\Delta S°/J\ K^{-1}\ mol^{-1} = - 70.039$

T/K	Mol Fraction X_1 x 10^3	$\Delta G°/J\ mol^{-1}$
278.15	3.04	13,403
283.15	2.90	13,754
288.15	2.78	14,104
293.15	2.66	14,454
298.15	2.55	14,804
303.15	2.45	15,154
308.15	2.35	15,505
313.15	2.27	15,855

The solubility values were adjusted to a partial pressure of krypton of 101.325 kPa (1 atm) by Henry's law.
The Bunsen coefficients were calculated by the compiler.

AUXILIARY INFORMATION

METHOD:

A. Degasser (1). B. Absorption of gas in a thin film of liquid (2,3)

APPARATUS/PROCEDURE:

Degassing. The solvent is sprayed into an evacuated chamber of an all glass apparatus; it is stirred and heated until the pressure drops to the vapor pressure of the liquid.
Solubility Determination. The degassed liquid passes in a thin film down a glass spiral tube at a total pressure of one atm of solute gas plus solvent vapor. The gas absorbed is measured in the attached buret system, and the solvent is collected in a tared flask and weighed.

SOURCE AND PURITY OF MATERIALS:

1. Krypton. Air Products & Chemicals. Research grade. Greater than 99 mol %.

2. 2-Methyl-1-propanol. Fisher Scientific Co. Certified (99 mol %).

ESTIMATED ERROR:

$\delta T/K = 0.03$
$\delta P/mmHg = 0.5$
$\delta X_1/X_1 = 0.015$

REFERENCES:
1. Battino, R.; Evans, F.D. Anal. Chem. 1966, 38, 1627.
2. Morrison, T. J.; Billet, F. J. Chem. Soc. 1948, 2033.
3. Clever, H. L.; Battino, R.; Saylor, J. H.; Gross, P. M. J. Phys. Chem. 1957, 61, 1078.

COMPONENTS:	ORIGINAL MEASUREMENTS:
1. Krypton; Kr; 7439-90-9 2. 1-Pentanol (Amyl Alcohol); $C_5H_{12}O$; 71-41-0	van Liempt, J. A. M.; van Wijk, W. Rec. Trav. Chim. Pays-Bas 1937, 56 632-634.

VARIABLES:	PREPARED BY:
T/K: 296.15 P/kPa: 101.325 (1 atm)	H. L. Clever

EXPERIMENTAL VALUES:

T/K	Mol Fraction X_1 x 10^3	Bunsen Coefficient α	Ostwald Coefficient L
296.15	3.18	0.66	0.72

The Ostwald coefficient was calculated by the compiler.

AUXILIARY INFORMATION

METHOD:	SOURCE AND PURITY OF MATERIALS:
The apparatus appears to be similar to the Winkler type apparatus used by Körösy (1).	1. Krypton. Source not given. Contained 5 % xenon. 2. Amyl Alcohol. No information.

APPARATUS/PROCEDURE:	ESTIMATED ERROR: $\delta X_1/X_1$ = 0.10 (Compiler)
	REFERENCES: 1. Körösy, F. Trans. Faraday Soc. 1937, 33, 416.

COMPONENTS:	ORIGINAL MEASUREMENTS:
1. Krypton; Kr; 7439-90-9 2. 1-Octanol; $C_8H_{18}O$; 111-87-5	Wilcock, R. J.; Battino, R.; Danforth, W. F.; Wilhelm, E. J. Chem. Thermodyn. 1978, 10, 817 - 822.

VARIABLES:	PREPARED BY:
T/K: 298.06 P/kPa: 101.325 (1 atm)	A.L. Cramer

EXPERIMENTAL VALUES:

T/K	Mol Fraction $X_1 \times 10^3$	Bunsen Coefficient α	Ostwald Coefficient L
298.06	3.773	0.5349	0.5837

A preliminary account of this work appeared in Conf. Int. Thermodyn. Chim. (C.R.) 4th 1975, 6, 122-128.

The solubility value was adjusted to a krypton partial pressure of 101.325 kPa (1 atm) by Henry's law.

The Bunsen coefficient was calculated by the compiler.

AUXILIARY INFORMATION

METHOD /APPARATUS/PROCEDURE:

The solubility apparatus is based on the design of Morrison and Billett (1) and the version used is described by Battino, Evans, and Danforth (2). The degassing apparatus is that described by Battino, Banzhof, Bogan, and Wilhelm (3).
See Krypton + Octane data sheet for more detail.

SOURCE AND PURITY OF MATERIALS:

1. Krypton. The Matheson Co., Inc., or Air Products and Chemicals. Purest commercially available.

2. 1-Octanol. Eastman Organic Chemicals. Distilled until refractive index equals the value recommended by Wilhoit, R.C.; Zwolinski,B. J. J. Phys. Chem. Ref. Data 1973, 2 (Suppl. No. 1), 1-212 and 1-278.

ESTIMATED ERROR:
$$\delta T/K = 0.03$$
$$\delta P/mmHg = 0.5$$
$$\delta X_1/X_1 = 0.005$$

REFERENCES:
1.Morrison, T.J.; Billett, F. J. Chem. Soc. 1948, 2033.
2.Battino, R.;Evans,F.D.;Danforth,W.F. J.Am.Oil Chem. Soc. 1968, 45, 830.
3.Battino, R.; Banzhof, M.; Bogan, M.; Wilhelm, E. Anal. Chem. 1971, 43, 806.

COMPONENTS:	ORIGINAL MEASUREMENTS:

COMPONENTS:

1. Krypton; Kr; 7439-90-9

2. 1-Decanol; $C_{10}H_{22}O$; 112-30-1

ORIGINAL MEASUREMENTS:

Wilcock, R. J.; Battino, R.;
 Danforth, W. F.; Wilhelm, E.

J. Chem. Thermodyn. 1978, 10, 817 - 822.

VARIABLES:

T/K: 298.06
P/kPa: 101.325 (1 atm)

PREPARED BY:

A.L. Cramer

EXPERIMENTAL VALUES:

T/K	Mol Fraction $X_1 \times 10^3$	Bunsen Coefficient α	Ostwald Coefficient L
298.06	4.250	0.4984	0.5439

A preliminary account of this work appeared in Conf. Int. Thermodyn. Chim. (C.R.) 4th 1975, 6, 122-128.

The solubility value was adjusted to a krypton partial pressure of 101.325 kPa by Henry's law.

The Bunsen coefficient was calculated by the compiler.

AUXILIARY INFORMATION

METHOD/APPARATUS/PROCEDURE:

   The solubility apparatus is based on the design of Morrison and Billett (1) and the version used is described by Battino, Evans, and Danforth (2). The degassing apparatus is that described by Battino, Banzhof, Bogan, and Wilhelm (3).
   See Krypton + Octane data sheet for more detail.

SOURCE AND PURITY OF MATERIALS:

1. Krypton. The Matheson Co., Inc., or Air Products and Chemicals. Purest commercially available.

2. 1-Decanol. Eastman Organic Chemicals. Distilled until refractive index value equals the value recommended by Wilhoit, R.C. Zwolinski, B. J. J. Phys. Chem. Ref. Data 1973, 2 (Suppl. No. 1), 1-212 and 1-278.

ESTIMATED ERROR:
$\delta T/K = 0.03$
$\delta P/mmHg = 0.5$
$\delta X_1/X_1 = 0.005$

REFERENCES:

1. Morrison, T.J.; Billett, F. J. Chem. Soc. 1948, 2033.
2. Battino,R.;Evans,F.D.;Danforth,W.F. J.Am.Oil Chem.Soc. 1968, 45, 830.
3. Battino,R.;Banzhof,M.;Bogan, M.; Anal. Chem. 1971, 43, 806.

COMPONENTS:	ORIGINAL MEASUREMENTS:
1. Krypton; Kr; 7439-90-9 2. 1,2,3-Propanetriol (Glycerol); $C_3H_8O_3$; 56-81-5	Körösy, F. Trans. Faraday Soc. 1937, 33, 416-425.
VARIABLES: T/K: 293.15 P/kPa: 101.325 (1 atm)	PREPARED BY: H.L. Clever

EXPERIMENTAL VALUES:

T/K	Mol Fraction $X_1 \times 10^3$	Bunsen Coefficient α	Ostwald Coefficient L
293.15	0.2	0.06	0.06

van Liempt and van Wijk (2) report a Bunsen coefficient of 0.01 for this system at 295.15 K and 101.325 kPa. Both the Körösy and the van Liempt and van Wijk solubility values appear to be of little more than qualitative significance.

AUXILIARY INFORMATION

METHOD:
The apparatus and method of Winkler (1) were used. However, the apparatus was usually not thermostated, and degassing was by evacuating and shaking the solvent, not by evacuating and boiling the solvent as was done by Winkler.

SOURCE AND PURITY OF MATERIALS:
1. Krypton. Source not given. The gas contained 5% xenon and 1% non-inert gases.

2. 1,2,3-Propantriol. No information.

APPARATUS/PROCEDURE:

ESTIMATED ERROR:

$$\delta X_1/X_1 = 0.05$$

REFERENCES:
1. Winkler, L.W.
Ber. 1891, 24, 89.

2. van Liempt, J.A.M.; van Wijk, W.
Rec. Trav. Chim. Pays-Bas 1937,
56,632.

COMPONENTS:	ORIGINAL MEASUREMENTS:
1. Krypton; Kr; 7439-90-9 2. Cyclohexanol; $C_6H_{12}O$; 108-93-0	Körösy, F. Trans. Faraday Soc. 1937, 33, 416-425.

VARIABLES:
    T/K: 295.15
    P/kPa: 101.325 (1 atm)
	PREPARED BY:
	
	    H.L. Clever

EXPERIMENTAL VALUES:

T/K	Mol Fraction $X_1 \times 10^3$	Bunsen Coefficient α	Ostwald Coefficient L
295.15	1.68	0.37	0.40

The mole fraction solubility and the Bunsen coefficient were calculated by the compiler. It was assumed that gas behavior is ideal, the Ostwald coefficient is independent of pressure, and that Henry's law is obeyed.

AUXILIARY INFORMATION

METHOD:	SOURCE AND PURITY OF MATERIALS:
The apparatus and method of Winkler (1) were used. However, the apparatus was usually not thermostated, and degassing was by evacuating and shaking the solvent, not by evacuating and boiling the solvent as was done by Winkler.	1. Krypton. Source not given. The gas contained 5% xenon and 1% non-inert gases. 2. Cyclohexanol. No information.

APPARATUS/PROCEDURE:

ESTIMATED ERROR:

$$\delta X_1/X_1 = 0.05$$

REFERENCES:

1. Winkler, L.W.
   Ber. 1891, 24, 89.

COMPONENTS:	ORIGINAL MEASUREMENTS:
1. Krypton; Kr; 7439-90-9 2. 2-Propanone (Acetone); C_3H_6O; 67-64-1	Körösy, F. Trans. Faraday Soc. 1937, 33, 416-425.

VARIABLES:	PREPARED BY:
T/K: 292.15 - 293.15 P/kPa: 101.325 (1 atm)	H.L. Clever

EXPERIMENTAL VALUES:

T/K	Mol Fraction $X_1 \times 10^3$	Bunsen Coefficient α	Ostwald Coefficient L
Technical Acetone			
292.15	-	-	0.83
Dried Acetone			
293.15	3.20	0.98	1.05

The mole fraction solubility and the Bunsen coefficient were calculated by the compiler. It was assumed that gas behavior is ideal, the Ostwald coefficient is independent of pressure, and that Henry's law is obeyed.

AUXILIARY INFORMATION

METHOD:	SOURCE AND PURITY OF MATERIALS:
The apparatus and method of Winkler (1) were used. However, the apparatus was usually not thermostated, and degassing was by evacuating and shaking the solvent, not by evacuating and boiling the solvent as was done by Winkler.	1. Krypton. Source not given. The gas contained 5% xenon and 1% non-inert gases. 2. 2-Propanone. Source not given. Technical grade and dried technical grade. No indication of the amount of water in the technical grade.

APPARATUS/PROCEDURE:	ESTIMATED ERROR:
	$\delta X_1/X_1 = 0.05$
	REFERENCES: 1. Winkler, L.W. Ber. 1891, 24, 89.

COMPONENTS:	ORIGINAL MEASUREMENTS:
1. Krypton; Kr; 7439-90-9 2. Acetic Acid, Glacial; $C_2H_4O_2$; 64-19-7	Körösy, F. Trans. Faraday Soc. 1937, 33, 416-425.

| VARIABLES:
   T/K: 295.15
   P/kPa: 101.325 (1 atm) | PREPARED BY:

   H.L. Clever |

EXPERIMENTAL VALUES:

T/K	Mol Fraction $X_1 \times 10^3$	Bunsen Coefficient α	Ostwald Coefficient L
295.15	1.11	0.44	0.47

The mole fraction solubility and the Bunsen coefficient were calculated by the compiler. It was assumed that gas behavior is ideal, the Ostwald coefficient is independent of pressure, and that Henry's law is obeyed.

AUXILIARY INFORMATION

METHOD:	SOURCE AND PURITY OF MATERIALS:
The apparatus and method of Winkler (1) were used. However, the apparatus was usually not thermostated, and degassing was by evacuating and shaking the solvent, not by evacuating and boiling the solvent as was done by Winkler.	1. Krypton. Source not given. The gas contained 5% xenon and 1% non-inert gases. 2. Acetic acid. Source not given. Glacial acetic acid.

APPARATUS/PROCEDURE:	ESTIMATED ERROR: $\delta X_1/X_1 = 0.05$
	REFERENCES: 1. Winkler, L. W. Ber. 1891, 24, 89.

COMPONENTS:	ORIGINAL MEASUREMENTS:
1. Krypton; Kr; 7439-90-9 2. Esters	Körösy, F. Trans. Faraday Soc. 1937, 33, 416-425.

VARIABLES:
     T/K: 293.15
     P/kPa: 101.325 (1 atm)
	PREPARED BY:
	
	     H.L. Clever

EXPERIMENTAL VALUES:

T/K	Mol Fraction $X_1 \times 10^3$	Bunsen Coefficient α	Ostwald Coefficient L
Butyl Acetate; $C_6H_{12}O_2$; 123-86-4			
293.15	4.6	0.79	0.85
Dibutyl Ester of 1,2 Benzene Dicarboxylic Acid (Dibutyl phthalate); $C_{16}H_{22}O_4$; 84-74-2			
293.15	5.2	0.44	0.47

The mole fraction solubility and the Bunsen coefficient were calculated by
the compiler. It was assumed that gas behavior is ideal, the Ostwald
coefficient is independent of pressure, and that Henry's law is obeyed.

AUXILIARY INFORMATION

METHOD: The apparatus and method of Winkler (1) were used. However, the apparatus was usually not thermostated, and degassing was by evacuating and shaking the solvent, not by evacuating and boiling the solvent as was done by Winkler.	SOURCE AND PURITY OF MATERIALS: 1. Krypton. Source not given. The gas contained 5% xenon and 1% non-inert gases. 2. Esters. Source not given. Technical grade.
APPARATUS/PROCEDURE:	ESTIMATED ERROR: $\delta X_1/X_1 = 0.05$
	REFERENCES: 1. Winkler, L. W. Ber. 1891, 24, 89.

COMPONENTS:	ORIGINAL MEASUREMENTS:
1. Krypton; Kr; 7439-90-9 2. Undecafluoro(trifluoromethyl)-cyclohexane (Perfluoromethyl-cyclohexane); C_7F_{14}; 355-02-2	Clever, H.L.; Saylor, J.H.; Gross, P.M. J. Phys. Chem. 1958, 62, 89-91.

VARIABLES:	PREPARED BY:
T/K: 289.15 - 316.25 Total P/kPa: 101.325 (1 atm)	P.L. Long

EXPERIMENTAL VALUES:

T/K	Mol Fraction X_1 x 10^3	Bunsen Coefficient α	Ostwald Coefficient L
289.15	9.06	1.058	1.12
303.15	8.08	0.925	1.027
316.25	7.77	0.872	1.01

Smoothed Data: $\Delta G^O/J\ mol^{-1} = - RT\ ln\ X_1 = -4302.3 + 54.078\ T$

Std. Dev. $\Delta G^O = 46.1$, Coef. Corr. = 0.9980

$\Delta H^O/J\ mol^{-1} = -4302.3$, $\Delta S^O/J\ K^{-1}\ mol^{-1} = -54.078$

T/K	Mol Fraction X_1 x 10^3	$\Delta G^O/J\ mol^{-1}$
288.15	9.02	11,280
293.15	8.75	11,551
298.15	8.49	11,821
303.15	8.25	12,091
308.15	8.03	12,362
313.15	7.81	12,632
318.15	7.61	12,903

The solubility values were adjusted to a partial pressure of krypton of 101.325 kPa (1 atm) by Henry's law.

The Bunsen coefficients were calculated by the compiler.

AUXILIARY INFORMATION

METHOD:
    Volumetric. The apparatus (1) is a modification of that used by Morrison and Billett (2). Modifications include the addition of a spiral solvent storage tubing, a manometer for constant reference pressure, and an extra gas buret for highly soluble gases.

SOURCE AND PURITY OF MATERIALS:
1. Krypton. Matheson Co., Inc. Both standard and research grades were used.

2. Perfluoromethylcyclohexane. du Pont FCS-326, shaken with concentrated H_2SO_4, washed, dried over Drierite and distilled.
b.p. 75.95 to 76.05° at 753 mm., lit. b.p. 76.14 at 760 mm.

APPARATUS/PROCEDURE:
    (a) Degassing. 700 ml of solvent is shaken and evacuated while attached to a cold trap, until no bubbles are seen; solvent is then transferred through a 1 mm capillary tubing, released as a fine mist into a continuously evacuated flask. (b) Solvent is saturated with gas as it flows through 8 mm x 180 cm of tubing attached to a gas buret. Pressure is maintained at 1 atm as the gas is absorbed.

ESTIMATED ERROR:
$\delta T/K = 0.05$
$\delta P/mmHg = 3$
$\delta X_1/X_1 = 0.03$

REFERENCES:
1. Clever, H.L.; Battino, R.; Saylor, J.H.; Gross, P.M. J. Phys. Chem. 1957, 61, 1078.

2. Morrison, T.J.; Billett, F. J. Chem. Soc. 1948, 2033; ibid. 1952, 3819.

COMPONENTS:	ORIGINAL MEASUREMENTS:

COMPONENTS:

1. Krypton; Kr; 7439-90-9

2. Dichlorodifluoromethane
   (Freon-12); CCl_2F_2; 75-71-8

ORIGINAL MEASUREMENTS:

Steinberg, M.; Manowitz, B.;
Pruzansky, J.

US AEC BNL-542 (T-140).
Chem. Abstr. 1959, 53, 21242g.

VARIABLES:

T/K: 190.15 - 273.15

PREPARED BY:

H.L. Clever

EXPERIMENTAL VALUES:

T/K	Absorption Coefficient	Henry's Constant K/atm	Mol Fraction $X_1 \times 10^2$	Bunsen Coefficient α	Ostwald Coefficient L
190.15	13.9	--	4.32	13.2	9.2
193.15	12.6	--	3.76	11.9	8.4
197.65	11.3	--	3.42	10.7	7.7$_5$
203.15	--	32.5	3.08	--	--$_5$
244.15	5.1	--	1.72	4.8	4.3
260.85	--	86	1.16	--	--
273.15	2.4	108	{0.925 / 0.89	2.3	2.3

Smoothed Data: $\Delta G^o/J\ mol^{-1} = - RT\ ln\ X_1 = -7772.6 + 67.346\ T$

Std. Dev. $\Delta G^o = 69.5$, Coef. Corr. = 0.9997

$\Delta H^o/J\ mol^{-1} = -7,772.6$, $\Delta S^o/J\ K^{-1}\ mol^{-1} = 0.9997$

T/K	Mol Fraction $X_1 \times 10^2$	$\Delta G^o/J\ mol^{-1}$
193.15	3.84	5,235.4
203.15	3.02	5,908.8
213.15	2.44	6,582.3
223.15	2.00	7,225.8
233.15	1.67	7,929.2
243.15	1.41	8,602.7
253.15	1.22	9,276.2
263.15	1.06	9,949.7
273.15	0.93	10,623

AUXILIARY INFORMATION

METHOD/APPARATUS/PROCEDURE:

Dynamic tracer technique (1). The Henry's constant is

$K = (P/atm)/X_1$.

The Henry's constants are probably from data smoothed by the authors.

The report is discussed further in a later paper (2).

SOURCE AND PURITY OF MATERIALS:

1. Krypton.

2. Dichlorodifluoromethane.

ESTIMATED ERROR:

$\delta X/X = 0.03 - 0.05$

(Compiler)

REFERENCES:

1. Steinberg, M.; Manowitz, B.
   Ind. Eng. Chem. 1959, 51, 47.

2. Steinberg, M.
   US AEC TID-7593, 1959, 217-218.
   Chem. Abstr. 1961, 55, 9083e.

COMPONENTS:	ORIGINAL MEASUREMENTS:
1. Krypton; Kr; 7439-90-9 2. Halogenated Methanes	Körösy, F. Trans. Faraday Soc. 1937, 33, 416-425.

VARIABLES:	PREPARED BY:
T/K: 273.15 - 295.15 P/kPa: 101.325 (1 atm)	H.L. Clever

EXPERIMENTAL VALUES:

T/K	Mol Fraction $X_1 \times 10^3$	Bunsen Coefficient α	Ostwald Coefficient L
Trichloromethane (Chloroform); $CHCl_3$; 67-66-3			
273.15	3.4	0.97	0.97
294.15	3.35	0.938	1.01
Tetrachloromethane (Carbon Tetrachloride); CCl_4; 56-23-5			
273.15	5.02	1.20	1.20
294.15	5.22	1.22	1.31
Tribromomethane (Bromoform); $CHBr_3$; 75-25-2			
295.15	3.01	0.43	0.46

The mole fraction solubility and the Bunsen coefficient were calculated by the compiler. It was assumed that gas behavior is ideal, the Ostwald coefficient is independent of pressure, and that Henry's law is obeyed.

AUXILIARY INFORMATION

METHOD:
   The apparatus and method of Winkler (1) were used. However, the apparatus was usually not thermostated, and degassing was by evacuating and shaking the solvent, not by evacuating and boiling the solvent as was done by Winkler.

SOURCE AND PURITY OF MATERIALS:
1. Krypton. Source not given. The gas contained 5% xenon and 1% non-inert gases.

2. Solvents. No information.

APPARATUS/PROCEDURE:

ESTIMATED ERROR:

$$\delta X_1/X_1 = 0.05$$

REFERENCES:
1. Winkler, L.W. Ber. 1891, 24, 89.

COMPONENTS:	ORIGINAL MEASUREMENTS:
1. Krypton; Kr; 7439-90-9 2. Fluorobenzene; C_6H_5F; 462-06-6	Saylor, J.H.; Battino, R. J. Phys. Chem. 1958, 62, 1334-1337.

VARIABLES: T/K: 288.15 - 328.15 P/kPa: 101.325 (1 atm)	PREPARED BY: H.L. Clever, A.L. Cramer

EXPERIMENTAL VALUES:

T/K	Mol Fraction $X_1 \times 10^3$	Bunsen Coefficient α	Ostwald Coefficient L
288.15	3.47	0.834	0.880
298.15	3.36	0.798	0.871
313.15	3.13	0.731	0.838
328.15	2.96	0.678	0.814

Smoothed Data: $\Delta G^O/J\ mol^{-1} = - RT \ln X_1 = -3210.3 + 58.186\ T$

Std. Dev. ΔG^O = 13.2, Coef. Corr. = .9999

$\Delta H^O/J\ mol^{-1} = -3210.3$, $\Delta S^O/J\ K^{-1}\ mol^{-1} = -58.186$

T/K	Mol Fraction $X_1 \times 10^3$	$\Delta G^O/J\ mol^{-1}$
288.15	3.49	13,556
293.15	3.41	13,847
298.15	3.33	14,138
303.15	3.26	14,429
308.15	3.20	14,720
313.15	3.13	15,011
318.15	3.07	15,302
323.15	3.02	15,593
328.15	2.96	15,884

Solubility values were adjusted to a partial pressure of krypton of 101.325 kPa (1 atm) by Henry's law. Bunsen coefficients were calculated by the compiler.

AUXILIARY INFORMATION

METHOD /APPARATUS/PROCEDURE:	SOURCE AND PURITY OF MATERIALS:
The solvent was degassed by evacuating the space above it, shaking, and then passing it as a fine mist into another evacuated container. The degassed liquid was saturated as it passed as a thin film inside a glass helix which contained the solute gas plus solvent vapor at a total pressure of 1 atm (1,2). The volume of liquid and the volume of gas absorbed are determined in a system of burets.	1. Krypton. Linde Air Products Co. 2. Fluorobenzene. Eastman Kodak Co., white label. Dried over P_4O_{10}, distilled, b.p. 84.28 - 84.68°C.

ESTIMATED ERROR:

$\delta T/K = 0.03$

$\delta P/mmHg = 1.0$

$\delta X_1/X_1 = 0.005$ (authors)

REFERENCES:

1. Morrison, T.J.; Billett, F. J. Chem. Soc. 1948, 2033.

2. Clever, H.L.; Battino, R.; Saylor, J.H.; Gross, P.M. J. Phys. Chem. 1957, 61, 1078.

COMPONENTS:	ORIGINAL MEASUREMENTS:
1. Krypton; Kr; 7439-90-9 2. Hexafluorobenzene; C_6F_6; 392-56-3	Evans, F. D.; Battino, R. J. Chem. Thermodyn. 1971, 3, 753-760.
VARIABLES: T/K: 282.91 - 297.92 P/kPa: 101.325 (1 atm)	PREPARED BY: H. L. Clever

EXPERIMENTAL VALUES:

T/K	Mol Fraction $X_1 \times 10^3$	Bunsen Coefficient α	Ostwald Coefficient L
282.91	6.45	1.286	1.332
283.09	6.42	1.279	1.326
297.76	5.92	1.155	1.259
297.92	5.89	1.148	1.252

Smoothed Data: $\Delta G°/J \ mol^{-1} = - RT \ ln \ X_1 = -4,063.6 + 56.302 \ T$

Std. Dev. $\Delta G° = 4.9$, Coef. Corr. = 0.9999

$\Delta H°/J \ mol^{-1} = -4063.6$, $\Delta S°/J \ K^{-1} \ mol^{-1} = -56.302$

T/K	Mol Fraction $X_1 \times 10^3$	$\Delta G°/J \ mol^{-1}$
278.15	6.64	11,597
283.15	6.44	11,878
288.15	6.25	12,160
293.15	6.07	12,441
298.15	5.90	12,723

The solubility values were adjusted to a partial pressure of krypton of 101.325 kPa (1 atm) by Henry's law.

The Bunsen coefficients were calculated by the compiler.

AUXILIARY INFORMATION

METHOD /APPARATUS/PROCEDURE:

The solubility apparatus is based on the design of Morrison and Billett (1) and the version used is described by Battino, Evans, and Danforth (2). The degassing apparatus is that described by Battino, Banzhof, Bogan, and Wilhelm (3).

Degassing. Up to 500 cm³ of solvent is placed in a flask of such size that the liquid is about 4 cm deep. The liquid is rapidly stirred and the vacuum is applied intermittently through a liquid N_2 trap until the permanent gas residual pressure drops to 5 microns.

Solubility Determination. The degassed solvent is passed in a thin film down a glass spiral tube containing solute gas plus the solvent vapor at a total pressure of one atm. The volume of gas absorbed is found by difference between the initial and final volumes in the buret system. The solvent is collected in a tared flask and weighed.

SOURCE AND PURITY OF MATERIALS:

1. Krypton. Either Air Products & Chemicals or Matheson. Better than 99 mol % (usually 99.9+)

2. Hexafluorobenzene. Imperial Smelting Co., Avonmouth, U.K. GC purity 99.7%, density at 25° C 1.60596 g cm⁻³. Purified by method in Anal. Chem. 1968, 40, 224.

ESTIMATED ERROR:
$\delta T/K = 0.03$
$\delta P/mmHg = 0.5$
$\delta X_1/X_1 = 0.015$

REFERENCES:
1. Morrison, T. J.; Billett, F. J. Chem. Soc. 1948, 2033.
2. Battino, R.; Evans, F. D.; Danforth, W. F. J. Am. Oil Chem. Soc. 1968, 45, 830.
3. Battino, R.; Banzhof, M.; Bogan, M.; Wilhelm, E. Anal. Chem. 1971, 43, 806.

COMPONENTS:	ORIGINAL MEASUREMENTS:
1. Krypton; Kr; 7439-90-9	Saylor, J.H.; Battino, R.
2. Chlorobenzene; C_6H_5Cl; 108-90-7	J. Phys. Chem. 1958, 62, 1334-1337.

VARIABLES:	PREPARED BY:
T/K: 288.15 - 328.15 P/kPa: 101.325 (1 atm)	H.L. Clever, A.L. Cramer

EXPERIMENTAL VALUES:

T/K	Mol Fraction X_1 x 10^3	Bunsen Coefficient α	Ostwald Coefficient L
288.15	2.84	0.629	0.664
298.15	2.75	0.604	0.659
313.15	2.56	0.553	0.634
328.15	2.44	0.520	0.625

Smoothed Data:

$$\Delta G^o/J\ mol^{-1} = -RT\ ln\ X_1 = -3079.8 + 59.412\ T$$

Std. Dev. ΔG^o = 13.8, Coef. Corr. = .9999

$$\Delta H^o/J\ mol^{-1} = -3079.8, \quad \Delta S^o/J\ K^{-1}\ mol^{-1} = -59.412$$

T/K	Mol Fraction X_1 x 10^3	$\Delta G^o/J\ mol^{-1}$
288.15	2.85	14,040
293.15	2.79	14,337
298.15	2.73	14,634
303.15	2.67	14,931
308.15	2.62	15,228
313.15	2.57	15,525
318.15	2.53	15,822
323.15	2.48	16,119
328.15	2.44	16,416

Solubility values were adjusted to a partial pressure of krypton of 101.325 kPa (1 atm) by Henry's law. Bunsen coefficients were calculated by the compiler.

AUXILIARY INFORMATION

METHOD /APPARATUS/PROCEDURE:

The solvent was degassed by evacuating the space above it, shaking, and then passing it as a fine mist into another evacuated container. The degassed liquid was saturated as it passed as a thin film inside a glass helix which contained the solute gas plus solvent vapor at a total pressure of 1 atm (1,2). The volume of the liquid and the volume of gas absorbed were determined in a system of burets.

SOURCE AND PURITY OF MATERIALS:

1. Krypton. Linde Air Products Co.

2. Chlorobenzene. Eastman Kodak Co., white label. Dried over P_4O_{10}, distilled, b.p. 131.67 - 131.71°C.

ESTIMATED ERROR:
$$\delta T/K = 0.03$$
$$\delta P/mmHg = 1.0$$
$$\delta X_1/X_1 = 0.005\ (authors)$$

REFERENCES:

1. Morrison, T.J.; Billett, F. J. Chem. Soc. 1948, 2033.

2. Clever, H.L.; Battino, R.; Saylor, J.H.; Gross, P.M. J. Phys. Chem. 1957, 61, 1078.

COMPONENTS:	ORIGINAL MEASUREMENTS:
1. Krypton; Kr; 7439-90-9 2. 1,4-Dimethylbenzene (p-Xylene); C_8H_{10}; 106-42-3 3. 1,4-Dichlorobenzene; $C_6H_4Cl_2$; 106-46-7	Clever, H. L. J. Phys. Chem. 1957, 61, 1082 - 1083.

VARIABLES:	PREPARED BY:
T/K: 303.15 P/kPa: 101.325 (1 atm) 1,4-Dichlorobenzene/X_3: 0 - 0.455	C. E. Eddelman A. L. Cramer

EXPERIMENTAL VALUES:

T/K	1,4-Dichloro-benzene Mol Fraction X_3	Mol Fraction $X_1 \times 10^3$	Bunsen Coefficient α	Ostwald Coefficient L
303.15	0.0	3.81	0.687	0.762
	0.170	3.56	0.652	0.724
	0.310	3.32	0.566	0.628
	0.455	3.10	0.583	0.647

The mole fraction solubility of krypton correlates well in a linear function with the mole fraction of 1,4-dichlorobenzene in the mixed solvent.

$$X_1 \times 10^3 = 3.815 - 1.573 X_3 \quad (r = 0.9996)$$

The solubility values were adjusted to a partial pressure of krypton of 101.325 kPa (1 atm) by Henry's law.

The Bunsen coefficients were calculated by the compiler.

AUXILIARY INFORMATION

METHOD:	SOURCE AND PURITY OF MATERIALS:
Volumetric. The solvent is saturated with gas as it flows through an 8 mm x 180 cm glass helix attached to a gas buret. The total pressure of solute gas plus solvent vapor is maintained at 1 atm as the gas is absorbed	1. Krypton. Linde Air Products Co. 2. 1,4-Dimethylbenzene. Eastman Kodak white label. Distilled. 3. 1,4-Dichlorobenzene. Eastman Kodak white label. Recrytalized twice from methanol, dried in air.

APPARATUS/PROCEDURE:	ESTIMATED ERROR:
The apparatus is a modification of the apparatus of Morrison and Billett (1). The modifications include the addition of a helical storage for the solvent, a manometer for a reference pressure, and an extra buret for highly soluble gases. The solvent is degassed by a modification of the method of Baldwin and Daniel (2).	$\delta T/K = 0.05$ $\delta P/mmHg = 3$ $\delta X_1/X_1 = 0.03$
	REFERENCES: 1. Morrison, T.J.; Billett, F. J. Chem. Soc. 1948, 2033; Ibid. 1952, 3819. 2. Baldwin, R.R.; Daniel, S.G. J. Appl. Chem. 1952, 2, 161.

COMPONENTS:	ORIGINAL MEASUREMENTS:
1. Krypton; Kr; 7439-90-9 2. Bromobenzene; C_6H_5Br; 108-86-1	Saylor, J.H.; Battino, R. J. Phys. Chem. 1958, 62, 1334-1337.

VARIABLES:	PREPARED BY:
T/K: 288.15 - 328.15 P/kPa: 101.325 (1 atm)	H.L. Clever, A.L. Cramer

EXPERIMENTAL VALUES:

T/K	Mol Fraction $X_1 \times 10^3$	Bunsen Coefficient α	Ostwald Coefficient L
288.15	2.38	0.510	0.538
298.15	2.29	0.487	0.532
313.15	2.14	0.448	0.514
328.15	2.04	0.423	0.508

Smoothed Data: $\Delta G^O/J\ mol^{-1} = -RT\ \ln X_1 = -3085.7 + 60.924\ T$

Std. Dev. ΔG^O = 9.1, Coef. Corr. = .9999

$\Delta H^O/J\ mol^{-1} = -3085.7$, $\Delta S^O/J\ K^{-1}\ mol^{-1} = -60.924$

T/K	Mol Fraction $X_1 \times 10^3$	$\Delta G^O/J\ mol^{-1}$
288.15	2.38	14,470
293.15	2.33	14,774
298.15	2.28	15,079
303.15	2.24	15,383
308.15	2.19	15,688
313.15	2.15	15,993
318.15	2.11	16,297
323.15	2.07	16,602
328.15	2.04	16,907

Solubility values were adjusted to a partial pressure of krypton of 101.325 kPa (1 atm) by Henry's law. Bunsen coefficients were calculated by the compiler.

AUXILIARY INFORMATION

METHOD/APPARATUS/PROCEDURE:	SOURCE AND PURITY OF MATERIALS:
The solvent was degassed by evacuating the space above it, shaking, and then passing it as a fine mist into another evacuated container. The degassed liquid was saturated as it passed as a thin film inside a glass helix which contained the solute gas plus solvent vapor at a totalpressure of 1 atm (1,2).The volume of liquid and the volume of gas absorbed are determined in a system of burets.	1. Krypton. Linde Air Products Co. 2. Bromobenzene. Eastman Kodak Co., white label. Dried over P_4O_{10}, distilled, b.p. 155.86 - 155.90°C.

ESTIMATED ERROR:

$\delta T/K = 0.03$
$\delta P/mmHg = 1.0$
$\delta X_1/X_1 = 0.005$ (authors)

REFERENCES:

1. Morrison, T.J.; Billett, F. J. Chem. Soc. 1948, 2033.

2. Clever, H.L.; Battino, R.; Saylor, J.H.; Gross, P. M. J. Phys. Chem. 1957, 61, 1078.

COMPONENTS:	ORIGINAL MEASUREMENTS:
1. Krypton; Kr; 7439-90-9 2. 1,4-Dimethylbenzene (p-Xylene); C_8H_{10}; 106-42-3 3. 1,4-Dibromobenzene; $C_6H_4Br_2$; 106-37-6	Clever, H.L. J. Phys. Chem. 1957, 61, 1082-1083.

VARIABLES: T/K: 303.15 P/kPa: 101.325 (1 atm) 1,4-Dibromobenzene/X_3: 0-0.255	PREPARED BY: C.E. Eddleman A.L. Cramer

EXPERIMENTAL VALUES:

T/K	1,4-Dibromo-benzene Mol Fraction X_3	Mol Fraction $X_1 \times 10^3$	Bunsen Coefficient α	Ostwald Coefficient L
303.15	0.0	3.81	0.687	0.762
	0.130	3.41	0.618	0.686
	0.255	3.13	0.569	0.631

The solubility values were adjusted to a partial pressure of krypton of 101.325 kPa (1 atm) by Henry's law.

The Bunsen coefficients were calculated by the compiler.

AUXILIARY INFORMATION

METHOD:

Volumetric. The solvent is saturated with gas as it flows through an 8 mm x 180 cm glass helix attached to a gas buret. The total pressure of solute gas plus solvent vapor is maintained at 1 atm as the gas is absorbed.

SOURCE AND PURITY OF MATERIALS:

1. Krypton. Linde Air Products Co.

2. 1,4-Dimtheylbenzene. Eastman Kodak white label. Distilled.

3. 1,4-Dibromobenzene. Eastman Kodak white label. Recrystallized twice and dried in air.

APPARATUS/PROCEDURE:

The apparatus is a modification of the apparatus of Morrison and Billett (1). The modifications include the addition of a helical storage for the solvent, a manometer for a reference pressure, and an extra buret for highly soluble gases. The solvent is degassed by a modification of the method of Baldwin and Daniel (2).

ESTIMATED ERROR:
$$\delta T/K = 0.05$$
$$\delta P/mmHg = 3$$
$$\delta X_1/X_1 = 0.03$$

REFERENCES:

1. Morrison, T.J.; Billett, F. J. Chem. Soc. 1948, 2033; ibid. 1952, 3819.

2. Baldwin, R.R.; Daniel, S.G. J. Appl. Chem. 1952, 2, 161.

COMPONENTS:	ORIGINAL MEASUREMENTS:
1. Krypton; Kr; 7439-90-9 2. Iodobenzene; C_6H_5I; 591-50-4	Saylor, J.H.; Battino, R. J. Phys. Chem. 1958, 62, 1334-1337.
VARIABLES: T/K: 288.15 - 328.15 P/kPa: 101.325 (1 atm)	PREPARED BY: H.L. Clever, A.L. Cramer

EXPERIMENTAL VALUES:

T/K	Mol Fraction X_1 x 10^3	Bunsen Coefficient α	Ostwald Coefficient L
288.15	1.73	0.349	0.368
298.15	1.70	0.339	0.370
313.15	1.63	0.322	0.369
328.15	1.58	0.309	0.371

Smoothed Data: ΔG^O/J mol^{-1} = - RT ln X_1 = -1841.9 + 59.241 T

Std. Dev. ΔG^O = 8.8, Coef. Corr. = .9999

ΔH^O/J mol^{-1} = -1841.9, ΔS^O/J K^{-1} mol^{-1} = -59.241

T/K	Mol Fraction X_1 x 10^3	ΔG^O/J mol^{-1}
288.15	1.74	15,228
293.15	1.71	15,525
298.15	1.69	15,821
303.15	1.67	16,117
308.15	1.65	16,413
313.15	1.63	16,709
318.15	1.61	17,006
323.15	1.60	17,302
328.15	1.58	17,598

Solubility values were adjusted to a partial pressure of krypton of 101.325 kPa (1 atm) by Henry's law. Bunsen coefficients were calculated by the compiler.

AUXILIARY INFORMATION

METHOD /APPARATUS/PROCEDURE:

The solvent was degassed by evacuating the space above it, shaking, and then passing it as a fine mist into another evacuated container. The degassed liquid was saturated as it passed as a thin film inside a glass helix which contained the solute gas plus solvent vapor at a total pressure of 1 atm (1,2). The volume of liquid and the volume of gas absorbed are determined in a system of burets.

SOURCE AND PURITY OF MATERIALS:

1. Krypton. Linde Air Products Co.

2. Iodobenzene. Eastman Kodak white label. Shaken with aq. $Na_2S_2O_3$, dried over P_4O_{10}, distilled. b.p. 77.40 - 77.60°C (20 mmHg).

ESTIMATED ERROR:
$\delta T/K$ = 0.03
δP/mmHg = 1.0
$\delta X_1/X_1$ = 0.005 (authors)

REFERENCES:

1. Morrison, T.J.; Billett, F. J. Chem. Soc. 1948, 2033.

2. Clever, H.L.; Battino, R.; Saylor, J.H.; Gross, P.M. J. Phys. Chem. 1957, 61, 1078.

COMPONENTS:	ORIGINAL MEASUREMENTS:
1. Krypton; Kr; 7439-90-9 2. 1,4-Dimethylbenzene (p-Xylene); C_8H_{10}; 106-42-3 3. 1,4-Diiodobenzene; $C_6H_4I_2$; 624-38-4	Clever, H.L. J. Phys. Chem. 1957, 61, 1082-1083.

VARIABLES:	PREPARED BY:
T/K: 303.15 P/kPa: 101.325 (1 atm) 1,4-Diiodobenzene/X_3: 0-0.078	C.E. Eddleman A.L. Cramer

EXPERIMENTAL VALUES:

T/K	1,4-Diiodo-benzene Mol Fraction X_3	Mol Fraction X_1 x 10^3	Bunsen Coefficient α	Ostwald Coefficient L
303.15	0.0	3.81	0.687	0.762
	0.078	3.48	0.626	0.695

The solubility values were adjusted to a partial pressure of krypton of 101.325 kPa (1 atm) by Henry's law.

The Bunsen coefficients were calculated by the compiler.

AUXILIARY INFORMATION

METHOD:

Volumetric. The solvent is saturated with gas as it flows through an 8 mm x 180 cm glass helix attached to a gas buret. The total pressure of solute gas plus solvent vapor is maintained at 1 atm as the gas is absorbed.

SOURCE AND PURITY OF MATERIALS:

1. Krypton. Linde Air Products Co.
2. 1,4-Dimethylbenzene. Eastman Kodak white label. Distilled.
3. 1,4-Diiodobenzene. Eastman Kodak white label. Recrystallized twice and dried in air.

APPARATUS/PROCEDURE:

The apparatus is a modification of the apparatus of Morrison and Billett (1). The modifications include the addition of a helical storage for the solvent, a manometer for a reference pressure, and an extra buret for highly soluble gases. The solvent is degassed by a modification of the method of Baldwin and Daniel (2).

ESTIMATED ERROR:
$$\delta T/K = 0.05$$
$$\delta P/mmHg = 3$$
$$\delta X_1/X_1 = 0.03$$

REFERENCES:

1. Morrison, T.J.; Billett, F. J. Chem. Soc. 1948, 2033; ibid. 1952, 3819.
2. Baldwin, R.R.; Daniel, S.G. J. Appl. Chem. 1952, 2, 161.

97

COMPONENTS:	ORIGINAL MEASUREMENTS:
1. Krypton; Kr; 7439-90-9 2. 1,1,2,2,3,3,4,4,4-nonafluoro-N, N-bis (nonafluorobutyl)-1-butanamine (Perfluorotributyl-amine); $(C_4F_9)_3N$; 311-89-7	Powell, R.J. $\underline{J}$. $\underline{Chem}$. $\underline{Eng}$. $\underline{Data}$ 1972, $\underline{17}$, 302-304.
VARIABLES: T/K: 298.15 P/kPa: 101.325 (1 atm)	PREPARED BY: P.L. Long

EXPERIMENTAL VALUES:

T/K	Mol Fraction X_1 x 10^3	Bunsen Coefficient α	Ostwald Coefficient L	$R\frac{\Delta \log x_1}{\Delta \log T} = N$
298.15	11.15	0.708	0.773	-3.30

The author implies that solubility measurements were made between 288.15 and 318.15 K, but only the solubility at 298.15 was given in the paper. The slope $R(\Delta \log x_1/\Delta \log T)$ was given. The smoothed data below were calculated by the compiler from the slope in the form:

$$\log x_1 = \log(11.15 \times 10^{-3}) + (-3.30/R)\log(T/298.15)$$

with $R = 1.9872$ cal K^{-1} mol^{-1}.

Smoothed Data:

T/K	Mol Fraction X_1 x 10^3
288.15	11.81
293.15	11.47
298.15	11.15
303.15	10.85
308.15	10.55
313.15	10.28
318.15	10.01

The Bunsen and Ostwald Coefficients were calculated by the compiler.

AUXILIARY INFORMATION

METHOD /APPARATUS/PROCEDURE:	SOURCE AND PURITY OF MATERIALS:
Solvent is degassed by freezing and pumping, then boiling under reduced pressure. The Dymond and Hildebrand (1) apparatus, with all glass pumping system, is used to spray slugs of degassed solvent into the krypton. Amount of gas dissolved is calculated from the initial and final gas pressure.	1. Krypton. No source. Manufacturer's research grade, dried over $CaCl_2$ before use. 2. Perfluorotributylamine. Minnesota Mining & Mfg. Co. Column distilled, used portion with b.p. = 447.85 - 448.64K, & single peak GC.

ESTIMATED ERROR:

$$\delta N /cal\ K^{-1}\ mol^{-1} = 0.1$$
$$\delta X_1/X_1 = 0.002$$

REFERENCES:

1. Dymond, J.; Hildebrand, J.H. $\underline{Ind}$. $\underline{Eng}$. $\underline{Chem}$. $\underline{Fundam}$. 1967, $\underline{6}$, 130.

COMPONENTS:	ORIGINAL MEASUREMENTS:
1. Krypton; Kr; 7439-90-9 ^{85}Kr; 13983-27-2 2. Carbon Dioxide; CO_2; 124-38-9	Notz, K. J.; Meservey, A. B. ORNL-5121, June 1976 Chem. Abstr. 1977, 86, 61170c, 79549t.
VARIABLES: T/K: 223.15 - 301.15	PREPARED BY: A. L. Cramer H. L. Clever

EXPERIMENTAL VALUES:

T/K	Mol Fraction $X_1 \times 10^3$	Bunsen Coefficient α	Ostwald Coefficient L
223.15	5.72	3.370	2.75
233.15	5.48	3.126	2.67
243.15	5.26	2.894	2.57
253.15	5.26	2.771	2.57
263.15	5.38	2.701	2.60
273.15	5.61	2.655	2.66
283.15	6.00	2.642	2.74
293.15	6.51	2.564	2.75
297.15	6.97	2.567	2.79
301.15	8.46	2.808	3.10

The mole fraction solubility at a krypton partial pressure of 101.325 kPa (1 atm) was calculated by the compiler.

A smoothed data fit with thermodynamic values for the transfer of one mole of krypton from the gas at 101.325 kPa to the hypothetical unit mole fraction krypton liquid is on the next page. The mole fraction solubility value at 301.15 K was not included in the smoothed data fit.

Another report (1) on this system gives values lower by a factor of 2. The results are thought to be in error because of a systematic sampling error.

AUXILIARY INFORMATION

METHOD:	SOURCE AND PURITY OF MATERIALS:
Tracer technique. Collimated counter with equilibrated gas-liquid samples. Krypton gas was 5 per cent ^{85}Krypton and 95 per cent stable Kr. The total pressure of the system was the equilibrium pressure of liquid CO_2 + the Kr pressure.	1. Krypton. Cryogenic Rare Gas Labs. Ultra high purity grade. Krypton-85. Isotopes Div., ORNL. 2. Carbon dioxide. Matheson Co., Inc. Research grade.
APPARATUS/PROCEDURE:	ESTIMATED ERROR: $\delta L/L$ = 0.006 at 220 K, less at higher temperatures.
	REFERENCES: 1. Laser, M.; Barnert-Wiemer, H.; Beanjean, H.; Merz, E.; Vygen, H. Proc. 13th AEC Air Cleaning Conf., 1974. CONF-740807, 1, 246-262.

COMPONENTS:	ORIGINAL MEASUREMENTS:

COMPONENTS:
1. Krypton; Kr; 7439-90-9
   ^{85}Kr; 13983-27-2
2. Carbon Dioxide; CO_2; 124-38-9

ORIGINAL MEASUREMENTS:
Notz, K.J.; Meservey, A. B.

ORNL-5121, June 1976
Chem. Abstr. 1977, 86, 61170c,79548t.

VARIABLES:

T/K: 223.15 - 301.15

PREPARED BY:
A.L. Cramer
H.L. Clever

EXPERIMENTAL VALUES:

Smoothed Data: $\ln X_1 = 41.5327 - 65.4529/(T/100) - 67.2463 \ln (T/100) + 16.4072 (T/100)$

Std. error about regression line 5.429×10^{-3}.

T/K	Mol Fraction $X_1 \times 10^3$	$\Delta G^{\circ}/kJmol^{-1}$	$\Delta H^{\circ}/kJmol^{-1}$	$\Delta S^{\circ}/JK^{-1}mol^{-1}$	$\Delta Cp^{\circ}/JK^{-1}mol^{-1}$
223.15	5.73	9.578	-2.416	-53.75	49.71
233.15	5.45	10.10	-1.782	-50.98	77.00
243.15	5.30	10.59	-0.876	-47.18	104.3
253.15	5.26	11.04	+0.302	-42.43	131.6
263.15	5.36	11.44	1.755	-36.81	158.8
273.15	5.60	11.78	3.480	-30.38	186.1
283.15	6.00	12.05	5.478	-23.20	213.4
293.15	6.60	12.24	7.748	-15.32	240.7
298.15	6.99	12.30	8.986	-11.13	254.3

The mol fraction solubility at 301.15 K was not used in the smoothed data fit.

AUXILIARY INFORMATION

METHOD:
See preceeding page.

SOURCE AND PURITY OF MATERIALS:
See preceeding page.

APPARATUS/PROCEDURE:

ESTIMATED ERROR:
See preceeding page.

REFERENCES:
See preceeding page.

COMPONENTS:	ORIGINAL MEASUREMENTS:
1. Krypton; Kr; 7439-90-9 2. Carbon Disulfide; CS_2; 75-15-0	Powell, R.J. J. Chem. Eng. Data 1972, 17, 302-304.

VARIABLES:
    T/K: 298.15
    P/kPa: 101.325 (1 atm)
	PREPARED BY:
	
	    P.L. Long

EXPERIMENTAL VALUES:

T/K	Mol Fraction $X_1 \times 10^3$	Bunsen Coefficient α	Ostwald Coefficient L	$R\frac{\Delta \log x_1}{\Delta \log T} = N$
298.15	1.756	0.650	0.710	-7.30

The author implies that solubility measurements were made between 273.15 and 308.15 K, but only the solubility at 298.15 was given in the paper. The slope $R(\Delta \log x_1/\Delta \log T)$ was given. The smoothed data below were calculated by the compiler from the slope in the form:

$$\log x_1 = \log(1.756 \times 10^3) + (-7.30/R)\log(T/298.15)$$

with $R = 1.9872$ cal K^{-1} mol^{-1}.

Smoothed Data:

T/K	Mol Fraction $X_1 \times 10^3$
273.15	2.422
278.15	2.266
283.15	2.122
288.15	1.980
293.15	1.868
298.15	1.756
303.15	1.652

The Bunsen and Ostwald Coefficients were calculated by the compiler.

AUXILIARY INFORMATION

METHOD/APPARATUS/PROCEDURE:

Solvent is degassed by freezing and pumping, then boiling under reduced pressure. The Dymond and Hildebrand (1) apparatus, with all glass pumping system, is used to spray slugs of degassed solvent into the krypton. Amount of gas dissolved is calculated from the initial and final gas pressures.

SOURCE AND PURITY OF MATERIALS:

1. Krypton. No source. Manufacturer's research grade, dried over $CaCl_2$ before use.

2. Carbon disulfide. No source given. Manufacturer's spectrochemical grade.

ESTIMATED ERROR:

$$\delta N /cal\ K^{-1}mol^{-1} = 0.1$$
$$\delta X_1/X_1 = 0.002$$

REFERENCES:
1. Dymond, J.; Hildebrand, J.H. Ind. Eng. Chem. Fundam. 1967, 6, 130.

COMPONENTS:	ORIGINAL MEASUREMENTS:
1. Krypton, Kr; 7439-90-9 2. Sulfinylbismethane (Dimethyl Sulfoxide); C_2H_6OS (CH_3SOCH_3); 67-68-5	Dymond, J. H. J. Phys. Chem. 1967, 71, 1829-1831.

VARIABLES:	PREPARED BY:
T/K: 298.15 P/kPa: 101.325 (1 atm)	M. E. Derrick

EXPERIMENTAL VALUES:

T/K	Mol Fraction $X_1 \times 10^3$	Bunsen Coefficient α	Ostwald Coefficient L
298.15	0.446	0.140	0.153

The Bunsen and Ostwald coefficients were calculated by the compiler.

AUXILIARY INFORMATION

METHOD /APPARATUS/PROCEDURE:	SOURCE AND PURITY OF MATERIALS:
The liquid is saturated with the gas at a gas partial pressure of 1 atm. The apparatus is that described by Dymond and Hildebrand (1). The apparatus uses an all-glass pumping system to spray slugs of degassed solvent into the gas. The amount of gas dissolved is calculated from the initial and final gas pressure.	1. Krypton. Matheson Co. Dried. 2. Dimethyl Sulfoxide. Matheson, Coleman, and Bell Co. Spectroquality reagent, dried, and a fraction frozen out. Melting pt. 18.37° C.
	ESTIMATED ERROR:
	REFERENCES: 1. Dymond, J.; Hildebrand, J. H. Ind. Eng. Chem. Fundam. 1967, 6, 130.

COMPONENTS:	ORIGINAL MEASUREMENTS:
1. Krypton; Kr; 7439-90-9 2. Nitrous Oxide; N_2O; 10024-97-2	Steinberg, M.; Manowitz, B.; Pruzansky, J. US AEC BNL-542 (T-140). Chem. Abstr. 1959, 53, 21242g.

VARIABLES:	PREPARED BY:
T/K: 190.15 - 243.65	H. L. Clever

EXPERIMENTAL VALUES:

T/K	Absorption Coefficient	Henry's Constant K/atm	Mol Fraction X_1 x 10^2	Bunsen Coefficient α	Ostwald Coefficient L
190.15	8.7	–	1.30	8.2	5.7
193.15	8.5	–	1.30	8.1	5.7
201.15	–	88	1.14	–	–
215.15	5.0	–	0.79$_5$	4.7	3.7
236.15	3.8	–	0.64^5	3.6	3.1
240.15	–	155	0.64$_5$	–	–
243.65	–	165	0.60$_5$	–	–

Smoothed Data: $\Delta G^o/J\ mol^{-1} = -\ RT\ ln\ X_1 = -5,741.0 + 66.134\ T$

Std. Dev. ΔG^o = 85.4, Coef. Corr. = 0.9984

$\Delta H^o/J\ mol^{-1} = -5,741.0$, $\Delta S^o/J\ K^{-1}\ mol^{-1} = -66.134$

T/K	Mol Fraction X_1 x 10^2	$\Delta G^o/J\ mol^{-1}$
193.15	1.25	7,032.8
203.15	1.05	7,694.1
213.15	0.896	8,355.4
223.15	0.775	9,016.8
233.15	0.679	9,678.1
243.15	0.601	10,339

AUXILIARY INFORMATION

METHOD/APPARATUS/PROCEDURE:	SOURCE AND PURITY OF MATERIALS:
Dynamic tracer technique (1). The Henry's constant is $K = (P/atm)/X_1$ The Henry's constants may be from data smoothed by the authors. The report is discussed further in a later paper (2).	1. Krypton. 2. Nitrous oxide. No information on either component.

ESTIMATED ERROR:

$\delta X/X = 0.03 - 0.05$

(Compiler)

REFERENCES:

1. Steinberg, M.; Manowitz, B. Ind. Eng. Chem. 1959, 51, 47.

2. Steinberg, M. US AEC TID-7593, 1959, 217-218. Chem. Abstr. 1961, 55, 9083e.

COMPONENTS:	ORIGINAL MEASUREMENTS:
1. Krypton; Kr; 7439-90-9 2. Nitromethane; CH_3NO_2; 75-52-5	Friedman, H.L. J. Am. Chem. Soc. 1954, 76, 3294-3297.

| VARIABLES:
 T/K: 298.00
 P/kPa: 101.325 (1 atm) | PREPARED BY:

 P.L. Long |

EXPERIMENTAL VALUES:

T/K	Mol Fraction $X_1 \times 10^3$	Bunsen Coefficient α	Ostwald Coefficient L
298.00			0.381 0.378
	0.838	0.348	0.380 av.

The author reports Ostwald coefficients measured at about 700 mmHg. The Bunsen coefficient and the mole fraction solubility at 101.325 kPa (1 atm) were calculated by the compiler with the assumptions that the gas is ideal, that Henry's law is obeyed, and that the Ostwald coefficient is independent of pressure.

AUXILIARY INFORMATION

METHOD:

Gas absorption. The method was essentially that employed by Eucken and Herzberg (1). Modifications included a magnetic stirring device instead of shaking the saturation vessel, and balancing the gas pressure against a column of mercury with electrical contacts instead of balancing the gas pressure against the atmosphere.

APPARATUS/PROCEDURE: The solvent was degassed by vacuum. The procedure, repeated 5-10 times, was to alternate 5-15 s evacuation and rapid stirring to produce cavitation. In the solubility measurement, gas, pre-saturated with solvent vapor, was brought into contact with about 80 ml of solvent in the saturation vessel. Initial conditions were established by a time extrapolation. Solubility equilibrium was approached from both under- and supersaturation by varying the rate.

SOURCE AND PURITY OF MATERIALS:

1. Krypton. Air Reduction Co. Reagent grade, 99.8 per cent pure by mass spectroscopy.

2. Nitromethane. Source not given. Distilled, dried by filtering at 253 K.

ESTIMATED ERROR:
$$\delta T/K = 0.05$$
$$\delta P/mmHg = 0.3$$
$$\delta L/L = 0.03$$

REFERENCES:

1. Euken, A.; Herzberg, G. Z. Phys. Chem. 1950, 195, 1.

COMPONENTS:	ORIGINAL MEASUREMENTS:
1. Krypton; Kr; 7439-90-9 2. Nitrobenzene; $C_6H_5NO_2$; 98-95-3	Saylor, J.H.; Battino, R. J. Phys. Chem. 1958, 62, 1334-1337.

VARIABLES:
   T/K: 288.15 - 328.15
   P/kPa: 101.325 (1 atm)
	PREPARED BY:
	
	   H.L. Clever, A.L. Cramer

EXPERIMENTAL VALUES:

T/K	Mol Fraction $X_1 \times 10^3$	Bunsen Coefficient α	Ostwald Coefficient L
288.15	1.37	0.301	0.318
298.15	1.42	0.310	0.338
313.15	1.37	0.295	0.338
328.15	1.36	0.290	0.348

Smoothed Data: $\Delta G^O/J\ mol^{-1} = -RT\ \ln X_1 = -418.82 + 56.124\ T$

Std. Dev. $\Delta G^O = 43.0$, Coef. Corr. = 0.9990

$\Delta H^O/J\ mol^{-1} = -418.82$, $\Delta S^O/J\ K^{-1}\ mol^{-1} = -56.124$

T/K	Mol Fraction $X_1 \times 10^3$	$\Delta G^O/J\ mol^{-1}$
288.15	1.39	15,753
293.15	1.39	16,034
298.15	1.39	16,315
303.15	1.38	16,595
308.15	1.38	16,876
313.15	1.37	17,157
318.15	1.37	17,437
323.15	1.37	17,718
328.15	1.36	17,998

Solubility values were adjusted to a partial pressure of krypton of 101.325 kPa (1 atm) by Henry's law. Bunsen coefficients were calculated by the compiler.

AUXILIARY INFORMATION

METHOD/APPARATUS/PROCEDURE:	SOURCE AND PURITY OF MATERIALS:
The solvent was degassed by evacuating the space above it, shaking, and then passing it as a fine mist into another evacuated container. The degassed liquid was saturated as it passed as a thin film inside a glass helix which contained the solute gas plus solvent vapor at a total pressure of 1 atm (1,2). The volume of the liquid and the volume of gas absorbed were determined in a system of burets.	1. Krypton. Linde Air Products Co. 2. Nitrobenzene. Eastman Kodak white label. Distilled from P_4O_{10}, b.p. 81.0 - 81.2°C (10 mmHg).
	ESTIMATED ERROR: $\delta T/K = 0.03$ $\delta P/mmHg = 1.0$ $\delta X_1/X_1 = 0.005$ (authors)
	REFERENCES: 1. Morrison, T.J.; Billett, F. J. Chem. Soc. 1948, 2033. 2. Clever, H.L.; Battino, R.; Saylor, J.H.; Gross, P.M. J. Phys. Chem. 1957, 61, 1078.

COMPONENTS:	ORIGINAL MEASUREMENTS:
1. Krypton; Kr; 7439-90-9 2. Ammonia; NH$_3$; 7664-41-7	Michels, A., Dumoulin, E. and Van Dijk, J. J. Th., *Physica*, 1959, *25*, 840.

VARIABLES:	PREPARED BY:
Temperature, pressure	C. L. Young

EXPERIMENTAL VALUES:

		Mole fraction of krypton in liquid, x_{Kr}
T/K	P/bar	x_{Kr}
253.05	25.414	0.0066
	50.702	0.0120
	101.421	0.0182
	203.103	0.0239
268.05	25.447	0.0067
	50.741	0.0147
	101.421	0.0259
	177.405	0.0330

AUXILIARY INFORMATION

METHOD /APPARATUS/PROCEDURE:

One pass flow method. Ammonia added to equilibrium cell. Krypton passed through liquid ammonia for many hours. Sample of gas and liquid analysed at atmospheric pressure by adsorbing ammonia in sulfuric acid. Details in source.

SOURCE AND PURITY OF MATERIALS:

No details given.

ESTIMATED ERROR:
$\delta T/K = \pm0.1$; $\delta P/bar = \pm0.005$;
$\delta x_{Ar} = \delta y_{Ar} = \pm0.5\%$.

REFERENCES:

COMPONENTS:	ORIGINAL MEASUREMENTS:

COMPONENTS:

1. Krypton; Kr; 7439-90-9

2. Phosphoric Acid, Tris(methyl phenyl) Ester (Tricresyl Phosphate); $C_{21}H_{21}O_4P$; 1330-78-5

ORIGINAL MEASUREMENTS:

Körösy, F.

Trans. Faraday Soc. 1937, 33, 416-425.

VARIABLES:
 T/K: 295.15
 P/kPa: 101.325 (1 atm)

PREPARED BY:
 H. L. Clever

EXPERIMENTAL VALUES:

T/K	Mol Fraction $X_1 \times 10^3$	Bunsen Coefficient α	Ostwald Coefficient L
295.15	3.3	0.21	0.23

The mole fraction solubility and the Bunsen coefficient were calculated by the compiler. It was assumed that gas behavior is ideal, the Ostwald coefficient is independent of pressure, and that Henry's law is obeyed.

An average density of the tricresyl phosphate isomers was used for the mole fraction solubility calculation.

AUXILIARY INFORMATION

METHOD:

 The apparatus and method of Winkler (1) were used. However, the apparatus was usually not thermostated, and degassing was by evacuating and shaking the solvent, not by evacuating and boiling the solvent as was done by Winkler.

SOURCE AND PURITY OF MATERIALS:

1. Krypton. Source not given. The gas contained 5 % xenon and 1 % non-inert gases.

2. Tris (methylphenyl) ester of phosphoric acid. Source not given. Technical grade.

APPARATUS/PROCEDURE:

ESTIMATED ERROR:
 $\delta X_1/X_1 = 0.05$

REFERENCES:
1. Winkler, L. W. Ber. 1891, 24, 89.

COMPONENTS:	ORIGINAL MEASUREMENTS:
1. Krypton; Kr; 7439-90-9 2. Octamethylcyclotetrasiloxane; $C_8H_{24}O_4Si_4$; 556-67-2	Wilcock, R.J.; McHale, J.L.; Battino, B.; Wilhelm, E. Fluid Phase Equil. 1978, 2, 225-230.

VARIABLES:	PREPARED BY:
T/K: 298.35 P/kPa: 101.325 (1 atm)	H.L. Clever

EXPERIMENTAL VALUES:

T/K	Mol Fraction $X_1 \times 10^3$	Bunsen Coefficient α	Ostwald Coefficient L
298.35	12.43	0.9019	0.9851

The solubility value was adjusted to a gas partial pressure of 101.325 kPa by Henry's law.

The Bunsen coefficient was calculated by the compiler.

AUXILIARY INFORMATION

METHOD /APPARATUS/PROCEDURE:

The apparatus is based on the design of Morrison and Billett (1), and the version used is described by Battino, Evans, and Danforth (2). The degassing apparatus and procedure are described by Battino, Banzhof, Bogan, and Wilhelm (3).

Degassing. Up to 500 cm³ of solvent is placed in a flask of such size that the liquid is about 4 cm deep. The liquid is rapidly stirred, and vacuum is applied intermittently through a liquid N_2 trap until the permanent gas residual pressure drops to 5 microns.

Solubility Determination. The degassed solvent is passed in a thin film down a glass spiral tube containing the solute gas plus the solvent vapor at a total pressure of one atm. The volume of gas absorbed is found by difference between the initial and final volumes in the buret system. The solvent is collected in a tared flask and weighed.

SOURCE AND PURITY OF MATERIALS:

1. Krypton. Matheson Co., Inc. Minimum per cent purity 99.995.

2. Octamethylcyclotetrasiloxane. General Electric Co. Distilled density of 298.15 K was 0.9500 g cm⁻³.

ESTIMATED ERROR:
$$\delta T/K = 0.03$$
$$\delta P/mmHg = 0.5$$
$$\delta X_1/X_1 = 0.01$$

REFERENCES·
1. Morrison,T.J.; Billett,F. J. Chem. Soc. 1948, 2033.
2. Battino, R.;Evans,F.D.;Danforth,W.F. J.Am.Oil Chem.Soc. 1968, 45, 830.
3. Battino,R.;Banzhof,M.;Bogan, M.; Wilhelm, E. Anal. Chem. 1971, 43, 806.

COMPONENTS:	EVALUATOR:
1. Krypton; Kr; 7439-90-9 2. Olive Oil	H. L. Clever Chemistry Department Emory University Atlanta, GA 30322 U.S.A. August 1978

CRITICAL EVALUATION:

The solubility of krypton in olive oil was measured by Lawrence, Loomis, Tobias and Turpin (1) at 295.15 and 310.15 K, by Yeh and Peterson (2) at 298.15, 303.15, 310.15 and 318.15 K, and at 310.15 K by Masson and Taylor (3) and by Kitani (4).

The data were converted to a mole fraction solubility at a partial pressure of krypton of 101.325 kPa (1 atm) assuming that olive oil is 1,2,3-propanetriyl ester of Z-9-octadecenoic acid, or triolein, of molecular weight 885.46. The data from the four laboratories shows considerable scatter. Yeh and Peterson made direct volumetric measurements at atmospheric pressure while the other workers used radiochemical techniques at low krypton partial pressures in the presence of air which can be subject to greater errors than the volumetric method. The data of Yeh and Peterson are internally self-consistent. It was decided to accept the Yeh and Peterson data as the basis of a tentative set of solubility data for krypton in olive oil.

The Yeh and Peterson data were used in a linear regression of a Gibbs energy equation linear in temperature. The tentative values of the thermodynamic changes for the transfer of one mole of krypton from the gas at 101.325 kPa (1 atm) to the hypothetical unit mole fraction solution are

$$\Delta G^{o}/J\ mol^{-1} = -\ RT\ ln\ X_1 = -4,690.4 + 48.782\ T$$

$$Std.\ Dev. = 5.4,\ Coef.\ Corr. = 0.9999$$

$$\Delta H^{o}/J\ mol^{-1} = -4,690.4,\ \Delta S^{o}/J\ K^{-1}\ mol^{-1} = -48.782$$

A table of tentative mole fraction solubility and Gibbs energy values as a function of temperature appears below.

TABLE 1. The solubility of krypton in olive oil. The tentative values of the mole fraction solubility at a krypton partial pressure of 101.325 kPa (1 atm) and the Gibbs energy change as a function of temperature.

T/K	Mol Fraction $X_1 \times 10^2$	$\Delta G^{o}/J\ mol^{-1}$
293.15	1.94	9,610.0
295.15	1.91	9,707.6
298.15	1.88	9,854.0
303.15	1.82	10,098
308.15	1.76_5	10,341
310.15	1.74_5	10,439
313.15	1.71_5	10,586
318.15	1.67^5	10,830

Continued on next page.

109

COMPONENTS:	EVALUATOR:
1. Krypton; Kr; 7439-90-9 2. Olive Oil	H. L. Clever Chemistry Department Emory University Atlanta, GA 30322 U.S.A. August 1978

CRITICAL EVALUATION:

Figure 1 shows the per cent deviation of all of the reported mole fraction
solubility data from the smoothed data of Yeh and Peterson. Since olive
oil is a natural product that may vary somewhat in composition and thus
show variation in its solvent capacity, the data sheets on all the workers
are included.

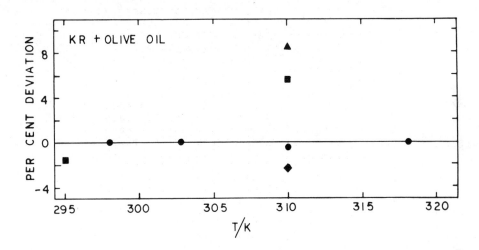

FIGURE 1.   Solubility of krypton in olive oil. Percent deviation from
regression equation fitted to data of Yeh and Peterson.
● Yeh and Peterson,   ■ Lawrence et al.,
▲ Masson and Taylor, and ◆ Kitani

REFERENCES

1. Lawrence, J.H.; Loomis, W.F.; Tobias, C.A.; Turpin, F.H.
    J. Physiol 1946, 105, 197.

2. Yeh, S.-Y.; Peterson, R.E.   J. Pharm. Sci. 1963, 52, 453.

3. Masson, M.B.R.; Taylor, K.   Phys. Med. Biol. 1967, 12, 93.

4. Kitani, K.   Scand. J. Clin. Lab. Invest. 1972, 29, 167.

COMPONENTS:	ORIGINAL MEASUREMENTS:
1. Krypton; Kr; 7439-90-9 2. Olive oil	Lawrence, J. H.; Loomis, W. F.; Tobias, C. A.; Turpin, F. H. J. Physiol. 1946, 105, 197-204.

VARIABLES:	PREPARED BY:
T/K: 295.15 - 310.15	H. L. Clever A. L. Cramer

EXPERIMENTAL VALUES:

T/K	Mol Fraction X_1 x 10^2	Bunsen Coefficient α	Ostwald Coefficient L
295.15	1.88	0.44	0.47_5
310.15	1.85	0.43	0.49_5

The mole fraction solubility and Ostwald coefficients were calculated by the compiler.

The molecular weight of olive oil was taken to be 885 and the density was calculated from $\rho = 0.9152 - 0.000468t/^{\circ}C$ (1) for the mole fraction calculation.

See the evaluation sheet for the solubility in olive oil for more information.

AUXILIARY INFORMATION

METHOD /APPARATUS/PROCEDURE:	SOURCE AND PURITY OF MATERIALS:
Radiochemical method. No details of the method given, but authors state they used an isotope of half life 34 hours. Possibly the isotope was krypton-79.	No information given.

ESTIMATED ERROR:

$\delta\alpha/\alpha = 0.05$ (Compiler)

REFERENCES:

1. Battino, R.; Evans, F. D.;
     Danforth, W. F.
     J. Am. Oil Chem. Soc. 1968,
     45, 830.

COMPONENTS:	ORIGINAL MEASUREMENTS:
1. Krypton; Kr; 7439-90-9 2. Olive Oil (1)	Yeh, S.Y.; Peterson, R.E. J. Pharm. Sci. 1963, 52, 453-458.

VARIABLES:
    T/K: 298.15 - 318.15
    P/kPa: 101.325 (1 atm)
	PREPARED BY:
	
	    H.L. Clever

EXPERIMENTAL VALUES:

T/K	Mol Fraction X_1 x 10^3	Bunsen Coefficient $\alpha \pm$ Std. Dev.	Ostwald Coefficient L
298.15	18.8	0.4376 + 0.0010	0.4746
303.15	18.2	0.4225 + 0.0025	0.4688
310.15	17.4	0.4031 + 0.0025	0.4581
318.15	16.7	0.3844 + 0.0021	0.4477

The Bunsen coefficients are the average of three measurements. The Ostwald coefficients were fitted by the method of least squares to the equation log L = A/T + B by the authors. The same line fitted olive oil and the fats. From the slope and intercept they obtained

$$\Delta H^{O} = (-1185 \pm 46) \text{ cal mol}^{-1} \text{ and } \Delta S^{O} = (-5.6 \pm 0.1) \text{ cal K}^{-1} \text{ mol}^{-1}$$

The thermodynamic values are for the standard state transfer of one mole of krypton from the gas phase at a concentration of one mole dm^{-3} to the solution at a concentration of one mole dm^{-3}. See the evaluation of the krypton + olive oil system on pages 108-109 for the thermodynamic values of the standard state tranfer on one mole of krypton from the gas phase at a partial pressure of one atm to the hypothetical unit mole fraction solution.

AUXILIARY INFORMATION

METHOD/APPARATUS/PROCEDURE:	SOURCE AND PURITY OF MATERIALS:
Oil was dried and degassed by stirring under vacuum at 80°C for about 12 hr. A 50 ml. sample was placed in an absorption flask attached to a Geffken gas buret (1). The oil was constantly stirred and equilibrated with increments of gas until no change was observed in a differential oil manometer for ½ hr. Difference between initial and final buret readings indicated amount of gas absorbed. Absorption at successively lower temperatures was determined. The authors also measured the viscosity and surface tension of the liquid.	1. Krypton. Matheson Co., Research grade, maximum impurity 0.04 mole per cent N_2 and Xe. 2. Olive Oil. Magnus, Mabee and Raynard Co., U.S.P.

ESTIMATED ERROR:
$$\delta T/K = 0.05$$
$$\delta P/mmHg = 0.5$$
$$\delta\alpha/\alpha = 0.005$$

REFERENCES:
1. Geffken, G. Z. Physik Chem. 1904, 49, 257.

COMPONENTS:	ORIGINAL MEASUREMENTS:
1. Krypton-85; ^{85}Kr; 13983-27-2 2. Olive oil	Masson, M. B. R.; Taylor, K. Phys. Med. Biol. 1967, 12, 93-98.

VARIABLES:	PREPARED BY:
T/K: 310.15	A.L. Cramer

EXPERIMENTAL VALUES:

T/K	Mol Fraction $X_1 \times 10^2$	Bunsen Coefficient $\alpha \pm$ Std Dev	Number of Determinations	Ostwald Coefficient L
310.15	1.90	0.440 ± 0.004	14	0.500

The mole fraction solubility and Ostwald coefficient were calculated by the compiler.

The molecular weight of olive oil was taken to be 885 and the density was calculated from $\rho = 0.9152 - 0.000468t/^{\circ}C$ (1) for the mole fraction calculation.

See the evaluation sheet for the solubility in olive oil for more information.

AUXILIARY INFORMATION

METHOD /APPARATUS/PROCEDURE:	SOURCE AND PURITY OF MATERIALS:
The olive oil was sprayed into a circulating mixture of krypton-85 and air. The difference between the count rate before and after the solvent was sprayed equals the amount absorbed by the olive oil. The measurement was checked by replacing the radioactive gas by air, recirculating the olive oil and measurement of the increase of radioactivity in the gas phase, which was compared with the original decrease. The sensor was a thallium activated sodium iodide crystal, behind a thin beryllium window.	1. Krypton-85. Radiochemical Centre, Amersham, UK. 2. Olive oil. No information given.
	ESTIMATED ERROR: $\delta\alpha/\alpha = 0.03$ (Compiler)
	REFERENCES: 1. Battino, R.; Evans, F.D. Danforth, W.F. J. Am. Oil Chem. Soc. 1968, 45, 830.

COMPONENTS:	ORIGINAL MEASUREMENTS:
1. Krypton-85; ^{85}Kr; 13983-27-2 2. Olive Oil	Kitani, K. Scand. J. Clin. Lab. Invest. 1972, 29, 167-172.
VARIABLES: T/K: 310.15	PREPARED BY: P.L. Long A.L. Cramer

EXPERIMENTAL VALUES:

T/K	Mol Fraction $X_1 \times 10^2$	Bunsen Coefficient α	Ostwald Coefficient $L \pm$ Std Dev	Number of Determinations
310.15	1.71	0.397	0.451 $\pm$ 0.015	19

The mole fraction solubility and Bunsen coefficient were calculated by the compiler.

The molecular weight of olive oil was taken to be 885, and the density (1) was calculated from $\rho = 0.9152 - 0.000468t/^{O}C$ for the mole fraction calculation.

See the evaluation sheet on olive oil for more information.

AUXILIARY INFORMATION

METHOD /APPARATUS/PROCEDURE:

Glass cuvettes were filled with the solvent. One-third was replaced by the radioactive gas and air. The sealed cuvette was rotated for two hours in a water bath at 37°C. The total pressure was adjusted to one atm by means of a thin needle. Samples from both the liquid and gas phase were counted by a scintillation detector and a Phillips pulse-height analyzer at the energy peak. Corrections for self absorption and scatter were made.

SOURCE AND PURITY OF MATERIALS:

1. Krypton-85. Radiochemical Centre, Amersham, UK.

2. Olive oil. Commercial sample.

ESTIMATED ERROR:

$$\delta L/L = 0.03$$

REFERENCES:
1. Battino, R.; Evans, F.D.; Danforth, W.F. J. Am. Oil Chem. Soc. 1968, 45, 830.

COMPONENTS:	EVALUATOR:
1. Krypton; Kr; 7439-90-9 2. Biological Systems	H. L. Clever Chemistry Department Emory University Atlanta, GA 30322 U.S.A. September 1978

CRITICAL EVALUATION:

The solubility of krypton in biological fluids and tissues.

There are several factors that make it difficult to compare and evaluate
the solubility of krypton in biological systems. First, the material from
biological specimens may show a natural variation in properties which
affects the solubility. Second, workers have used quite different experimental techniques to measure the solubility. Some have used classical
volumetric methods with the krypton at a partial pressure near atmospheric
pressure, many have used radiochemical techniques either with natural
krypton tagged with radioactive krypton at a total pressure near atmospheric
or with a small unknown partial pressure of radioactive krypton. In
neither of the radiochemical techniques is it necessary to know the total
pressure to obtain an Ostwald coefficient. However, to compare the results
of krypton solubility determinations by the volumetric method and by the
radiochemical techniques one must assume the Ostwald coefficient is independent of pressure. This may not be true, especially if the gas associates
with one or more components of the biological fluid. In these systems the
solubility data are classed as tentative. Below are comments, which
compare rather than evaluate, the solubility data in several types of
biological systems.

Fat. Yeh and Peterson (1) found little difference in the solubility of
krypton in olive oil, human fat, dog fat and rat fat by a volumetric
method. Masson and Taylor (2) used a radiochemical method and obtained
solubility values that were eight per cent higher in olive oil and three
per cent higher in human fat than Yeh and Peterson's results. The Masson
and Taylor method, although elegant, appears to be more subject to error
than the Yeh and Peterson technique. The results of the two laboratories
probably agree within experimental error. Kirk, Parrish and Morken's (3)
results with guinea pig fat are reported on a weight basis. They appear
to be of similar magnitude to the solubility found in other fats.

Blood and blood components. Several studies showed that the solubility of
krypton and other gases varies linearly with the percent of either red
blood cells, serum albumin, or hemoglobin in blood and in other fluids.
The results of these studies are sometimes extrapolated to 100% blood
component to obtain the solubility of the gas in that component. Both
Hardewig, Rochester and Briscoe (4) and Kitani (5) report a value of the
Ostwald coefficient of krypton in human red blood cells. The values
agree within 5 per cent. Yeh and Peterson (1) show that 0.075 mole of
krypton associates with a mole of human hemoglobin and that 0.2 to 0.05
mole of krypton associate with a mole of human serum albumin at temperatures
of 298.15 and 310.15 K respectively. Muehlbaecher, DeBon and Featherstone
(6) carried out qualitative experiments that showed that bovine hemoglobin
associates with krypton while bovine gamma-globulin and serum albumin
apparently do not associate. Kirk, Parrish and Morken (3) reported the
solubility of krypton in dog, cat, hamster and guinea pig blood and Strang
(7) reported the solubility of krypton in rabbit blood.

Tissues. The distribution of krypton between the gas phase and the tissue
can be measured either with homogenized tissue in a saline or other solution, with solid tissue, or with tissue from an animal sacrificed after
breathing air containing krypton. It is likely that the structured tissue
interacts somewhat differently with the gas than does the homogenized
tissue, but there is not enough data to draw any conclusions about the
effect of tissue structure and "solubility". Both Yeh and Peterson (1)
and Strang (7) reported on work with rabbit tissues. Kitani (5) reported
results with human liver tissues. Kirk, Parrish and Morken (3) reported
on a range of guinea pig tissues and guinea pig brain and Yeh and
Peterson (1) reported results on beef brain tissues.

115

COMPONENTS:	EVALUATOR:
1. Krypton; Kr; 7439-90-9 2. Biological Systems	H. L. Clever Chemistry Department Emory University Atlanta, GA 30322 U.S.A. September 1978

CRITICAL EVALUATION:

In addition there are whole blood tissue coefficients for the partition of
krypton between blood and tissue determined incidental to regional blood
flow studies. These results could be converted into gas-tissue coefficients using the gas-blood Ostwald coefficients. The tissue-blood partition coefficients for krypton are not given in this compilation. Many are
summarized by Kirk, Parrish and Morken (3) from the work of Lassen and
Munck (8), Lassen and Ingvar (9), Lassen (10, 11), Ingvar and Lassen (12),
Glass and Harper (13), Thornburn, Kopald, Herd, Hollenberg, O'Morchoe, and
Barger (14), Friedman, Kopald and Smith (15), Bell and Harper (16),
Hollenberg and Dougherty (17), Setchell, Waites and Thorburn (18), and
Bell and Battersby (19).

REFERENCES

1. Yeh, S.Y.; Peterson, R.E. J. Pharm. Sci. 1963, 52, 453;
   J. Appl. Physiol. 1965, 20, 1041.
2. Masson, M.B.R.; Taylor, K. Phys. Med. Biol. 1967, 12, 93.
3. Kirk, W.P.; Parish, P.W.; Morken, D.A. Health Phys. 1975, 28, 249.
4. Hardewig, A.; Rochester, D.F.; Briscoe, W.A. J. Appl. Physiol. 1960,
   15, 723.
5. Kitani, K. Scand. J. Clin. Lab. Invest. 1972, 29, 167.
6. Muehlbaecher, C.; DeBon, F.Z.; Featherstone, R.M. Inst. Anesth.
   Clinics 1963, 1, 937.
7. Strang, R. Phys. Med. Biol. 1975, 20, 1025.
8. Lassen, N.A.; Munck, O. Acta physiol scand. 1955, 33, 30.
9. Lassen, N.A.; Ingvar, D.H. Experientia 1961, 17, 42.
10. Lassen, N.A. Minerva Nucl. 1964, 8, 211.
11. Lassen, N.A. in Radioisotopes in Medical Diagnosis, Belcher, E.H.;
    Vetter, H. Editors, Appleton-Century-Crofts, 1971.
12. Ingvar, D.H.; Lassen, N.A. Acta physiol scand. 1962, 54, 325.
13. Glass, H.I.; Harper, A.M. Phys. Med. Biol. 1962, 7, 335.
14. Thorburn, G.D.; Kopald, H.H.; Herd, J.A.; Hollenberg, M.;
    O'Morchoe, C.C.C.; Barger, A.C. Circulation Res. 1963, 13, 290.
15. Friedman, E.; Kopald, H.H.; Smith, T.R. Invest. Opthalmol. 1964, 3,
    539.
16. Bell, G.; Harper, A.M. J. Surg. Res. 1965, 5, 382.
17. Hollenberg, M.; Dougherty, J. Am. J. Physiol 1966, 210, 926.
18. Setchell, B.P.; Waites, G.M.H.; Thorburn, G.D. Circulation Res 1966,
    18, 755.
19. Bell, P.R.F.; Battersby, A.C. Surgery 1967, 62, 468.

COMPONENTS:	ORIGINAL MEASUREMENTS:
1. Krypton; Kr; 7439-90-9 2. Human Fat	Yeh, S.Y.; Peterson, R.E. J. Pharm. Sci. 1963, 52, 453-458.

VARIABLES:
        T/K: 298.15 - 318.15
        P/kPa: 101.325 (1 atm)
	PREPARED BY:
	
	        H.L. Clever

EXPERIMENTAL VALUES: Human fat 1

T/K	Bunsen Coefficient $\alpha \pm$ Std. Dev.	Ostwald Coefficient L	T/K	Bunsen Coefficient $\alpha \pm$ Std. Dev.	Ostwald Coefficient L
298.15	0.4412 ± 0.0061	0.4816	298.15	0.4404 ± 0.0015	0.4807
303.15	0.4258 ± 0.0012	0.4725	303.15	0.4247 ± 0.0013	0.4713
310.15	0.4071 ± 0.0005	0.4626	310.15	0.4062 ± 0.0004	0.4617
318.15	0.3878 ± 0.0015	0.4516	318.15	0.3875 ± 0.0012	0.4513

(right block header: Human fat 2)

The Bunsen coefficients are the average of three measurements. The Ostwald coefficients were fitted by the method of least squares to the equation log L = A/T + B by the authors. The same line fitted olive oil and the fats. From the slope and intercept they obtained

$$\Delta H^{o} = (-1185 \pm 46) \text{ cal mol}^{-1} \text{ and } \Delta S^{o} = (-5.6 \pm 0.1) \text{ cal K}^{-1} \text{ mol}^{-1}$$

The thermodynamic values are for the transfer of one mole of krypton from the gas phase at a concentration of one mole dm^{-3} to the solution at a concentration of one mole dm^{-3}.

AUXILIARY INFORMATION

METHOD /APPARATUS/PROCEDURE:	SOURCE AND PURITY OF MATERIALS:
Fat was dried and degassed by stirring under vacuum at 80°C for about 12 hr. A 50 ml. sample was placed in an absorption flask attached to a Geffken gas buret (1). The fat was constantly stirred and equilibrated with increments of gas until no change was observed in a differential oil manometer for ½ hr. Difference between initial and final buret readings indicated amount of gas absorbed. Absorption at successively lower temperatures was determined. The authors also measured the viscosity and surface tension of the liquid.	1. Krypton. Matheson Co. Research grade, maximum impurity 0.04 mole per cent N_2 and Xe. 2. Human omental fats obtained from two deceased patients (1 and 2). Extracted with petroleum ether (b.p. 309-338 K). The ether was evaporated at 353 K under vacuum for several hours. Stored under refrigeration until use. Fat 2 appeared to have more stearine precipitate than fat 1 at 296 K.

	ESTIMATED ERROR:
	        $\delta T/K$ = 0.05
	        $\delta P/mmHg$ = 0.5
	        $\delta\alpha/\alpha$ = 0.005

	REFERENCES: 1. Geffken, G. Z. Physik Chem. 1904, 49, 257.

COMPONENTS:	ORIGINAL MEASUREMENTS:
1. Krypton-85; $^{85}_{36}$Kr; 13983-27-2 2. Human Fat	Masson, M.B.R.; Taylor, K. Phys. Med. Biol. 1967, 12, 93-98.

VARIABLES:	PREPARED BY:
T/K: 310.15 Total P/kPa: 101.325 (1 atm)	A.L. Cramer

EXPERIMENTAL VALUES:

T/K	Bunsen Coefficient $\alpha \pm$ Std. Dev.	Number of Determinations
Pooled Fat		
310.15	0.420 ± 0.006	16
Abdominal Fat		
310.15	0.414 ± 0.006	6

AUXILIARY INFORMATION

METHOD /APPARATUS/PROCEDURE:

Liquid fat was sprayed into a circulating mixture of ^{85}Kr and air. Difference between count rate in air before and after spraying equals amount absorbed by fat. Then, radioactive gas mixture was replaced by air. The spray pump was re-connected, and increase of count rate in air was compared with original decrease. Sensor was a B-sensitive thallium activated sodium iodide crystal, behind a thin (0.008 in) beryllium window.

SOURCE AND PURITY OF MATERIALS:

1. Krypton-85. Radiochemical Centre, Amersham, U.K.

2. Fat. Abdominal, perirenal and omental. Heated 50-60°C for 2-3 days; pure lipid filtered off and stored under refrigeration.

ESTIMATED ERROR:

REFERENCES:

COMPONENTS:	ORIGINAL MEASUREMENTS:
1. Krypton; Kr; 7439-90-9 2. Rat-pooled Fat	Yeh, S.Y.; Peterson, R.E. J. Pharm. Sci. 1963, 52, 453-458.

VARIABLES:
    T/K: 298.15 - 318.15
    P/kPa: 101.325 (1 atm)
	PREPARED BY:
	
	    H.L. Clever

EXPERIMENTAL VALUES:

T/K	Bunsen Coefficient $\alpha \pm$ Std. Dev.	Ostwald Coefficient L
298.15	0.4363 ± 0.0060	0.4762
303.15	0.4219 ± 0.0011	0.4755
310.15	0.4037 ± 0.0013	0.4588
318.15	0.3847 ± 0.0001	0.4481

The Bunsen coefficients are the average of three measurements. The Ostwald coefficients were fitted by the method of least squares to the equation $\log L = A/T + B$ by the authors. The same line fitted olive oil and the fats. From the slope and intercept they obtained

$$\Delta H^{o} = (-1185 \pm 46) \text{ cal mol}^{-1} \text{ and } \Delta S^{o} = (-5.6 \pm 0.1) \text{ cal K}^{-1} \text{ mol}^{-1}$$

The thermodynamic values are for the transfer of one mole of krypton from the gas phase at a concentration of one mole dm^{-3} to the solution at a concentration of one mole dm^{-3}.

AUXILIARY INFORMATION

METHOD/APPARATUS/PROCEDURE:

Fat was dried and degassed by stirring under vacuum at 80°C for about 12 hr. A 50 ml. sample was placed in an absorption flask attached to a Geffken gas buret (1). The fat was constantly stirred and equilibrated with increments of gas until no change was observed in a differential oil manometer for ½ hr. Difference between initial and final buret readings indicated amount of gas absorbed. Absorption at successively lower temperatures was determined. The authors also measured the viscosity and surface tension of the liquid.

SOURCE AND PURITY OF MATERIALS:

1. Krypton. Matheson Co. Research grade, maximum impurity 0.04 mole per cent N_2 and Xe.

2. Rat retroperitoneal, mesenteric, omental, and hair clipped skin was cut into about 2.5 cm squares, dried at 353 K under vacuum, coarsely crushed and then extracted with petroleum ether (b.p. 308-338 K) in a Soxhlet extractor. The ether was evaporated at 353 K for several hours under vacuum. Stored under refrigeration until use.

ESTIMATED ERROR: $\delta T/K = 0.05$
$\delta P/mmHg = 0.5$
$\delta\alpha/\alpha = 0.005$

REFERENCES:

1. Geffken, G. Z. Physik Chem. 1904, 49, 257.

COMPONENTS:	ORIGINAL MEASUREMENTS:
1. Krypton; Kr; 7439-90-9 2. Dog Fat	Yeh, S.Y.; Peterson, R.E. J. Pharm. Sci. 1963, 52, 453-458.
VARIABLES: T/K: 298.15 - 318.15 P/kPa: 101.325 (1 atm)	PREPARED BY: H.L. Clever

EXPERIMENTAL VALUES:

T/K	Bunsen Coefficient $\alpha \pm$ Std. Dev.	Ostwald Coefficient L
298.15	0.4364 $\pm$ 0.0006	0.4764
303.15	0.4225 $\pm$ 0.0013	0.4721
310.15	0.4031 $\pm$ 0.0007	0.4581
318.15	0.3853 $\pm$ 0.0004	0.4426

The Bunsen coefficients are the average of three measurements. The Ostwald coefficients were fitted by the method of least squares to the equation log L = A/T + B by the authors. The same line fitted olive oil and the fats. From the slope and intercept they obtained

$$\Delta H^O = (-1185 \pm 46) \text{ cal mol}^{-1} \text{ and } \Delta S^O = (-5.6 \pm 0.1) \text{ cal K}^{-1} \text{ mol}^{-1}$$

The thermodynamic values are for the transfer of one mole of krypton from the gas phase at a concentration of one mole dm^{-3} to the solution phase at a concentration of one mole dm^{-3}.

AUXILIARY INFORMATION

METHOD/APPARATUS/PROCEDURE:

Fat was dried and degassed by stirring under vacuum at 80°C for about 12 hr. A 50 ml. sample was placed in an absorption flask attached to a Geffken gas buret (1). The fat was constantly stirred and equilibrated with increments of gas until no change was observed in a differential oil manometer for ½ hr. Difference between initial and final buret readings indicated amount of gas absorbed. Absorption at successively lower temperatures was determined. The authors also measured the viscosity and surface tension of the liquid.

SOURCE AND PURITY OF MATERIALS:

1. Krypton. Matheson Co. Research grade, maximum impurity 0.04 mole per cent N_2 and Xe.

2. Dog perineal, mesenteric, omental, and other adipose fats were extracted with petroleum ether (b.p. 309-338 K). The ether was evaporated at 353 K under vacuum for several hours. Stored under refrigeration until use.

ESTIMATED ERROR:
$\delta T/K = 0.05$
$\delta P/mmHg = 0.5$
$\delta\alpha/\alpha = 0.005$

REFERENCES:

1. Geffken, G. Z. Physik Chem. 1904, 49, 257.

COMPONENTS:	ORIGINAL MEASUREMENTS:
1. Krypton-85; $^{85}_{36}$Kr; 13983-27-2 2. Paraffin Oil 3. Lecithin	Kitani, K. Scand. J. Clin. Lab. Invest. 1972, 29, 167-172.

VARIABLES: T/K: 310.15 P/kPa: 101.325 (1 atm)	PREPARED BY: P.L. Long A.L. Cramer

EXPERIMENTAL VALUES:

T/K	Bunsen Coefficient α	Ostwald Coefficient L ± Std. Dev.	Replications
Paraffin Oil			
310.15	0.424	0.481 ± 0.019	9
	-	0.492*	-
Lecithin			
310.15	-	0.375*	-

*Extrapolated values. See equation below.

The Bunsen coefficients were calculated by the compiler assuming that the Ostwald coefficient was independent of pressure.

The solubility value for krypton in lecithin is a value extrapolated from the solubility of the gas in mixtures of paraffin oil and lecithin. The coefficient of solubility of the gas is linear in lecithin per cent (graph in the paper).

$$L = 0.492 - (0.00127) \text{ (Lecithin \%w/v)}$$

AUXILIARY INFORMATION

METHOD /APPARATUS/PROCEDURE:	SOURCE AND PURITY OF MATERIALS:
A glass cuvette is filled with the liquid. One-third of the liquid is replaced with radioactive gas in air. The sealed cuvette is placed in a thermostated bath for 2 hours. The pressure is adjusted to 1 atm by means of a thin needle. The radioactive assay, corrected for self absorption and scatter, is made by a scintillation detector and a Philips pulse-height analyzer.	1. Krypton. Radiochemical Centre, Amersham, England. 2. Paraffin Oil. Commercial quality. 3. Egg lecithin. Purified twice with ether and acetone.
	ESTIMATED ERROR: See standard deviations above.
	REFERENCES:

COMPONENTS:	ORIGINAL MEASUREMENTS:
1. Krypton-85; $^{85}_{36}$Kr; 13983-27-2 2. Water; H_2O; 7732-18-5 3. Blood and Blood components	Hardewig, A.; Rochester, D.F.; Briscoe, W.A. J. Appl. Physiol. 1960, 15, 723-725.

VARIABLES:	PREPARED BY:
T/K: 310.15	A.L. Cramer

EXPERIMENTAL VALUES:

T/K	Ostwald Coefficient L $\pm$ Std. Dev.	Number of Determinations
Water; H_2O; 7732-18-5		
310.15	0.0499 $\pm$ 0.0013	10
Plasma		
310.15	0.0510 $\pm$ 0.0010	6
Red Blood Cells		
310.15	0.0677*	

*Extrapolated value. See equation below.

Twenty measurements of the krypton-85 solubility in whole blood were made at hematocrit of 30 to 65 percent. The data were given in a graph. The regression equation

$$L = 0.05199 + (0.0001573)(Hcf \%)$$

fitted the data with a regression coefficient of 0.65.

AUXILIARY INFORMATION

METHOD:	SOURCE AND PURITY OF MATERIALS:
In vitro equilibration at 37 °C (310.15 K), 2 hr. Lassen and Munck counting (1).	1. Krypton-85. Oak Ridge. 10-15 mc cm^{-3} diluted with air to 30 mc cm^{-3}.

APPARATUS/PROCEDURE:	ESTIMATED ERROR:
	Standard error of the mean is 0.004 for both water and plasma.
	REFERENCES:
	1. Lassen, T.A.; Munck, O. Acta. physiol. Scandinav. 1955, 33, 30.

COMPONENTS:	ORIGINAL MEASUREMENTS:
1. Krypton; Kr; 7439-90-9 2. Water; H_2O; 7732-18-5 3. Sodium Phosphate Buffer 4. Bovine Blood Components	Muehlbaecher, C.; DeBon, F.L.; Featherstone, R.M. Inst. Anesth. Clinics 1963, 1, 937-952.

VARIABLES:
    T/K: 310.15
	PREPARED BY:
	    H.L. Clever

EXPERIMENTAL VALUES:

The Bunsen coefficients were presented on large scale graphs. The krypton Bunsen coefficient in sodium phosphate buffer was about 0.056. Comments about the individual systems follow:

T/K	Comments
	Bovine Gamma-Globulin + Phosphate Buffer (pH 6.3-6.5, Ionic Strength 0.16)
310.15	No apparent change in krypton Bunsen coefficient as bovine gamma-globulin was increased from 0-8 per cent.
	Bovine Serum Albumin + Phosphate Buffer (pH 5.6-6.3, Ionic Strength 0.16)
310.15	The change in the krypton Bunsen coefficient was within the 8 per cent uncertainty as the bovine serum albumin was increased from 0 to 20 per cent.
	Bovine Hemoglobin + Phosphate Buffer (pH 6.3-6.6, Ionic Strength 0.16)
310.15	The krypton Bunsen coefficient appears to increase about 12 per cent as the bovin hemoglobin was increased from 0 to 20 per cent.

AUXILIARY INFORMATION

METHOD /APPARATUS/PROCEDURE:	SOURCE AND PURITY OF MATERIALS:
Gas chromatography	No information
	ESTIMATED ERROR: $\delta\alpha/\alpha = 0.08$
	REFERENCES:

COMPONENTS:	ORIGINAL MEASUREMENTS:
1. Krypton; Kr; 7439-90-9 2. Water; H_2O; 7732-18-5 3. Human Whole Blood and Blood Components	Yeh, S-Y.; Peterson, R.E. J. Appl. Physiol. 1965, 20, 1041
VARIABLES: T/K: 310.15 Total P/kPa: 101.325 (1 atm)	PREPARED BY: A.L. Cramer

EXPERIMENTAL VALUES:

T/K	Absorption Coefficient mean $\beta \pm$ Std. Dev.	Number of Determinations	Absorption Coefficient* g^{-1} Hemoglobin
Hemoglobin Solution (7.5% hemoglobin, 92.27% water, 1.72 mg cm^{-3} lipid)			
310.15	0.0444	1	0.0346
Hemoglobin Solution (15.39% hemoglobin, 84.60% water, 2.50 mg cm^{-3} lipid)			
310.15	0.0420 ± 0.0018	3	0.0214
Human Blood (425 ml whole blood + 120 ml 1.32% sodium citrate solution)			
310.15	0.0455 ± 0.0044	6	-

*The authors give a weighted average of 0.0247 cm^3 (STP) gas g^{-1} hemoglobin which is equivalent to 0.075 mole Kr mole $^{-1}$ hemoglobin (mol. wt. = 68,000). The values are for a partial pressure of krypton of (1-solution vapor pressure) atm. At 1 atm krypton the values would be about 6% higher.

AUXILIARY INFORMATION

METHOD /APPARATUS/PROCEDURE:

In Yeh and Peterson (1) modification of Geffcken (2) apparatus, 45 ml. of liquid was frozen, evacuated, and melted repeatedly until no bubbles appeared in liquid under vacuum. Equilibration with gas and measurement of solubility followed (1).

SOURCE AND PURITY OF MATERIALS:

1. Krypton. The Matheson Co. Research grade, maximum impurity 0.04% nitrogen and/or oxygen.
2. Human blood. 425 ml. from a normal donor, mixed with 120 ml. 1.32% sodium citrate solution. Frozen and thawed to hemolyze red blood cells.
3. Hemoglobin. From centrifuged citrated human blood.

ESTIMATED ERROR:
$\delta T/K = 0.05$
$\delta P/mmHg = 0.2$
$\delta L/L = 0.0015$

REFERENCES:

1. Yeh, S-Y.; Peterson, R.E. J. Pharm. Sci. 1963, 52, 453.

2. Geffcken, G. Z. Physik. Chem. 1904, 49, 257.

124

COMPONENTS:	ORIGINAL MEASUREMENTS:
1. Krypton-85; $^{85}_{36}$Kr; 13983-27-2 2. Water, Saline Solution, Plasma, and Human Red Blood Cells	Kitani, K. Scand. J. Clin. Lab. Invest. 1972, 29, 167-172.

VARIABLES:	PREPARED BY:
T/K: 310.15 P/kPa: 101.325 (1 atm)	P.L. Long A.L. Cramer

EXPERIMENTAL VALUES:

T/K	Bunsen Coefficient α	Ostwald Coefficient L $\pm$ Std. Dev.	Replications
Water; H_2O; 7732-18-5			
310.15	0.0424	0.0481 $\pm$ 0.0022	17
Water + Sodium Chloride; NaCl; 7647-14-5 0.9 per cent Saline			
310.15	0.403	0.0458 $\pm$ 0.0020	16
Plasma (3 Samples)			
310.15	0.0442	0.0502 $\pm$ 0.0002	8
	0.0435	0.0494 $\pm$ 0.0003	8
	0.0440	0.0500 $\pm$ 0.0006	9
Human Red Blood Cells			
310.15	–	0.0718	–

The Bunsen coefficients were calculated by the compiler assuming that the Ostwald coefficient was independent of pressure.

The solubility value for red blood cells is a value extrapolated from the solubility of the gas in mixtures of plasma and heparinized blood from healthy donors. The coefficient of solubility is linear in hematocrit per cent (graph in paper). L = 0.04961 + (0.0002223) (Hematocrit %v/v)(r=0.975)

AUXILIARY INFORMATION

METHOD/APPARATUS/PROCEDURE:

A glass cuvette is filled with the liquid. One-third of the liquid is replaced with radioactive gas in air. The sealed cuvette is placed in a thermostated bath for 2 hours. The pressure is adjusted to 1 atm by means of a thin needle.

The radioactive assay, corrected for self absorption and scatter, is made by a scintillation detector and a Philips pulse-height analyzer.

SOURCE AND PURITY OF MATERIALS:

1. Krypton. Radiochemical Centre, Amersham, England.

2. Water and 0.9 per cent saline were prepared according to the criteria for purity in the Nordic Pharmacopeia. Heparinized blood from healthy donors was used.

ESTIMATED ERROR:

See standard deviations above.

REFERENCES:

125

COMPONENTS:	ORIGINAL MEASUREMENTS:

COMPONENTS:

1. Krypton-85; $^{85}_{36}$Kr; 13983-27-2

2. Blood of Various Animals

ORIGINAL MEASUREMENTS:

Kirk, W.P.; Parish, P. W.; Morken,D.A.

Health Phys. 1975, 28, 249 - 261.

VARIABLES:
     T/K: 308.65 - 312.15
     P/kPa: 101.325 (1 atm)

PREPARED BY:

A. L. Cramer

EXPERIMENTAL VALUES:

T/K	Ostwald Coefficient L ± Std. Dev.	Number of Determinations
Dog blood		
308.65	0.0691 ± 0.0054	9
Chinese hamster blood		
310.15	0.0822 ± 0.0014	10
Cat blood		
309.15	0.0595 ± 0.0065	7
312.15	0.0559 ± 0.0040	8
Guinea pig blood		
309.15	0.0546 ± 0.0076	16
310.15	0.0538 ± 0.0033	8
312.25	0.0513 ± 0.0041	8

N,N'-1,2-ethanediyl-bis(N-carboxymethyl)glycine
(Ethylenediaminetetraacetic acid or EDTA);$C_{10}H_{16}N_2O_8$;
60-00-4. A 10 % solution in water.

| 310.15 | 0.0406 ± 0.0028 | 8 |

Sodium chloride; NaCl; 7647-14-5.Aqueous soln.0.9 wt.%

| 310.15 | 0.0472 ± 0.0012 | 4 |

The authors conversion of their experimental data to the Ostwald coefficient assumed that Henry's law is obeyed and the Ostwald coefficient is independent of pressure.

AUXILIARY INFORMATION

METHOD:
   The blood was obtained by veni-puncture (cats, dogs),cardiac puncture (guinea pigs) or intraorbital sinus puncture (Chinese hamsters). The blood was mixed with a "known volume" of 10 % EDTA to prevent clotting. The samples were equilibrated with a ^{85}Kr + air atmosphere for 24 hours; samples were counted in a NaI well system. The solubility coefficients were corrected for the presence of aqueous saline and EDTA solution used as diluents.

SOURCE AND PURITY OF MATERIALS:

1. Krypton. No information given. The ^{85}Kr + air mixture had a count of 0.9 μCi cm^{-3}.

2. Guinea pigs. Duncan-Hartley random bred. 700-1000 g, from commercial suppliers or bred in laboratory.

APPARATUS/PROCEDURE:

   The authors fitted the Ostwald coefficients in Guinea pig blood to the equation

   $\log L = -4.1378 + 889.2/T$
   (r = 0.994)

ESTIMATED ERROR:

$\delta L/L = 0.02 - 0.14$

REFERENCES:

COMPONENTS:	ORIGINAL MEASUREMENTS:
1. Krypton; Kr; 7439-90-9 2. Water; H_2O; 7732-18-5 3. Sodium Chloride; NaCl; 7647-14-5 4. Human Albumin; 9048-46-8	Yeh, S.-Y.; Peterson, R. E. J. Appl. Physiol. 1965, 20, 1041-1047.
VARIABLES: T/K: 298.15 - 310.15 P/kPa: 101.325 (1 atm) Albumin/wt %: 5.2 - 25.3	PREPARED BY: A. L. Cramer H. L. Clever

EXPERIMENTAL VALUES:

T/K	Albumin wt %	Mol Fraction $X_1 \times 10^3$	Bunsen Coefficient α	Ostwald Coefficient L ± Std. Dev.	Number of Determinations.
298.15	5.2		0.0503	0.0549 ± 0.0008	6
303.15			0.0446	0.0495 ± 0.0008	6
310.15			0.0380	0.0431 ± 0.0008	5
298.15	15.4		0.0504	0.0550 ± 0.0023	5
303.15			0.0437	0.0485 ± 0.0015	5
310.15			0.0358	0.0406 ± 0.0018	5
298.15	25.3		0.0507	0.0553 ± 0.0006	3
303.15			0.0427	0.0474 ± 0.0009	3
310.15			0.0336	0.0381 ± 0.0007	3
298.15	100	161.1	0.0572	0.0624	
303.15		108.7	0.0371	0.0412	
310.15		48.5	0.0172	0.0195	

The Ostwald coefficients in 100 % human albumin were values extrapolated
from the values at lower concentration by the authors. The Bunsen coefficients and the mole fraction solubilities were calculated by the compiler. The
authors used an albumin molecular weight of 69,000 and a density of 1.0 to
calculate the mole ratio of krypton to albumin of 0.192, 0.122, and 0.051
at 298.15, 303.15, and 310.15 K, respectively.

AUXILIARY INFORMATION

METHOD:	SOURCE AND PURITY OF MATERIALS:
A 45 cm^3 sample of albumin in standard saline solution was frozen, evacuated, and melted repeatedly until no bubbles appeared in the liquid under vacuum in the Yeh and peterson modification (1) of the Geffcken apparatus (2). Equil-ibration with gas and measurement of the solubility followed.	1. Krypton. The Matheson Co. Research grade with maximum impurity of 0.04 % N_2 and O_2. 2. Serum albumin solution. Cutter Laboratories. A 25 % solution stabilized with 0.02 M sodium caprylate and 0.02 M sodium acetyltryptophanate.
APPARATUS/PROCEDURE:	ESTIMATED ERROR: $\delta T/K = 0.05$ $\delta P/mmHg = 0.2$ $\delta L/L = 0.0015$
	REFERENCES: 1. Yeh, S.-Y.; Peterson, R. E. J. Pharm. Sci. 1963, 52, 453. 2. Geffcken, G. Z. Phys. Chem. 1904, 49, 257.

COMPONENTS:	ORIGINAL MEASUREMENTS:
1. Krypton-85; $^{85}_{36}$Kr; 13983-27-2 2. Rabbit Tissues	Strang, R. Phys. Med. Biol. 1975, 20, 1025-1028.

VARIABLES:	PREPARED BY:
T/K: 310.15 P/kPa: 101.325 (1 atm) Time/hr: 0.5 - 24	A.L. Cramer

EXPERIMENTAL VALUES:

T/K	Tissue	Bunsen Coefficient α	Ostwald Coefficient L $\pm$ Std. Dev.	Number of Replications
310.15	Erythrocytes	0.094	0.107 $\pm$ 0.009	32
	Plasma	0.052	0.059 $\pm$ 0.005	32
	Whole Blood	0.067	0.076 $\pm$ 0.007	32
	Vitreous Body	0.052	0.059 $\pm$ 0.008	28
	Retina	0.055	0.062 $\pm$ 0.011	35
	Choroid Layer	0.026	0.030 $\pm$ 0.008	44
	Sclera	0.024	0.028 $\pm$ 0.006	46

The Bunsen coefficients were calculated by the compiler.

AUXILIARY INFORMATION

METHOD:

The method of Ladefoged and Anderson (1), modified to exclude rubber stoppers, was used. Tissue was equilibrated with ^{85}Kr at 310.15 K for 0.5, 1, 2, 3, 4, 5, and 24 hours. Radioactivity was determined in a well scintillation counter for tissue samples alone and for sample + gas at the end of equilibration. The uptake of gas did not vary significantly with the time of equilibration. Transfer loss was less than 10%.

APPARATUS/PROCEDURE:

SOURCE AND PURITY OF MATERIALS:

1. Krypton. No information given.

2. Choroid, retina, and sclera. Freshly enucleated eyes dissected under saline to obtain 0.015 - 0.05 g samples.

ESTIMATED ERROR:

See standard deviations above.

REFERENCES:

1. Ladefoged, J.; Anderson, A. Phys. Med. Biol. 1967, 12, 353.

COMPONENTS:	ORIGINAL MEASUREMENTS:
1. Krypton; Kr; 7439-90-9 2. Water; H_2O; 7732-18-5 3. Sodium chloride; NaCl; 7647-14-5 4. Rabbit leg muscle	Yeh, S-Y.; Peterson, R.E. J. Appl. Physiol. 1965, 20, 1041-1047.

VARIABLES:	PREPARED BY:
T/K: 298.15 - 310.15 P/kPa: 101.325 (1 atm)	A.L. Cramer

EXPERIMENTAL VALUES:

T/K	Absorption Coefficient mean β ± Std. Dev.	Number of Determinations	cm^3 Kr g^{-1} muscle*
298.15	0.0549 ± 0.0015	4	–
303.15	0.0501 ± 0.0010	4	–
310.15	0.0439 ± 0.0010	4	0.0405 - 0.0423

*Calculated by the authors. They corrected for dilution with saline solution and assumed a muscle specific gravity of 1.07. The range of values reflects the range of lipid, water, and protein found by analysis of their samples.

AUXILIARY INFORMATION

METHOD /APPARATUS/PROCEDURE:

In Yeh & Peterson (1) modification of Geffcken (2) apparatus, 45 ml of liquid is frozen, evacuated, and melted repeatedly until no bubbles appeared in liquid under vacuum.

Equilibration with gas and measurement of solubility followed (1).

SOURCE AND PURITY OF MATERIALS:

1. Krypton. The Matheson Co. Research grade, maximum impurity 0.04% nitrogen and/or oxygen.
2. Rabbit leg muscle, homogenized with 4x its volume normal saline, and 0.05% "mercury chloride" added as microbial poison.

ESTIMATED ERROR:
$$\delta T/K = 0.005$$
$$\delta P/mmHg = 0.2$$
$$\delta L/L = 0.05$$

REFERENCES:
1. Yeh, S-Y.; Peterson, R.E. J. Pharm. Sci. 1963, 52, 453.
2. Geffcken, G. Z. Physik Chem. 1904, 49, 257.

COMPONENTS:	ORIGINAL MEASUREMENTS:
1. Krypton; ^{85}Kr; 13983-27-2 2. Guinea Pig Tissues	Kirk, W.P.; Parish, P.W.; Morken, D.A. Health Phys. 1975, 28, 249-261.

VARIABLES:
 	
     T/K: 310.15
     P/kPa: 101.325 (1 atm)
	PREPARED BY:
	
	     A.L. Cramer

EXPERIMENTAL VALUES: ^{85}Kr

TABLE. In vivo ^{85}Kr weight partition coefficients for guinea pig tissues.

T/K	Tissue	Number of Samples	Whole Tissue $L_w \pm$ Std. Dev.	Bloodless Tissue $L_w \pm$ Std. Dev.
310.15	Omental fat	17	0.391 ± 0.059	0.421 ± 0.064
	Subcutaneous fat	16	0.380 ± 0.081	0.405 ± 0.090
	Lungs (see note (1))	4	0.240 ± 0.088	0.274 ± 0.166
	Thymus	6	0.228 ± 0.114	0.259 ± 0.132
	Lymph nodes	8	0.126 ± 0.064	0.138 ± 0.075
	Bone marrow	15	0.113 ± 0.093	0.134 ± 0.119
	Adrenals	18	0.0939 ± 0.0318	0.102 ± 0.038
	Thyroid	9	0.0799 ± 0.0313	0.0829 ± 0.0373
	Liver	9	0.0713 ± 0.0214	0.0768 ± 0.0281
	Large intestine	9	0.0734 ± 0.0402	0.0744 ± 0.0423
	Small intestine	9	0.0715 ± 0.0384	0.0722 ± 0.0391

Note: (1) L_w is not a valid measurement in lungs because of ^{85}Kr in

        air spaces.

AUXILIARY INFORMATION

METHOD/APPARATUS/PROCEDURE:

The guinea pig is enclosed in monitored air with ^{85}Kr for 6 hr (1); 15 min before sacrifice, ^{131}I-tagged albumin (2) is injected to indicate the amount of blood in the tissue. Sacrifice, necropsy, and radioactive count at ^{85}Kr level and ^{131}I level determine the uptake of gas in the tissue.

SOURCE AND PURITY OF MATERIALS:

1. Krypton. No information given. The ^{85}Kr + air mixture had a count of 0.9 µCi cm^{-3}.

2. Guinea pigs. Duncan-Hartley random bred. 700-1000 g, from commercial suppliers or bred in laboratory.

ESTIMATED ERROR:

    $\delta L/L = 0.13$ (blood) - 0.89 (bone marrow)

REFERENCES:

1. Glass, H.I.; Harper, A.M. Phys. Med. Biol. 1962, 7, 335.

2. McCall, M.S.; Camp, M.F. J. Lab. Clin. Med. 1961, 58, 772.

COMPONENTS:	ORIGINAL MEASUREMENTS:
1. Krypton; ^{85}Kr; 13983-27-2 2. Guinea Pig Tissues	Kirk, W.P.; Parish, P.W.; Morken, D.A. Health Phys. 1975, 28, 249-261.

VARIABLES:	PREPARED BY:
T/K: 310.15 P/kPa: 101.325 (1 atm)	A.L. Cramer

EXPERIMENTAL VALUES:
TABLE.   In vivo ^{85}Kr weight partition coefficients for guinea pig tissues.

T/K	Tissue	Number of Samples	Whole Tissue L_w ± Std. Dev.	Bloodless Tissue L_w ± Std. Dev.
310.15	Testes	10	0.0578± 0.0154	0.0579 ± 0.0160
	Ovaries	9	0.0571± 0.0129	0.0574 ± 0.0137
	Blood (see note (2))	10(23)	0.0567± 0.0073	–
	Kidneys	14	0.0455± 0.0160	0.0430 ± 0.0188
	Uterus	4	0.0426± 0.0125	0.0418 ± 0.0127
	Stomach	10	0.0420± 0.0191	0.0415 ± 0.0206
	Brain	19	0.0410± 0.0162	0.0405 ± 0.0167
	Eyes (whole)	8	0.0404± 0.0127	0.0402 ± 0.0128
	Muscle	15	0.0402± 0.0169	0.0396 ± 0.0173
	Seminal vesicles	2	0.0374± 0.0066	0.0369 ± 0.0068
	Spleen	9	0.0352± 0.0144	0.0365 ± 0.0090
	Heart	8	0.0347± 0.0129	0.0300 ± 0.0148

Note: (2)   Weighted average. The number of samples per animal ranged from 1 to 3. The total number of samples is in parentheses.

AUXILIARY INFORMATION

METHOD/APPARATUS/PROCEDURE:	SOURCE AND PURITY OF MATERIALS:
See preceeding page.	See preceeding page.
	ESTIMATED ERROR: See preceeding page.
	REFERENCES: See preceeding page.

COMPONENTS:	ORIGINAL MEASUREMENTS:
1. Krypton-85; $^{85}_{36}$Kr; 13983-27-2 2. Human Liver Tissue	Kitani, K.; Winkler, K. Scand. J. Clin. Lab. Invest. 1972, 29, 173-176.
VARIABLES: T/K: 310.15	PREPARED BY: A.L. Cramer

EXPERIMENTAL VALUES:

Thirty three measurements of the solubility of krypton-85 were made in liver tissue with triglyceride content varying between 1 and 20 weight per cent. One measurement was made at 50 weight percent triglyceride. The results were given in a graph. The data fitted the regression equation

$$L = 0.04924 + (0.004072)(\text{triglyceride wt \%})$$

with a regression coefficient of 0.98.

The triglyceride weight per cent was calculated as tripalmitin. The tissue solubility was corrected for the water added to the sample. Below a liver triglyceride content of 5 per cent the standard deviation was 0.7 per cent, at higher triglyceride content the standard error increases, and reaches a value of 5 per cent at 25 per cent triglyceride.

Tripalmitin is hexadecanoic acid, 1,2,3-propanetriyl ester; $C_{51}H_{98}O_6$; 555-44-2.

AUXILIARY INFORMATION

METHOD /APPARATUS/PROCEDURE:

Five g of liver tissue was homogenized at 277 K for 15 m after addition of 5 cm^3 of water. A glass cuvette (1 x 1 x 10 cm) was 2/3 filled with the homogenate, and closed with a rubber stopper. The space above the homogenate was evacuated and filled with air containing krypton-85. The cuvette was rotated for 2 h in a water bath at 310 K. The radioactivity in the homogenate and in the air phase were determined by a scintillation counter placed in a thermostated box at 310 K (1).

The triglyceride was determined enzymatically in chloroform-methanol extracts.

SOURCE AND PURITY OF MATERIALS:

1. Krypton-85. Radiochemical Centre, Amersham, U.K.

2. Liver tissue. Autopsy material.

ESTIMATED ERROR:

REFERENCES:

1. Kitani, K.
   Scand. J. Clin. Lab. Invest. 1972, 29, 167.

COMPONENTS:	ORIGINAL MEASUREMENTS:
1. Krypton; Kr; 7439-90-9 2. Water; H_2O; 7732-18-5 3. Sodium chloride, NaCl; 7647-14-5 4. Beef brain	Yeh, S-Y; Peterson, R.E. J. Appl. Physiol. 1965, 20, 1041-1047.

VARIABLES:	PREPARED BY:
T/K: 298.15 - 310.15 P/kPa: 101.325 (1 atm)	A.L. Cramer

EXPERIMENTAL VALUES:

T/K	Absorption Coefficient mean β ± Std. Dev.	Number of Determinations	cm^3 Kr g^{-1} muscle*
298.15	0.0572 ± 0.0017	3	–
303.15	0.0517 ± 0.0020	3	–
310.15	0.0454 ± 0.0020	3	0.066 - 0.072

*Calculated by the authors. They corrected for dilution with saline solution and assumed a brain specific gravity of 1.04. The range of values reflects the range of lipid, water, and protein content found by analysis of their samples.

AUXILIARY INFORMATION

METHOD

In Yeh and Peterson (1) modification of Geffcken (2) apparatus, 45 ml of liquid homogenate is frozen, evacuated, and melted repeatedly until no bubbles appear in liquid under vacuum. Equilibration with gas and measurement of solubility follow (1).

SOURCE AND PURITY OF MATERIALS:

1. Krypton. The Matheson Co. Research grade; maximum impurity 0.04% nitrogen and/or oxygen.
2. Beef brain. Homogenized with 4 x volume normal saline (w/w); then mixed with 0.05% "mercury chloride" added as microbial poison.

ESTIMATED ERROR:
$\delta T/K = 0.05$
$\delta P/mmHg = 0.2$
$\delta L/L = 0.05$

REFERENCES:

1. Yeh, S-Y.; Peterson, R.E. J. Pharm. Sci. 1963, 52, 453.

2. Geffcken, G. Z. Physik Chem. 1904, 49, 257.

COMPONENTS:	ORIGINAL MEASUREMENTS:
1. Krypton; ^{85}Kr; 13983-27-2	Kirk, W.P.; Parish, P.W.; Morken,D.A.
2. Water; H_2O; 7732-18-5	
3. Sodium Chloride; NaCl; 7647-14-5	Health Phys. 1975, 28, 249-261.
4. Guinea Pig Brain	

VARIABLES:	PREPARED BY:
T/K: 310.15 P/kPa: 101.325 (1 atm)	A.L. Cramer

EXPERIMENTAL VALUES:

T/K	Solvent	Ostwald coefficient $L \pm$ Std. Dev.	Number of Determinations
310.15	Guinea pig brain	0.0494 + 0.0179	12
310.15	0.9% NaCl solution	0.0472 + 0.0012	4

The authors conversion of their experimental data to the Ostwald coefficient assumed that Henry's law is obeyed and the Ostwald coefficient is independent of pressure.

AUXILIARY INFORMATION

METHOD/APPARATUS/PROCEDURE:

Brain samples were obtained and homogenized under aseptic conditions and diluted 4:1 with sterile 0.9% saline. Penicillin (200 units/ml) and streptomycin (100µg/ml) were added as antibacterial agents. Solubility for 20% brain homogenate in normal saline was determined and Ostwald coefficient for brain was calculated from that value.

SOURCE AND PURITY OF MATERIALS:

1. Krypton. No information given. The ^{85}Kr + air mixture had a count of 0.9 µCi cm^{-3}.

2. Guinea pigs. Duncan-Hartley random bred. 700-1000 g, from commercial suppliers or bred in laboratory.

ESTIMATED ERROR:

$$\delta L/L = 0.36 \text{ (brain)}$$
$$\delta L/L = 0.025 \text{ (saline)}$$

REFERENCES:

COMPONENTS:	EVALUATOR:
1. Xenon; Xe; 7440-63-3 2. Water; H_2O; 7732-18-5	Rubin Battino, Department of Chemistry, Wright State University, Dayton, Ohio, 45435 U.S.A. July 1978

CRITICAL EVALUATION:

The experimental solubility data of three workers was considered to be of sufficient reliability to use in the smoothing equation. In the process of fitting the data to the smoothing equation any data points which differed from the smooth equation by about two standard deviations or more were rejected. The 20 points used for the final smoothing were obtained as follows (reference - number of data points used from that reference): 1-3; 2-4; 3-11; 4-1; 5-1. The fitting equation used was

$$\ln x_1 = A + B/(T/100K) + C \ln (T/100K) + DT/100K \tag{1}$$

Using T/100K as the variable rather than T/K gives coefficients of approximately equal magnitude. The best fit for the 20 data points was

$$\ln x_1 = -74.7398 + 105.210/(T/100K) + 27.4664 \ln (T/100K) \tag{2}$$

where x_1 is the mole fraction solubility of xenon at 101,325 Pa partial pressure of gas. The fit in $\ln x_1$ gave a standard deviation of 0.35% taken at the middle of the temperature range. Table 1 gives smoothed values of the mole fraction solubility at 101,325 Pa partial pressure of gas and the Ostwald coefficient at 5K intervals.

Table 1 also gives the thermodynamic functions $\Delta \overline{G}_1^\circ$, $\Delta \overline{H}_1^\circ$, $\Delta \overline{S}_1^\circ$, and $\Delta \overline{C}_{p_1}^\circ$ for the transfer of gas from the vapor phase at 101,325 Pa partial gas pressure to the (hypothetical) solution phase of unit mole fraction. These thermodynamic properties were calculated from the smoothing equation according to the following equations:

$$\Delta \overline{G}_1^\circ = - RAT - 100RB - RCT \ln (T/100) - RDT^2/100 \tag{3}$$

$$\Delta \overline{S}_1^\circ = RA + RC \ln (T/100) + RC + 2RDT/100 \tag{4}$$

$$\Delta \overline{H}_1^\circ = - 100RB + RCT + RDT^2/100 \tag{5}$$

$$\Delta \overline{C}_{p_1}^\circ = RC + 2RDT/100 \tag{6}$$

The results from five other laboratories were rejected for several reasons. Antropoff's data (6) were very high - being off by as much as 40%. On the other hand, König's data (7) were consistently low by 8 to 22 per cent. Steinberg and Manowitz (8) only crudely measured the solubility and their single value was 6% high. Eucken and Hertzberg's two measurements (9) were both very high (5 to 30%). Wood and Caputi's measurements (10) ranged from 12% high to 3% low although their stated precision was 0.5%.

Figure 1 shows the temperature dependence of the solubility of xenon in water. The curve was obtained from the smoothing equation. There is no minimum in the range of the measurements, but the solubility curve flattens out markedly at the higher temperatures. The ln X equation minimum is 383 K.

Alexander (11) made five calorimetric determinations of the xenon enthalpy of solution in water at 298.15 K. He obtained an enthalpy of solution of -17.2 ± 0.5 kJ mol^{-1} which is 11% below the value derived from the temperature coefficient of the solubility values. We have found no reports of the partial molal volume of xenon in water.

Ewing and Ionescu (12) determined Henry's constant of xenon in water at three temperatures. Their value at 288.15 K agrees with the recommended value, but their values at 278.15 and 298.15 K are 11.5 and 5 per cent high, respectively.

COMPONENTS:	EVALUATOR:
1. Xenon; Xe; 7440-63-3	Rubin Battino, Department of Chemistry, Wright State University, Dayton, Ohio, 45435 U.S.A.
2. Water; H_2O; 7732-18-5	
	July 1978

CRITICAL EVALUATION:

Table 1. Smoothed values of the solubility of xenon in water and thermodynamic functions* at 5K intervals using equation 2 at 101.325 kPa partial pressure of xenon.

T/K	Mole Fraction $X_1 \times 10^5$	Ostwald Coefficient $L \times 10^2$	$\Delta\bar{G}_1^\circ$/ Jmol^{-1}	$\Delta\bar{H}_1^\circ$/ Jmol^{-1}	$\Delta\bar{S}_1^\circ$/JK^{-1}mol^{-1}
273.15	17.991	22.384	19.58	-25.10	-163.6
278.15	14.816	18.773	20.39	-23.95	-159.4
283.15	12.393	15.981	21.18	-22.81	-155.4
288.15	10.519	13.795	21.94	-21.67	-151.4
293.15	9.051	12.066	22.69	-20.53	-147.4
298.15	7.890	10.684	23.42	-19.39	-143.6
303.15	6.961	9.571	24.13	-18.25	-139.8
308.15	6.212	8.668	24.82	-17.10	-136.0
313.15	5.604	7.932	25.49	-15.96	-132.4
318.15	5.107	7.329	26.14	-14.82	-128.7
323.15	4.698	6.833	26.78	-13.68	-125.2
328.15	4.362	6.426	27.39	-12.54	-121.7
333.15	4.084	6.093	27.99	-11.39	-118.2
338.15	3.854	5.822	28.58	-10.25	-114.8
343.15	3.666	5.602	29.14	-9.11	-111.5
348.15	3.511	5.429	29.69	-7.97	-108.2

*$\Delta\bar{C}_{P_1}^\circ$ was independent of temperature and has the value 228 J K^{-1} mol^{-1}.

cal$_{th}$ = 4.184 joule.

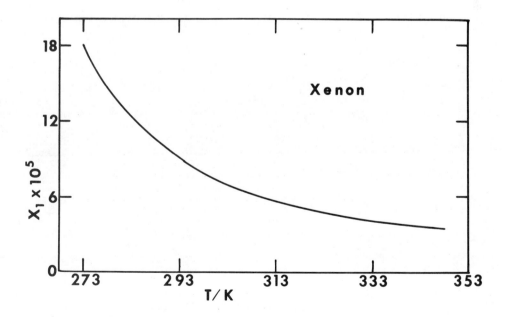

FIGURE 1. The mole fraction solubility of xenon in water at a xenon partial pressure of 101.325 kPa (1 atm).

COMPONENTS:	EVALUATOR:
1. Xenon; Xe; 7440-63-3 2. Water; H_2O; 7732-18-5	Rubin Battino, Department of Chemistry, Wright State University, Dayton, Ohio, 45435 U.S.A. July 1978

CRITICAL EVALUATION:

References

1. Yeh, S-H.; Peterson, R. E. J. Pharm. Sci. 1964, 53, 822.

2. Morrison, T. J.; Johnstone, N. B. J. Chem. Soc. 1954, 3441.

3. Benson, B. B.; Krause, D. J. Chem. Phys. 1976, 64, 689.

4. Kitani, K. Scand. J. Clin. Lab. Invest. 1972, 29, 167.

5. Ladefoged, J.; Anderson, A. M. Phys. Med. Biol. 1967, 12, 353.

6. Antropoff, A. Proc. Roy. Soc. (London) 1910, 83, 474; Z. Elektrochem. 1919, 25, 269.

7. König, H. Z. Naturforsch. 1963, 18a, 363.

8. Steinberg, M.; Manowitz, B. Ind. Eng. Chem. 1959, 51, 47.

9. Eucken, A.; Hertzberg, G. Z. Physik. Chem. 1950, 195, 1.

10. Wood, D.; Caputi, R. AD631557, Avail. CFSTI, 1966, 17 pp.

11. Alexander, D. M. J. Phys. Chem. 1959, 63, 994.

12. Ewing, G. J.; Ionescu, L. G. J. Chem. Eng. Data 1974, 19, 367.

COMPONENTS:	ORIGINAL MEASUREMENTS:

COMPONENTS:

1. Xenon; Xe; 7440-63-3

2. Water; H_2O; 7732-18-5

ORIGINAL MEASUREMENTS:

Morrison, T. J.; Johnstone, N. B.
J. Chem. Soc.
1954, 3441 - 3443.

VARIABLES:

  T/K: 285.85 - 344.85

PREPARED BY:

R. Battino

EXPERIMENTAL VALUES:

T/K	Mol Fraction $X_1 \times 10^5$	Kuenen Coefficient $S_o \times 10^3$
285.85	10.896	135.5
288.24	9.943	123.6
288.65	9.742	121.1
292.75	8.928	110.9
303.35	6.821	84.5
313.15	5.613*	69.3
313.65	5.493	67.8
321.05	4.827*	59.4
332.15	4.150*	50.8
344.85	3.604*	43.8

The mole fraction solubility at a partial pressure of xenon of 101.325 kPa was calculated by the compiler.

The solubility, So, is the cm^3 xenon at 273.15 K and 101.325 kPa per kg of water at a partial gas pressure of 101.325 kPa.

*Solubility values which were used in the final smoothing equation for the recommended solubility values given in the critical evaluation.

AUXILIARY INFORMATION

METHOD:

  Flowing the previously degassed liquid in a thin film through the gas in an absorption spiral. Volume changes are measured in burets. See ref. 1.

SOURCE AND PURITY OF MATERIALS:

1. Xenon-British Oxygen Company, Ltd. Contained about 1% krypton.

2. Water-no comments by authors.

APPARATUS/PROCEDURE:

  Apparatus of Morrison and Billett used. See ref. 1. The authors used the smoothing equation

$\log_{10}s_o = -60.836 + 3605/(T/K)$.

ESTIMATED ERROR:

REFERENCES:

1. Morrison, T. J.; Billett, F. J. Chem. Soc. 1952, 3819.

K.X.R.—L

COMPONENTS:	ORIGINAL MEASUREMENTS:
1. Xenon; Xe; 7440-63-3 2. Water; H_2O; 7732-18-5	Yeh, S-Y.; Peterson, R. E. J. Pharm. Sci. 1964, 53, 822 - 824.

VARIABLES:	PREPARED BY:
T/K: 298.15 - 318.15	R. Battino

EXPERIMENTAL VALUES:

T/K	Mol Fraction $X_1 \times 10^5$	Ostwald Coefficient L
298.15	7.887*	0.1068
303.15	6:968*	0.0958
310.15	5.893	0.0827
318.15	5.136*	0.0737

The mole fraction solubility at 101.325 Pa (1 atm) partial pressure of xenon was calculated by the compiler.

*Solubility values which were used in the final smoothing equation for the recommended solubility values given in the critical evaluation.

AUXILIARY INFORMATION

METHOD: Used a 125 cm^3 absorption flash in a water bath. The amount of gas absorbed by the degassed liquid is measured on a thermostated gas buret. Details are given in ref. 1.

SOURCE AND PURITY OF MATERIALS:

1. Xenon-Matheson Co. Maximum impurities of 0.02% nitrogen and 0.05% Krypton.

2. Water-redistilled from glass apparatus.

APPARATUS/PROCEDURE:

ESTIMATED ERROR:
Authors estimate error in solubility to be 1%.

REFERENCES:

1. Yeh, S-Y.; Peterson, R. E.
    J. Pharm. Sci. 1963, 52, 453.

COMPONENTS:	ORIGINAL MEASUREMENTS:
1. Xenon-133; $^{133}_{54}$Xe; 14932-42-4 2. Water; H_2O; 7732-18-5	Ladefoged, J.; Andersen, A. M. Phys. Med. Biol. 1967, 12, 353-8.

VARIABLES: T/K: 310.15	PREPARED BY: R. Battino

EXPERIMENTAL VALUES:

T/K	Mol Fraction X_1 x 10^5	Ostwald Coefficient L x 10^2
310.15	5.943*	8.34

The mole fraction solubility at 101.325 Pa (1 atm) was calculated by compiler.

*Solubility value which was used in the final smoothing equation for the recommended solubility values which are given in the critical evaluation.

AUXILIARY INFORMATION

METHOD: An airtight 2 ml vial was half filled with distilled water. Some ^{133}Xe was added to the air in the vial. After 24 hrs equilibration the activity in the aqueous and gas phases was measured in a well-type scintillation counter. The result cited above was the average of 107 observations. Solubility was also measured in saline, olive oil, liquid paraffin, solutions of albumin, and blood at 310.15 K.	SOURCE AND PURITY OF MATERIALS: 1. Xenon-133. Radiochemical Centre, Amersham. Impurity of other isotopes less than 2%. (Authors) 2. Water.Distilled.
	ESTIMATED ERROR: Result reported by authors as 0.0834 ± 0.0002 for 107 observations.
	REFERENCES:

COMPONENTS:	ORIGINAL MEASUREMENTS:
1. Xenon-133; $^{133}_{54}$Xe; 14932-42-4 2. Water; H_2O; 7732-18-5	Kitani, K. Scand. J. Clinical and Lab. Invest. 1972, 29, 167-172.
VARIABLES: T/K: 310.15	PREPARED BY: R. Battino

EXPERIMENTAL VALUES:

T/K	Mol Fraction $X_1 \times 10^5$	Ostwald Coefficient $L \times 10^2$
310.15	5.914*	8.30

The mole fraction solubility at 101.325 kPa (1 atm) was calculated by compiler.

*Solubility value which was used in the final smoothing equation for the recommended solubility values which are given in the critical evaluation.

AUXILIARY INFORMATION

METHOD: Liquid was placed in a septum sealed 1 x 1 x 10 cm curette filled two-thirds. The radioactive gas was in air in the other third. After 2 hrs equilibration the middle portion of the liquid and gas phases were counted using a scintillation counter. Solubilities were also measured in saline, lipids, and blood.

SOURCE AND PURITY OF MATERIALS:

1. Xenon-133. Radiochemical Centre, Amersham.

2. Water. Distilled.

ESTIMATED ERROR:
Temperature control to $\pm 0.5^{\circ}$C. Results reported by author as 0.0830 ± 0.0017 (S.D.) for 17 measurements.

REFERENCES:

141

COMPONENTS:	ORIGINAL MEASUREMENTS:
1. Xenon; Xe; 7440-63-3 2. Water; H_2O; 7732-18-5	Ewing, G. J.; Ionescu, L. G. J. Chem. Eng. Data 1974, 19, 367 -369.
VARIABLES: T/K: 278.15 298.15 P/kPa: 101.325 - 506.625 (1 - 5 atm)	PREPARED BY: H. L. Clever

EXPERIMENTAL VALUES:

T/K	Henry's Constant $K = (P/atm)/(m/mol\ Xe\ kg^{-1}\ H_2O)$	Mol Fraction $X_1 \times 10^5$
278.15	109 ± 10	16.5
288.15	171 ± 8	10.5
298.15	217 ± 6*	8.3

*An earlier report from the same laboratory gave a
value of 227 (1) at 298.15 K.

AUXILIARY INFORMATION

METHOD:

Deionized water was introduced into
the reaction vessel, degassed, and
then equilibrated with xenon at the
desired temperature. Equilibration
required up to five hours. The change
in pressure of the xenon gas was
measured. The moles of xenon absorbed
by the water was calculated from the
ideal gas equation from the pressure
decrease, the gas volume and the
temperature. Measurements were made
at several pressures up to five atm,
and the Henry's law slope determined.

SOURCE AND PURITY OF MATERIALS:

1. Xenon. J.T.Baker Chemical Co.
   Stated to contain the impurities
   Kr 10 ppm, N_2 50 ppm, O_2 5 ppm,
   H_2 5 ppm, hydrocarbons 15 ppm, and
   moisture 5 ppm.

2. Water. Deionized.

ESTIMATED ERROR:
$\delta T/K = 0.1$
$\delta P/mmHg = 0.5$

REFERENCES:

1. Ewing, G. J.; Maestas, S.
   J. Phys. Chem. 1970, 74, 2341.

COMPONENTS:	ORIGINAL MEASUREMENTS:
1. Xenon; Xe; 7440-63-3 2. Water; H_2O; 7732-18-5	Benson, B. B.; Krause, D. J. Chem. Phys. 1976, 64, 689 - 709.
VARIABLES: T/K: 274.150 - 318.405	PREPARED BY: R. Battino

EXPERIMENTAL VALUES:

T/K	Mol Fraction X_1 x 10^5	Bunsen Coefficient α x 10^3
274.150	17.458	214.578
278.157	14.837*	182.403
283.154	12.382*	152.208
288.150	10.498*	128.398
288.150	10.532*	129.398
293.129	9.041*	110.997
298.145	7.875*	96.573
298.148	7.885*	96.702
298.149	7.917*	97.092
303.154	6.966*	85.320
308.155	6.232*	76.219
313.150	5.620*	68.614
318.405	5.134*	62.584

The mole fraction solubility at 101.325 kPa (1 atm) partial pressure of xenon was calculated by the compiler.

*Solubility values which were used in the final smoothing equation for the recommended solubility values given in the critical evaluation.

AUXILIARY INFORMATION

METHOD/APPARATUS/PROCEDURE:	SOURCE AND PURITY OF MATERIALS:
Gas-free water and the pure gas are equilibrated, and volumetric samples of the liquid and gaseous phases are isolated. The gas dissolved in the water is extracted and the number of moles determined on a special mercury manometer. After removal of the water vapor, the number of moles in the sample of the gaseous phase is measured with the same manometer; from which the pressure (and fugacity) above the solution may be calculated. Real gas corrections are made. Predicted maximum error is 0.02%. No drawings of the apparatus are given in the original paper.	1. Xenon. No comment by authors. 2. Water. No comment by authors.
	ESTIMATED ERROR: Smoothed data were fit to 0.20% rms by the authors. Calculated error in the measurements was 0.02%.
	REFERENCES:

143

COMPONENTS:	ORIGINAL MEASUREMENTS:
1. Xenon; Xe; 7440-63-3 2. Water; H_2O; 7732-18-5	Stephan, E. L., Hatfield, N. S., Peoples, R. S. and Pray, H. A. H., *Battelle Memorial Institute Report BMI-1067*, <u>1956</u>.

VARIABLES:	PREPARED BY:
Temperature, pressure	C. L. Young

EXPERIMENTAL VALUES:

T/K	$P^\dagger$/bar	Mole fraction of xenon in liquid, x_{Xe}	T/K	$P^\dagger$/bar	Mole fraction of xenon in liquid, x_{Xe}
373.15	9.31	0.000286	533.15	11.8	0.000956
	10.0	0.000281		12.3	0.000932
	16.9	0.000514		14.5	0.00117
	17.9	0.000547		15.0	0.00124
	18.6	0.000555		15.4	0.00120
	20.0	0.000631	574.82	6.48	0.000900
435.93	13.4	0.000498		7.17	0.000932
	14.1	0.000490		7.65	0.000932
	15.4	0.000571		8.20	0.00103
	15.9	0.000579		9.24	0.00108
	16.5	0.000635			

$P^\dagger$ partial pressure of xenon.

AUXILIARY INFORMATION

METHOD/APPARATUS/PROCEDURE:	SOURCE AND PURITY OF MATERIALS:
Static equilibrium apparatus. Gas and liquid equilibrated for 18 hours. Pressure measured with Bourdon gauge and temperature measured with thermocouple. Composition of liquid estimated by volumetric method. Details in source. Partial pressure estimated by subtracting vapor pressure of water from total pressure.	No details given.
	ESTIMATED ERROR: $\delta T/K = \pm 0.6$; δP/bar $= \pm 0.3$; $\delta x_{Xe} = \pm 0.00002$ (estimated by compiler).
	REFERENCES:

COMPONENTS:	ORIGINAL MEASUREMENTS:
1. Xenon; Xe; [7440-63-3] 2. Water; H_2O; [7732-18-5]	Franck, E. U.; Lentz, H.; Welsch, H., Z. Phys. Chem., N.F., <u>1974</u>, 93, 95.
VARIABLES: Temperature, pressure	PREPARED BY: C. L. Young

EXPERIMENTAL VALUES:

T/K	P/bar	Mole fraction of xenon at phase boundary, x_{Xe}	T/K	P/bar	Mole fraction of xenon at phase boundary, x_{Xe}
627	210	0.030	603	2000	0.187
634	240	0.030	583	200	0.192
635	260	0.030	603	240	0.192
635	270	0.030	617	300	0.192
629	300	0.030	623	350	0.192
575	400	0.030	625	430	0.192
489	600	0.030	615	600	0.192
622	230	0.064	606	850	0.192
633	290	0.064	606	2000	0.192
628	330	0.064	609	2370	0.192
611	420	0.064	612	2700	0.192
573	625	0.064	503	155	0.300
513	1875	0.064	583	250	0.300
604	215	0.096	593	280	0.300
615	240	0.096	609	340	0.300
623	270	0.096	620	420	0.300
629	320	0.096	621	520	0.300
630	340	0.096	616	770	0.300
623	370	0.096	623	1300	0.300
611	500	0.096	527	150	0.495
603	265	0.187	538	175	0.495
617	305	0.187	557	220	0.495
623	340	0.187	571	300	0.495
618	520	0.187	583	370	0.495
603	820	0.187	599	650	0.495 (cont.)

AUXILIARY INFORMATION

METHOD/APPARATUS/PROCEDURE:	SOURCE AND PURITY OF MATERIALS:
Small volume static cell with sapphire window. Temperature measured with thermocouple and pressure measured with strain gauge. Phase boundary conditions obtained from discontinuity in isochoric temperature pressure curves for mixtures of fixed composition.	1. Deutsche l'Air Liquide sample, purity 99.998 mole per cent. 2. Triple distilled.
	ESTIMATED ERROR: $\delta T/K = \pm 2$; $\delta P/bar = \pm 1$; $\delta x_{Xe} = \pm 0.002$ (estimated by compiler).
	REFERENCES:

COMPONENTS:

1. Xenon; Xe; [7440-63-3]

2. Water; H_2O; [7732-18-5]

ORIGINAL MEASUREMENTS:

Franck, E. U.; Lentz, H.; Welsch, H.; *Z. Phys. Chem.*, *N.F.*, <u>1974</u>, *93*, 95.

EXPERIMENTAL VALUES:

T/K	P/bar	Mole fraction of xenon at phase boundary, x_{Xe}	T/K	P/bar	Mole fraction of xenon at phase boundary, x_{Xe}
621	1300	0.495	582	450	0.630
528	170	0.625	595	680	0.630
547	240	0.625	496	200	0.771
563	310	0.625	533	260	0.771
579	470	0.625	555	370	0.771
603	1200	0.625	573	520	0.771
613	2600	0.625	588	870	0.771
544	220	0.630	597	1600	0.771
551	260	0.630	598	1820	0.771
563	325	0.630	601	2460	0.771

COMPONENTS:	ORIGINAL MEASUREMENTS:
1. Xenon; Xe; 7440-63-3 2. Deuterium oxide; D_2O; 7789-20-0	Stephan, E. L., Hatfield, N. S., Peoples, R. S. and Pray, H. A. H., *Battelle Memorial Institute Report* *BMI-1067*, <u>1956</u>.

VARIABLES:	PREPARED BY:
Temperature, pressure	C. L. Young

EXPERIMENTAL VALUES:

T/K	$P^{\dagger}$/bar	Mole fraction of xenon in liquid, x_{Xe}	T/K	$P^{\dagger}$/bar	Mole fraction of xenon in liquid, x_{Xe}
435.93	10.3	0.000362	533.15	8.9	0.000830
	11.0	0.000436		11.0	0.000972
	12.7	0.000491		11.7	0.001008
	13.4	0.000509		12.3	0.001017
	14.1	0.000495	574.82	6.5	0.000874
533.15	5.8	0.000571		7.2	0.000946
	6.3	0.000589		7.5	0.001088
	6.8	0.000598		8.2	0.001133
	8.2	0.000776		8.9	0.001186

$P^{\dagger}$ partial pressure of xenon.

AUXILIARY INFORMATION

METHOD /APPARATUS/PROCEDURE:	SOURCE AND PURITY OF MATERIALS:
Gas and liquid equilibrated for 18 hours. Pressure measured with Bourdon gauge and temperature measured with thermocouple. Composition of liquid estimated by volumetric method. Details in source. Partial pressure estimated by subtracting vapor pressure of deuterium oxide from total pressure.	No details given.
	ESTIMATED ERROR: $\delta T/K = \pm 0.6$; $\delta P/bar = \pm 0.3$; $\delta x_{Xe} = \pm 2\text{-}3\%$ (estimated by compiler).
	REFERENCES:

COMPONENTS:	ORIGINAL MEASUREMENTS:
1. Xenon; Xe; 7440-63-3 2. Sea Water	König, H. Z. Naturforschg. 1963, 18a, 363-367.

VARIABLES:	PREPARED BY:
T/K: 273.15 - 298.15 P/kPa: 101.325 (1 atm)	H. L. Clever A. L. Cramer

EXPERIMENTAL VALUES:

T/K	Xenon $cm^3(STP)\ kg^{-1}$ water	Xenon $cm^3(STP)\ kg^{-1}$ sea water	Bunsen Coefficient α sea water
273.15	–	136	0.140
278.15	147	115	0.118
283.15	128	103	0.106
288.15	115	90.0	0.0922
293.15	–	80.0	0.0818
298.15	96.4	70.2	0.0717

The sea water chlorinity was 19.12‰ (Salinity S‰ = 34.54).

The Bunsen coefficients were calculated by the compiler using sea water densities from the International Critical Tables.

AUXILIARY INFORMATION

METHOD/APPARATUS/PROCEDURE:	SOURCE AND PURITY OF MATERIALS:
The apparatus is a modification of that of Morrison and Billett (1).	1. Xenon. Linde. Greater than 99.9 percent pure.

APPARATUS/PROCEDURE:	ESTIMATED ERROR: $\delta T/K = 1.0 \overset{<}{=} 283.15\ K$ $= 0.5 \overset{\geq}{} 298.15\ K$
	REFERENCES: 1. Morrison, T. J.; Billett, F. J. Chem. Soc. 1948, 2033; ibid, 1952, 3821.

COMPONENTS:	ORIGINAL MEASUREMENTS:
1. Xenon; Xe; 7440-63-3 2. Sea Water	Wood, D.; Caputi, R. U.S.N.R.D.L.-TR-988, 27 Feb. 1966 Chem. Abstr. 1967, 66, 118693u.

VARIABLES:	PREPARED BY:
T/K: 274.45 - 321.25 P/kPa: 101.325 (1 atm)	A. L. Cramer

EXPERIMENTAL VALUES:

T/K	Henry's Constant $K = (P_1/mmHg)/X_1$	Percent Error*	Number of Determinations	Mol Fraction X_1 x 10^2	Bunsen Coefficient α
		Water; H_2O; 7732-18-5			
274.45	0.425 x 10^7	0.5	3	0.0179	0.224
299.35	0.987 x 10^7	2.0	4	0.00770	0.0955
320.35	1.586 x 10^7	0.6	4	0.00479	0.0591
		Artificial Sea Water (1), S‰ = 34.727			
274.45	0.568 x 10^7	2.1	3	0.0134	0.167
300.15	1.267 x 10^7	0.0	2	0.00600	0.0744
321.25	1.848 x 10^7	2.5	4	0.00411	0.0506

*Percent error is the maximum spread in Henry's constant times 100 divided by average Henry's constant.

The mole fractions were calculated by the compiler from the average Henry's constant. The Bunsen coefficients were calculated by the compiler from the mole fractions using a solvent mean molecular weight of 18.4823 and sea water densities from the International Critical Tables.

AUXILIARY INFORMATION

METHOD /APPARATUS/PROCEDURE:

   Degassed water was introduced into an evacuated apparatus (< 50 µ Hg) and gas bled into burette. After the system was isolated, gas was admitted to equilibrium column and the water was circulated through the column, flowing over packing of 4 mm Berl saddles at 110 ml/min for 4-5 hr.

   Dissolved gas was reclaimed and measured by evacuating the system → < 1 µ Hg and allowing water to dis- till and condense in a cold trap. Water was melted and refrozen until all the gas was recovered. Gas was then transferred to a gas burette and the pressure was measured with a Hg manometer. Purity was checked by gas chromatography.

SOURCE AND PURITY OF MATERIALS:

1. Xenon. AIRCO, Certified 0.0042 impurity, Kr and N_2. Determined 0.0001 after experiment.

2. Water. Distilled three times be- fore degassing.
 Sea Water. Artificial, modified from (1).

ESTIMATED ERROR: $\delta T/K$ = 0.1, 0.005, 0.03
 (as T increases)
 $\delta P/P$ = 0.001
 $\delta H/H$ = 0.005
 (author's error analysis)

REFERENCES:

1. Lyman, J.; Fleming, R. H.
 J. Mar. Res. 1940, 3, 134.

COMPONENTS:	EVALUATOR:
1. Xenon; Xe; 7440-63-3	H. L. Clever
	Chemistry Department
2. Water; H_2O; 7732-18-5	Emory University
	Atlanta, GA 30322
3. Electrolyte	U.S.A.
	August 1978

CRITICAL EVALUATION:

The solubility of xenon in aqueous electrolyte solutions.

There are data on the solubility of xenon in aqueous NaCl, KI, and Na_3PO_4 solutions. The Setschenow salt effect parameters $k_s = (1/m)\log(S^o/S)$ and $k_{sx} = (1/m)\log(X^o/X)$, where S is volume of gas at STP dissolved kg^{-1} water and X is the gas mole fraction, are summarized in Table 1.

TABLE 1. Setschenow salt effect parameters for xenon dissolved in aqueous electrolyte solution.

Electrolyte	T/K	mol salt kg^{-1} H_2O	k_s	k_{sx}	Ref.
NaCl	273.15	0.76	0.182	0.197	2
		1.17	0.185	0.200	2
		2.68	0.175	0.190	2
	293.15	0.285	0.161	0.176	2
		1.32	0.161	0.176	2
		1.80	0.154	0.169	2
	298.15	0.155	0.237	0.252	3
		1.0	0.149	0.164	1
	303.15	0.155	0.172	0.187	3
	310.15	0.155	0.143	0.158	3
	318.15	0.155	0.185	0.200	3
Na_3PO_4*	298.15	0.066	0.499	--	3
	303.15	0.066	0.462	--	3
	310.15	0.066	0.368	--	3
	318.15	0.066	0.453	--	3
KI	298.15	1.0	0.113	0.128	1

*Buffer solution, 0.066 mol Na_3PO_4 dm^{-3} solution. The Setschenow parameter is $k_L = (1/m)\log(L^o/L)$.

The Setschenow parameters from the data of Eucken and Hertzberg (2) and of Morrisson and Johnstone (1) appear to form a self-consistent set of values for aqueous NaCl solutions and they are recommended. The Setschenow salt effect parameters for aqueous NaCl solutions calculated from the data of Yeh and Peterson (3) appear high and erratic. This is partly due to the experimental difficulties in making accurate gas solubility measurements in water and dilute aqueous salt solutions and it may indicate the salt effect parameter is of greater magnitude in the dilute NaCl solution (0.155 molal) than in solutions of 1.0 molal and greater concentration. The Setschenow parameters for xenon in aqueous Na_3PO_4 solution and aqueous KI solution are tentative.

REFERENCES

1. Morrison, T.J.; Johnstone, N.B.B. J. Chem. Soc. 1955, 3655.

2. Eucken, A.; Hertzberg, G. Z. phys. Chem. 1950, 195, 1.

3. Yeh, S.-Y.; Peterson, R.E. J. Pharm. Sci. 1964, 53, 822.

   Ladefoged and Anderson (4) report the solubility of xenon-133 in water and standard saline solution. The k_s value calculated from their data is 0.188 at 310.15 K. Isbister, Schofield and Torrance (5) report the solubility of xenon-133 in saline solution. The saline solution was not defined. The solubility value appears to be high and is not recommended.

4. Ladefoged, J.; Anderson, A. M. Phys. Med. Biol. 1967, 12, 353.

5. Isbister, W. H.; Schofield, P. F.; Torrance, H. B. Phys. Med. Biol. 1965, 10, 243.

COMPONENTS:	ORIGINAL MEASUREMENTS:

COMPONENTS:

1. Xenon; Xe; 7440-63-3

2. Dioxosulfatouranium (Uranyl
    Sulfate); UO_6S; 1314-64-3

3. Water; H_2O; 7732-18-5

ORIGINAL MEASUREMENTS:

Stephan, E. L., Hatfield, N. S.,
Peoples, A. S. and Pray, H. A. H.,
*Battelle Memorial Institute Report
BMI-1067*, <u>1956</u>.

VARIABLES:

Temperature, pressure, composition

PREPARED BY:

C. L. Young

EXPERIMENTAL VALUES:

T/K	g Uranium per liter	$P^\dagger$/bar	Solubility*	T/K	g Uranium per liter	$P^\dagger$/bar	Solubility*
373.15	40	12.1	0.432	435.93	40	18.9	0.756
		12.4	0.443			19.6	0.79
		12.8	0.465	533.15	40	17.5	1.51
435.93	40	11.7	0.47			18.2	1.53
		12.0	0.505			18.5	1.63
		12.3	0.52			18.9	1.66
		18.2	0.735				

$P^\dagger$ partial pressure of xenon.

*   ml of xenon at S.T.P. per g of solution.

AUXILIARY INFORMATION

METHOD/APPARATUS/PROCEDURE:

Static equilibrium apparatus. Gas
and liquid equilibrated for 18 hours.
Pressure measured with Bourdon gauge
and temperature measured with thermo-
couple. Composition of liquid
estimated by volumetric method.
Details in source.   Partial pressure
estimated by subtracting vapor pressure from total pressure.

SOURCE AND PURITY OF MATERIALS:

No details given.

ESTIMATED ERROR:
$\delta T/K = \pm0.6$; δP/bar $= \pm0.3$;
δ(solubility) $= \pm3\%$ (estimated by
compiler).

REFERENCES:

COMPONENTS:	ORIGINAL MEASUREMENTS:
1. Xenon; Xe; 7440-63-3 2. Water; H_2O; 7732-18-5 3. Alkali Halides	Morrison, T.J.; Johnstone, N.B.B. J. Chem. Soc. 1955, 3655-3659.

VARIABLES:
        T/K: 298.15
        P/kPa: 101.325 (1 atm)
	PREPARED BY:
	        T.D. Kittredge
	        H.L. Clever

EXPERIMENTAL VALUES:

T/K	k_s = (1/m) log (S^o/S)	k_{sX} = (1/m) log (X^o/X)
Sodium Chloride; NaCl; 7647-14-5		
298.15	0.149	0.164
Potassium Iodide; KI; 7681-11-0		
298.15	0.113	0.128

The values of the Setschenow salt effect parameter, k_s, were apparently determined from only two solubility measurements. They were the solubility of xenon in pure water, S^o, and the solubility of xenon in a near one equivalent of salt per kg of water solution, S. No solubility values are given in the paper. The S^o/S ratio was referenced to a solution containing one kg of water. The compiler calculated the salt effect parameter k_{sX} from the mole fraction solubility ratio X^o/X. The electrolytes were assumed to be 100 per cent dissociated and both cation and anion were used in the mole fraction calculation.

AUXILIARY INFORMATION

METHOD:	SOURCE AND PURITY OF MATERIALS:
Gas absorption in a flow system.	1. Xenon. British Oxygen Co. Ltd. 2. Water. No information given. 3. Electrolyte. No information given.

APPARATUS/PROCEDURE:
 	
   The previously degassed solvent flows in a thin film down an absorption spiral containing the gas plus solvent vapor at a total pressure of one atm. The volume of gas absorbed is measured in attached calibrated burets (1).
	ESTIMATED ERROR:
	
	        δk_s = 0.010

	REFERENCES: 1. Morrison, T.J.; Billett, F. J. Chem. Soc. 1952, 3819.

COMPONENTS:	ORIGINAL MEASUREMENTS:
1. Xenon; Xe; 7440-63-3	Eucken, A.; Hertzberg, G.
2. Water; H_2O; 7732-18-5	
3. Sodium Chloride; NaCl; 7647-14-5	Z. physik. Chem. 1950, 195, 1-23.

VARIABLES:	PREPARED BY:
T/K: 273.15 - 293.15 P/kPa: 101.325 (1 atm)	P.L. Long

EXPERIMENTAL VALUES:

T/K	Sodium Chloride mol kg^{-1} H_2O	Ostwald Coefficient L	Setschenow Constant k = (1/m) log (L^o/L)
273.15	0	0.222	
	0.76	0.1595	0.188
	1.17	0.1323	0.190
	2.68	0.0719	0.184
293.15	0	0.1188	
	0.285	0.1062	0.164
	1.32	0.0710	0.168
	1.80	0.0605	0.163

AUXILIARY INFORMATION

METHOD /APPARATUS/PROCEDURE:

Gas absorption. The apparatus consists of a gas buret and an adsorption flask connected by a capillary tube. The whole apparatus is shaken. The capillary tube is a 2m-long glass helix. An amount of gas is measured at STP and placed in the gas buret. After shaking, the difference from the original amount of gas placed in the gas buret is determined.

SOURCE AND PURITY OF MATERIALS:

Components. No information given.

ESTIMATED ERROR:

$\delta L/L$ = 0.01

REFERENCES:

COMPONENTS:	ORIGINAL MEASUREMENTS:
1. Xenon; Xe; 7440-63-3 2. Water; H_2O; 7732-18-5 3. Sodium Chloride; NaCl; 7647-14-5	Yeh, S.Y.; Peterson, R.E. J. Pharm. Sci., 1964, 53, 822-824.
VARIABLES: T/K: 298.15 - 318.15 P/kPa: 101.325 (1 atm)	PREPARED BY: H.L. Clever

EXPERIMENTAL VALUES:

T/K	Ostwald Coefficient L	ΔH^O/cal mol^{-1}	ΔS^O/cal K^{-1} mol^{-1}
298.15	0.0976	-4801	-27.0
303.15	0.0895	-4533	-26.1
310.15	0.0778	-4158	-24.9
318.15	0.0680	-3729	-26.6

For comparison, the authors' Ostwald coefficients in water were 0.1068, 0.0958, 0.0827 and 0.0737 at the four temperatures.

The sodium chloride solution was 0.9 weight percent which is about 0.155 molal.

Each solubility value is the average of three to four measurements, the standard deviation of each measurement closely approximated 1.0 per cent.

AUXILIARY INFORMATION

METHOD/APPARATUS/PROCEDURE:	SOURCE AND PURITY OF MATERIALS:
Freshly boiled solution was introduced into 125 ml. absorption flask of solubility apparatus (1), then frozen and boiled under vacuum three times. Water-saturated gas was introduced and equilibrated (2) and weight of solution was determined. Thermodynamic constants were calculated from equations (3): $\log S = A/T + B \log T - C$ $\Delta H^O = R(-2.3A + BT - T)$ $\Delta S^O = R(-B-B\ln T + 2.3C +$ $\ln(0.082T) + 1)$	1. Xenon. Matheson Co. 2. Water. Distilled from glass apparatus. 3. Sodium chloride. Analytical grade.

ESTIMATED ERROR:
$\delta T/K = 0.05$
$\delta P/mmHg = 0.2$
$\delta L/L = 0.002$

REFERENCES:
1. Geffken, G. Z. Physik Chem. 1904, 49, 257.
2. Yeh, S.Y.; Peterson, R.E. J. Pharm. Sci. 1963, 52, 453-8.
3. Eley, E.E. Trans. Faraday Soc. 1939, 35, 1281.

154

COMPONENTS:	ORIGINAL MEASUREMENTS:
1. Xenon; Xe; 7440-63-3	Yeh, S.Y.; Peterson, R.E.
2. Water; H_2O; 7732-18-5	
3. Sodium Phosphate (phosphate buffer); Na_3PO_4; 7601-54-9	J. Pharm. Sci., 1964, 53, 822-4.

VARIABLES:	PREPARED BY:
T/K: 298.15 - 318.15 P/kPa: 101.325 (1 atm)	H.L. Clever

EXPERIMENTAL VALUES:

T/K	Ostwald Coefficient L	ΔH^o/cal mol^{-1}	ΔS^o/cal K^{-1} mol^{-1}
298.15	0.0990	-4566	-26.3
303.15	0.0893	-4320	-25.4
310.15	0.0782	-3977	-24.3
318.15	0.0688	-3585	-23.1

For comparison, the authors' Ostwald coefficients in water were 0.1068 0.0958, 0.827 and 0.0737 at the four temperatures.

The sodium phosphate solution was 0.066 M at pH 7.0.

Each solubility value is the average of three to four measurements. The standard deviation of each measurement closely approximates 1.0 per cent.

AUXILIARY INFORMATION

METHOD /APPARATUS/PROCEDURE:

Freshly boiled solution was introduced into 125 ml. absorption flask of solubility apparatus (1), then frozen and boiled under vacuum three times. Water-saturated gas was introduced and equilibrated (2) and weight of solution was determined.

Thermodynamic constants were calculated from equations (3):

$$\log S = A/T + B \log T - C$$
$$\Delta H^o = R(-2.3A + BT - T)$$
$$\Delta S^o = R(-B-B\ln T + 2.3C + \ln(0.082T) +1)$$

SOURCE AND PURITY OF MATERIALS:

1. Xenon. Matheson Co.
2. Water. Distilled from glass apparatus.
3. Sodium phosphate. Analytical grade.

ESTIMATED ERROR:
$\delta T/K = 0.05$
$\delta P/mmHg = 0.2$
$\delta L/L = 0.002$

REFERENCES:
1. Geffken, G. Z. Physik Chem. 1904, 49, 257.
2. Yeh, S.Y.; Peterson, R.E. J. Pharm. Sci. 1963, 52, 453-8.
3. Eley, E.E. Trans. Faraday Soc. 1939, 35, 1281.

155

COMPONENTS:	ORIGINAL MEASUREMENTS:
1. Xenon; Xe; 7440-63-3 2. Pentane; C_5H_{12}; 109-66-0	Makranczy, J.; Megyery-Balog, K.; Rusz, L.; Patyi, L. Hung. J. Ind. Chem. 1976, 4, 269-280.

VARIABLES:
   T/K: 298.15
   P/kPa: 101.325 (1 atm)
	PREPARED BY:
	
	   S.A. Johnson

EXPERIMENTAL VALUES:

T/K	Mol Fraction X_1 x 10^2	Bunsen Coefficient α	Ostwald Coefficient L
298.15	2.731	5.420	5.916

The mole fraction and Bunsen coefficient were calculated by the compiler.

AUXILIARY INFORMATION

METHOD /APPARATUS/PROCEDURE:	SOURCE AND PURITY OF MATERIALS:
Volumetric method. The apparatus of Bodor, Bor, Mohai, and Sipos (1) was used.	Both the gas and liquid were analytical grade reagents of Hungarian or foreign origin. No further information.
	ESTIMATED ERROR: $\delta X_1/X_1 = 0.03$

	REFERENCES:
	1. Bodor, E.; Bor, Gy.; Mohai, B.;
	   Sipos, G.
	   Veszpremi Vegyip. Egy. Kozl.
	   1957, 1, 55.
	   Chem. Abstr. 1961, 55, 3175h.

COMPONENTS:	ORIGINAL MEASUREMENTS:
1. Xenon; Xe; 7440-63-3 2. Hexane; C_6H_{14}; 110-54-3	Clever, H.L. J. Phys. Chem. 1958, 62, 375-376.

VARIABLES:
     T/K: 289.15 - 316.25
     P/kPa: 101.325 (1 atm)
	PREPARED BY:
	
	     C.E.Edelman
	     A.L. Cramer

EXPERIMENTAL VALUES:

T/K	Mol Fraction $X_1 \times 10^2$	Bunsen Coefficient α	Ostwald Coefficient L
289.15	2.98	5.29	5.60
298.45	2.54	4.43	4.84
307.55	2.29	3.95	4.45
316.25	2.01	3.41	3.95

Smoothed Data: $\Delta G^o/J\ mol^{-1} = - RT\ ln\ X_1 = -10808 + 66.635\ T$

Std. Dev. ΔG = 27.6, Coef. Corr. = 0.9994

$\Delta H^o/J\ mol^{-1}$ = -10808 , $\Delta S^o/J\ K^{-1}\ mol^{-1}$ = -66.635

T/K	Mol Fraction $X_1 \times 10^2$	$\Delta G^o/J\ mol^{-1}$
288.15	3.01	8,393.1
293.15	2.79	8,726.3
298.15	2.59	9,059.4
303.15	2.41	9,392.6
308.15	2.25	9,725.8
313.15	2.10	10,059
318.15	1.97	10,392

Makranczy, J.; Megyery-Balog, K.; Rusz, L.; Patyi, L. Hung. J. Ind. Chem. 1976, 4, 269 report an Ostwald coefficient of 5.065 (mole fraction 2.652 $\times 10^{-2}$) at 298.15K. The value was not used in the smoothed data fit above.

AUXILIARY INFORMATION

METHOD:
    The solvent is saturated with gas as it flows through an 8 mm x 180 cm glass helix attached to a gas buret. The total pressure of solute gas plus solvent vapor is maintained at 1 atm as the gas is absorbed.

The solubility values were adjusted to a partial pressure of xenon of 101.325 kPa (1 atm) by Henry's law.

The Bunsen coefficients were calculated by the compiler.

APPARATUS/PROCEDURE:

    The apparatus is a modification of that of Morrison and Billett (1). The modifications include the addition of a spiral storage for the solvent, a manometer for a constant reference pressure, and an extra buret for highly soluble gases. The solvent is degassed by a modification of the method of Baldwin and Daniel (2).

SOURCE AND PURITY OF MATERIALS:

1. Xenon. Linde Air Products Co.

2. Hexane. Humphrey-Wilkinson, Inc. Shaken with conc. H_2SO_4, water washed, dried over Na, distilled, b.p. 68.23 - 68.52°C.

ESTIMATED ERROR:
$$\delta T/K = 0.05$$
$$\delta P/mmHg = 3$$
$$\delta X_1/X_1 = 0.01$$

REFERENCES:

1. Morrison, T.J.; Billett, F. J. Chem. Soc. 1948, 2033; ibid. 1952, 3819.

2. Baldwin, R.R.; Daniel, S.G. J. Appl. Chem. 1952, 2, 161.

COMPONENTS:	ORIGINAL MEASUREMENTS:
1. Xenon; Xe; 7440-63-3 2. Heptane; C_7H_{16}; 142-82-5	Makranczy, J.; Megyery-Balog, K.; Rusz, L.; Patyi, L. Hung. J. Ind. Chem. 1976, 4, 269-280.

VARIABLES:	PREPARED BY:
T/K: 298.15 P/kPa: 101.325 (1 atm)	S.A. Johnson H.L. Clever

EXPERIMENTAL VALUES:

T/K	Mol Fraction X_1 x 10^2	Bunsen Coefficient α	Ostwald Coefficient L
298.15	2.602	4.060	4.432

The mole fraction and Bunsen coefficient were calculated by the compiler.

Steinberg, M.; Manowitz, B. Ind. Eng. Chem. 1959, 51, 47 report an absorption coefficient of 3.27 (mole fraction 1.97 x 10^{-2}) at 294.65 K for this system. A comparison of the solubility of xenon in several closely related hydrocarbons with the values of the solubility of xenon in heptane indicates the Steinberg and Manowitz value is probably low. The Makranczy et al. value is consistent with other values and is accepted as the tentative solubility value.

AUXILIARY INFORMATION

METHOD/APPARATUS/PROCEDURE:	SOURCE AND PURITY OF MATERIALS:
Volumetric method. The apparatus of Bodor, Bor, Mohai, and Sipos (1) was used.	Both the gas and liquid were analytical grade reagents of Hungarian or foreign origin. No further information.
	ESTIMATED ERROR: $\delta X_1/X_1 = 0.03$
	REFERENCES: 1. Bodor, E.; Bor, Gy.; Mohai, B.; Sipos, G. Veszpremi Vegyip. Egy. Kozl. 1957, 1, 55. Chem. Abstr. 1961, 55, 3175h.

COMPONENTS:	ORIGINAL MEASUREMENTS:
1. Xenon; Xe; 7440-63-3 2. 2,2,4-Trimethylpentane (Isooctane); C_8H_{18}; 540-84-1	Clever, H.L. J. Phys. Chem. 1958, 62, 375-376.

VARIABLES:	PREPARED BY:
T/K: 289.15 - 316.15 P/kPa: 101.325 (1 atm)	C.E. Edelman A.L. Cramer

EXPERIMENTAL VALUES:

T/K	Mol Fraction $X_1 \times 10^2$	Bunsen Coefficient α	Ostwald Coefficient L
289.15	3.13	4.42	4.68
298.65	2.64	3.73	4.08
307.55	2.41	3.29	3.70
316.15	2.14	2.90	3.36

Smoothed Data: $\Delta G^o/J\ mol^{-1} = - RT \ln X_1 = -10,419 + 64.929\ T$

Std. Dev. ΔG = 36.7, Coef. Corr. = 0.9988

$\Delta H^o/J\ mol^{-1} = -10,419$, $\Delta S^o/J\ K^{-1}\ mol^{-1} = -64.929$

T/K	Mol Fraction $X_1 \times 10^2$	$\Delta G^o/J\ mol^{-1}$
288.15	3.14	8,289.9
293.15	2.92	8,614.6
298.15	2.72	8,939.2
303.15	2.53	9,263.9
308.15	2.37	9,588.5
313.15	2.22	9,913.2
318.15	2.08	10,238

AUXILIARY INFORMATION

METHOD:

The solvent is saturated with gas as it flows through an 8 mm x 180 cm glass helix attached to a gas buret. The total pressure of solute gas plus solvent vapor is maintained at 1 atm as the gas is absorbed.

The solubility values were adjusted to a partial pressure of xenon of 101.325 kPa (1 atm) by Henry's law.

The Bunsen coefficients were calculated by the compiler.

SOURCE AND PURITY OF MATERIALS:

1. Xenon. Linde Air Products Co.

2. 2,2,4-Trimethylpentane. Enjay Co. Used as received.

APPARATUS/PROCEDURE:

The apparatus is a modification of that of Morrison and Billett (1). The modifications include the addition of a spiral storage for the solvent, a manometer for a constant reference pressure, and an extra buret for highly soluble gases. The solvent is degassed by a modification of the method of Baldwin and Daniel (2).

ESTIMATED ERROR:

$\delta T/K = 0.05$
$\delta P/mmHg = 3$
$\delta X_1/X_1 = 0.01$

REFERENCES:

1. Morrison, T.J.; Billett, F. J. Chem. Soc. 1948, 2033; Ibid. 1952, 3819.

2. Baldwin, R.R.; Daniel, S.G. J. Appl. Chem. 1952, 2, 161.

COMPONENTS:	ORIGINAL MEASUREMENTS:
1. Xenon; Xe; 7440-63-3 2. C_8 to C_{11} Alkanes	Makranczy, J.; Megyery-Balog, K.; Rusz, L.; Patyi, L. Hung. J. Ind. Chem. 1976, 4, 269-280.
VARIABLES: T/K: 298.15 P/kPa: 101.325 (1 atm)	PREPARED BY: S.A. Johnson

EXPERIMENTAL VALUES:

T/K	Mol Fraction X_1 x 10^2	Bunsen Coefficient α	Ostwald Coefficient L
Octane; C_8H_{18}; 111-65-9			
298.15	2.538	3.570	3.897
Nonane; C_9H_{20}; 111-84-2			
298.15	2.502	3.200	3.493
Decane; $C_{10}H_{22}$; 124-18-5			
298.15	2.472	2.900	3.165
Undecane; $C_{11}H_{24}$; 1120-21-4			
298.15	2.465	2.670	2.914

The mole fraction and Bunsen coefficient were calculated by the compiler.

AUXILIARY INFORMATION

METHOD /APPARATUS/PROCEDURE:

Volumetric method. The apparatus of Bodor, Bor, Mohai, and Sipos (1) was used.

SOURCE AND PURITY OF MATERIALS:

Both the gas and liquid were analytical grade reagents of Hungarian or foreign origin. No further information.

ESTIMATED ERROR:

$$\delta X_1/X_1 = 0.03$$

REFERENCES:

1. Bodor, E.; Bor, Gy.; Mohai, B.; Sipos, G.
Veszpremi Vegyip. Egy. Kozl. 1957, 1, 55.
Chem. Abstr. 1961, 55, 3175h.

COMPONENTS:	ORIGINAL MEASUREMENTS:
1. Xenon; Xe; 7440-63-3 2. Dodecane; $C_{12}H_{26}$; 112-40-3	Clever, H.L. $\underline{J}$. $\underline{Phys}$. $\underline{Chem}$. 1958, $\underline{62}$, 375-376.

VARIABLES:	PREPARED BY:
T/K: 289.15 - 316.15 P/kPa: 101.325 (1 atm)	C.E. Edelman A.L. Cramer

EXPERIMENTAL VALUES:

T/K	Mol Fraction $X_1 \times 10^2$	Bunsen Coefficient α	Ostwald Coefficient L
289.15	3.48	3.57	3.78
298.15	3.04	3.07	3.35
307.55	2.74	2.74	3.08
316.15	2.47	2.45	2.84

Smoothed Data: $\Delta G^o / J\ mol^{-1} = -\ RT\ ln\ X_1 = -9,514.1 + 60.872\ T$

Std. Dev. $\Delta G = 17.2$, Coef. Corr. $= 0.9997$

$\Delta H^o / J\ mol^{-1} = -9,514.1$, $\Delta S^o / J\ K^{-1}\ mol^{-1} = -60.872$

T/K	Mol Fraction $X_1 \times 10^2$	$\Delta G^o / J\ mol^{-1}$
288.15	3.51	8,026.1
293.15	3.28	8,330.5
298.15	3.07	8,634.8
303.15	2.88	8,939.2
308.15	2.71	9,243.5
313.15	2.55	9,547.9
318.15	2.41	9,852.3

Makranczy, J.; Megyery-Balog, K.; Rusz, L.; Patyi, L. $\underline{Hung}$. $\underline{J}$. $\underline{Ind}$. $\underline{Chem}$. 1976, $\underline{4}$, 269 report an Ostwald coefficient of 2.685 (mole fraction 2.448 x 10^{-2}) at 298.15K. The value was not used in the smoothed data fot above.

The solubility values were adjusted to a partial pressure of xenon of 101.325 kPa (1 atm) by Henry's law.

AUXILIARY INFORMATION

METHOD:	SOURCE AND PURITY OF MATERIALS:
The solvent is saturated with gas as it flows through an 8 mm x 180 cm glass helix attached to a gas buret. The total pressure of solute gas plus solvent vapor is maintained at 1 atm as the gas is absorbed. The Bunsen coefficients were calculated by the compiler.	1. Xenon. Linde Air Products Co. 2. Dodecane. Humphrey-Wilkinson, Inc. Shaken with conc. H_2SO_4, water washed, dried over Na.

APPARATUS/PROCEDURE:	ESTIMATED ERROR:
The apparatus is a modification of that of Morrison and Billett (1). The modifications include the addition of a spiral storage for the solvent, a manometer for a constant reference pressure, and an extra buret for highly soluble gases. The solvent is degassed by a modification of the method of Baldwin and Daniel (2).	$\delta T/K = 0.05$ $\delta P/mmHg = 3$ $\delta X_1/X_1 = 0.01$
	REFERENCES: 1. Morrison, T.J.; Billett, F. $\underline{J}$. $\underline{Chem}$. $\underline{Soc}$. 1948, 2033; $\underline{Ibid}$. 1952, 3819. 2. Baldwin, R.R.; Daniel, S.G. $\underline{J}$. $\underline{Appl}$. $\underline{Chem}$. 1952, $\underline{2}$, 161.

COMPONENTS:	ORIGINAL MEASUREMENTS:
1. Xenon; Xe; 7440-63-3 2. C_{13} to C_{16} Alkanes	Makranczy, J.; Megyery-Balog, K.; Rusz, L.; Patyi, L. Hung. J. Ind. Chem. 1976, 4, 269-280.

VARIABLES:	PREPARED BY:
T/K: 298.15 P/kPa: 101.325 (1 atm)	S.A. Johnson

EXPERIMENTAL VALUES:

T/K	Mol Fraction X_1 x 10^2	Bunsen Coefficient α	Ostwald Coefficient L
Tridecane; $C_{13}H_{28}$; 629-50-5			
298.15	2.421	2.270	2.478
Tetradecane; $C_{14}H_{30}$; 629-59-4			
298.15	2.412	2.120	2.314
Pentadecane; $C_{15}H_{32}$; 629-62-9			
298.15	2.382	1.970	2.150
Hexadecane; $C_{16}H_{34}$; 544-76-3			
298.15	2.369	1.850	2.019

The mole fraction and Bunsen coefficient were calculated by the compiler.

AUXILIARY INFORMATION

METHOD/APPARATUS/PROCEDURE:	SOURCE AND PURITY OF MATERIALS:
Volumetric method. The apparatus of Bodor, Bor, Mohai, and Sipos (1) was used.	Both the gas and liquid were analytical grade reagents of Hungarian or foreign origin. No further information.

ESTIMATED ERROR:

$$\delta X_1/X_1 = 0.03$$

REFERENCES:

1. Bodor, E.; Bor, Gy.; Mohai, B.; Sipos, G.
   Veszpremi Vegyip. Egy. Kozl.
   1957, 1, 55.
   Chem. Abstr. 1961, 55, 3175h.

COMPONENTS:	EVALUATOR:
1. Xenon; Xe; 7440-63-3 2. Cyclohexane; C_6H_{12}; 110-82-7	H. L. Clever Chemistry Department Emory University Atlanta, GA 30322 U.S.A. August 1978

CRITICAL EVALUATION:

Both Clever (1) and Dymond (2) measured the solubility of xenon in cyclohexane at four temperatures in the room temperature range. Dymond's data is more self-consistent than Clever's data. The smoothed data differs by 5.4 per cent at 288.15 K and 2.1 per cent at 308.15 K with Dymond's mole fraction solubilities greater than Clever's.

Dymond's four values and Clever's values at 307.60 and 316.25 K were combined in a linear regression of a Gibbs energy equation linear in temperature to obtain the recommended equation.

The recommended thermodynamic values for the transfer of xenon from the gas at 101.325 kPa (1 atm) to the hypothetical unit mole fraction solution are

$$\Delta G^o/J\ mol^{-1} = -\ RT\ ln\ X_1 = -9,980.4 + 65.625\ T$$

Std. Dev. ΔG^o = 28.3, Coef. Corr. = 0.9987

$$\Delta H^o/J\ mol^{-1} = -9,980.4, \quad \Delta S^o/J\ K^{-1}\ mol^{-1} = -65.625$$

The recommended values of the mole fraction solubility at 101.325 kPa and the Gibbs energy of solution as a function of temperature are given in Table 1.

TABLE 1. Solubility of xenon in cyclohexane. Recommended mole fraction solubility and Gibbs energy of solution as a function of temperature.

T/K	Mol Fraction X_1 x 10^2	$\Delta G^o/J\ mol^{-1}$
288.15	2.41	8,929.6
293.15	2.24	9,257.7
298.15	2.09	9,585.9
303.15	1.96	9,914.0
308.15	1.84	10,242
313.15	1.73	10,570
318.15	1.62	10,898

REFERENCES

1. Clever, H.L. J. Phys. Chem. 1958, 62, 375.

2. Dymond, J.H. J. Phys. Chem. 1967, 71, 1829.

COMPONENTS:	ORIGINAL MEASUREMENTS:
1. Xenon; Xe; 7440-63-3 2. Cyclohexane; C_6H_{12}; 110-82-7	Clever, H.L. J. Phys. Chem. 1958, 62, 375-376.

| VARIABLES:
    T/K: 289.15 - 316.25
    P/kPa: 101.325 (1 atm) | PREPARED BY:
    C.E. Edelman
    A.L. Cramer |

EXPERIMENTAL VALUES:

T/K	Mol Fraction X_1 x 10^2	Bunsen Coefficient α	Ostwald Coefficient L
289.15	2.33	4.97	5.26
299.15	1.92	4.04	4.42
307.60	1.81	3.77	4.25
316.25	1.68	3.45	4.00

Smoothed Data: $\Delta G^o/J\ mol^{-1} = - RT\ ln\ X_1 = -8,847.8 + 62.092\ T$

Std. Dev. ΔG = 76.7, Coef. Corr. = 0.9944

See the evaluation of xenon + cyclohexane for the recommended Gibbs energy equation and smoothed solubility values.

The solubility values were adjusted to a partial pressure of xenon of 101.325 kPa (1 atm) by Henry's law.

The Bunsen coefficients were calculated by the compiler.

AUXILIARY INFORMATION

METHOD:

The solvent is saturated with gas as it flows through an 8 mm x 180 cm glass helix attached to a gas buret. The total pressure of solute gas plus solvent vapor is maintained at 1 atm as the gas is absorbed.

SOURCE AND PURITY OF MATERIALS:

1. Xenon. Linde Air Products Co.

2. Cyclohexane. Phillips Petroleum Co. Used as received.

APPARATUS/PROCEDURE:

The apparatus is a modification of that of Morrison and Billett (1). The modifications include the addition of a spiral storage for the solvent, a manometer for a constant reference pressure, and an extra buret for highly soluble gases. The solvent is degassed by a modification of the method of Baldwin and Daniel (2).

ESTIMATED ERROR:
$\delta T/K = 0.05$
$\delta P/mmHg = 3$
$\delta X_1/X_1 = 0.01$

REFERENCES:

1. Morrison, T.J.; Billett, F. J. Chem. Soc. 1948, 2033; Ibid. 1952, 3819.

2. Baldwin, R.R.; Daniel, S.G. J. Appl. Chem. 1952, 2, 161.

COMPONENTS:	ORIGINAL MEASUREMENTS:
1. Xenon; Xe; 7440-63-3 2. Cyclohexane; C_6H_{12}; 110-82-7	Dymond, J. H. J. Phys. Chem. 1967, 71, 1829 - 1831.

VARIABLES:
 	
     T/K: 292.32 - 309.20
     P/kPa: 101.325 (1 atm)
	PREPARED BY:
	          M. E. Derrick

EXPERIMENTAL VALUES:

T/K	Mol Fraction X_1 x 10^2	Bunsen Coefficient α	Ostwald Coefficient L
292.32	2.28	4.84	5.18
298.01	2.10	4.42	4.82
306.15	1.885	3.92	4.39
309.20	1.815	3.76	4.26

Smoothed Data: $\Delta G°/J\ mol^{-1} = - RT\ ln\ X_1 = -10{,}147 \quad + 66.159\ T$

Std. Dev. $\Delta G° = 3.0$, Coef. Corr. $= 0.9999$

See the evaluation of xenon + cyclohexane for the recommended Gibbs energy equation and the smoothed solubility values.

The Bunsen and Ostwald coefficients were calculated by the compiler.

AUXILIARY INFORMATION

METHOD:	SOURCE AND PURITY OF MATERIALS:
Saturation of liquid with gas at partial pressure of gas equal to 1 atm.	1. Xenon. Matheson Co., dried. 2. Cyclohexane. Matheson, Coleman, and Bell. Chromatoquality reagent. Dried and fractionally frozen. m.p. 6.45 ° C.
APPARATUS/PROCEDURE: Dymond-Hildebrand apparatus (1) using an all-glass pumping system to spray slugs of degassed solvent into the gas. Amount of gas dissolved calculated from initial and final gas pressures.	ESTIMATED ERROR: $\delta X_1/X_1 = 0.01$
	REFERENCES: 1. Dymond, J.; Hildebrand, J. H. Ind. Eng. Chem. Fundam. 1967, 6, 130.

COMPONENTS:	ORIGINAL MEASUREMENTS:
1. Xenon; Xe; 7440-63-3 2. Methylcyclohexane; C_7H_{14}; 108-87-2	Clever, H.L.; Saylor, J.H.; Gross, P.M. J. Phys. Chem. 1958, 62, 89-91.

VARIABLES:	PREPARED BY:
T/K: 289.15 - 316.25 Total P/kPa: 101.325 (1 atm)	P.L. Long

EXPERIMENTAL VALUES:

T/K	Mol Fraction $X_1 \times 10^2$	Bunsen Coefficient α	Ostwald Coefficient L
289.15	2.50	4.53	4.80
303.15	2.18	3.87	4.30
316.25	1.85	3.22	3.73

Smoothed Data:

$$\Delta G^{o}/\text{J mol}^{-1} = - RT \ln X_1 = -8453.0 + 59.833\, T$$

Std. Dev. $\Delta G^{o} = 36.8$, Coef. Corr. = 0.9990

$$\Delta H^{o}/\text{J mol}^{-1} = -8453.0, \quad \Delta S^{o}/\text{J K}^{-1}\text{mol}^{-1} = -59.833$$

T/K	Mol Fraction $X_1 \times 10^2$	$\Delta G/\text{J mol}^{-1}$
288.15	2.55	8,788.0
293.15	2.40	9,087.1
298.15	2.27	9,386.3
303.15	2.14	9,685.5
308.15	2.03	9,984.6
313.15	1.93	10,284
318.15	1.83	10,583

The solubility values were adjusted to a partial pressure of xenon of 101.325 kPa (1 atm) by Henry's law.

The Bunsen coefficients were calculated by the compiler.

AUXILIARY INFORMATION

METHOD:
   Volumetric. The apparatus (1) is a modification of that used by Morrison and Billett (2). Modifications include the addition of a spiral solvent storage tubing, a manometer for constant reference pressure, and an extra gas buret for highly soluble gases.

SOURCE AND PURITY OF MATERIALS:
1. Xenon. Matheson Co., Inc. Both standard and research grades were used.

2. Methylcyclohexane. Eastman Kodak Co., white label. Dried over Na and distilled; corrected b.p. 100.95 to 100.97°, lit. b.p. 100.93°C.

APPARATUS/PROCEDURE:
   (a) Degassing. 700 ml of solvent is shaken and evacuated while attached to a cold trap, until no bubbles are seen; solvent is then transferred through a 1 mm capillary tubing, released as a fine mist into a continuously evacuated flask. (b) Solvent is saturated with gas as it flows through 8 mm x 180 cm of tubing attached to a gas buret. Pressure is maintained at 1 atm as the gas is absorbed.

ESTIMATED ERROR:
$$\delta T/K = 0.05$$
$$\delta P/\text{mmHg} = 3$$
$$\delta X_1/X_1 = 0.03$$

REFERENCES:
1. Clever, H.L.; Battino, R.; Saylor, J.H.; Gross, P.M. J. Phys. Chem. 1957, 61, 1078.

2. Morrison, T.J.; Billett, F. J. Chem. Soc. 1948, 2033; Ibid. 1952, 3819.

COMPONENTS:	ORIGINAL MEASUREMENTS:
1. Xenon; Xe; 7440-63-3 2. Benzene; C_6H_6; 71-43-2	Clever, H.L. J. Phys. Chem. 1958, 62, 375-376.

VARIABLES:
    T/K: 289.15 - 316.25
    P/kPa: 101.325 (1 atm)
	PREPARED BY:
	        C.E. Edelman
	        A.L. Cramer

EXPERIMENTAL VALUES:

T/K	Mol Fraction $X_1 \times 10^2$	Bunsen Coefficient α	Ostwald Coefficient L
289.15	1.32	3.39	3.59
298.15	1.11	2.82	3.08
307.60	1.06	2.65	2.98
316.25	1.01	2.50	2.90

Smoothed Data: $\Delta G^O/\text{J mol}^{-1} = - RT \ln X_1 = -7,062.5 + 60.704\ T$

Std. Dev. $\Delta G = 91.5$, Coef. Corr. $= 0.9918$

$\Delta H^O/\text{J mol}^{-1} = -7,062.5$, $\Delta S^O/\text{J K}^{-1}\text{mol}^{-1} = -60.704$

T/K	Mol Fraction $X_1 \times 10^2$	$\Delta G^O/\text{J mol}^{-1}$
288.15	1.29	10,429
293.15	1.22	10,733
298.15	1.16	11,037
303.15	1.11	11,340
308.15	1.06	11,644
313.15	1.02	11,947
318.15	0.974	12,251

AUXILIARY INFORMATION

METHOD:

The solvent is saturated with gas as it flows through an 8 mm x 180 cm glass helix attached to a gas buret. The total pressure of solute gas plus solvent vapor is maintained at 1 atm as the gas is absorbed.

The solubility values were adjusted to a partial pressure of xenon of 101.325 kPa (1 atm) by Henry's law.

The Bunsen coefficients were calculated by the compiler.

SOURCE AND PURITY OF MATERIALS:

1. Xenon. Linde Air Products Co.

2. Benzene. Jones & Laughlin Steel Co. Shaken with conc. H_2SO_4, water washed, dried over Na, distilled, b.p. 80.03-80.04°C.

APPARATUS/PROCEDURE:

The apparatus is a modification of that of Morrison and Billett (1). The modifications include the addition of a spiral storage for the solvent, a manometer for a constant reference pressure, and an extra buret for highly soluble gases. The solvent is degassed by a modification of the method of Baldwin and Daniel (2).

ESTIMATED ERROR:
$\delta T/K = 0.05$
$\delta P/\text{mmHg} = 3$
$\delta X_1/X_1 = 0.01$

REFERENCES:

1. Morrison, T.J.; Billett, F. J. Chem. Soc. 1948, 2033; ibid. 1952, 3819.

2. Baldwin, R.R.; Daniel, S.G. J. Appl. Chem. 1952, 2, 161.

COMPONENTS:	ORIGINAL MEASUREMENTS:
1. Xenon; Xe; 7440-63-3 2. Methylbenzene (Toluene); C_7H_8; 108-88-3	Saylor, J. H.; Battino, R. $\underline{J}$. $\underline{Phys}$. $\underline{Chem}$. 1958, $\underline{62}$, 1334-1337.

| VARIABLES:
 T/K: 288.15 - 328.15
 P/kPa: 101.325 (1 atm) | PREPARED BY:

 H. L. Clever |

EXPERIMENTAL VALUES:

T/K	Mol Fraction $X_1 \times 10^2$	Bunsen Coefficient α	Ostwald Coefficient L
288.15	1.637	3.526	3.722
328.15	1.078	2.210	2.655

Smoothed Data: $\Delta G°/J\ mol^{-1} = -RT \ln X_1 = -8,257.7 + 62.829\ T$

$\Delta H°/J\ mol^{-1} = -8,257.7$, $\Delta S°/J\ K^{-1}\ mol^{-1} = -62.829$

T/K	Mol Fraction $X_1 \times 10^2$	$\Delta G°/J\ mol^{-1}$
288.15	1.64	9,846.4
298.15	1.46	10,475
308.15	1.31	11,102
318.15	1.19	11,731
328.15	1.08	12,360

The solubility values were adjusted to a partial pressure of xenon of 101.325 kPa (1 atm) by Henry's law.

The Bunsen coefficients were calculated by the compiler.

AUXILIARY INFORMATION

METHOD /APPARATUS/PROCEDURE:	SOURCE AND PURITY OF MATERIALS:
The solvent was degassed by evacuating the space above it, shaking, and then passing it as a fine mist into another evacuated container. The degassed liquid was saturated as it passed as a thin film inside a glass helix ehich contained the solute gas plus solvent vapor at a total pressure of 1 atm (1,2). The volume of liquid and the volume of gas absorbed are determined in a system of burets.	1. Xenon. Linde Air Products Co. 2. Toluene. Mallinckrodt. Reagent grade. Shaken over conc. H_2SO_4, water washed, dried over Drierite, distilled b.p. 110.40 - 110.60°C.

	ESTIMATED ERROR: $\delta T/K = 0.03$ $\delta P/torr = 1$ $\delta X_1/X_1 = 0.005$

| The smoothed data above are based on only two experimental points. They should be used with caution.

 Evaluator's Note: Steinberg, M.; Manowitz, B. $\underline{Ind}$. $\underline{Eng}$. $\underline{Chem}$. 1959, $\underline{51}$, 47 report an Absorption coefficient of 3.17 (mole fraction 1.41 x 10^{-2}) at 297.15 K for this system. | REFERENCES:

 1. Morrison, T. J.; Billett, F. J. Chem. Soc. 1948, 2033; $\underline{ibid}$, 1952, 3819.

 2. Baldwin, R. R.; Daniel, S. G. $\underline{J}$. $\underline{Appl}$. $\underline{Chem}$. 1952, $\underline{2}$, 161. |

COMPONENTS:	ORIGINAL MEASUREMENTS:
1. Xenon; Xe; 7440-63-3 2. Aromatic Hydrocarbons	Steinberg, M.; Manowitz, B. Ind. Eng. Chem. 1959, 51, 47-51.

VARIABLES:
     T/K: 273.15 - 305.15
     P/kPa: 101.325 (1 atm)
	PREPARED BY:
	
	     H.L. Clever
	     A.L. Cramer

EXPERIMENTAL VALUES:

T/K	Mol Fraction $X_1 \times 10^2$	Bunsen Coefficient α	Absorption Coefficient β
Toluene; C_7H_8; 108-88-3			
297.15	1.41	3.00	3.17
1,4-Dimethylbenzene; C_8H_{10}; 106-42-3			
273.15	1.67	3.11	3.28
302.15	1.55	2.80	2.95
Dimethylbenzene Isomer Mixture; C_8H_{10}; 1330-20-7			
305.15	0.743	1.38	1.46
1,3,5-Trimethylbenzene (Mesitylene); C_9H_{12}; 108-67-8			
293.15	1.45	2.39	2.52

The authors define the absorption coefficient as the volume of gas, corrected to 288.15 K and 101.325 kPa, absorbed under a total system pressure of 101.325 kPa per unit volume of solvent at 288.15 K.

The mole fraction solubilities and Bunsen coefficients were calculated by the compiler.

AUXILIARY INFORMATION

METHOD /APPARATUS/PROCEDURE:	SOURCE AND PURITY OF MATERIALS:
Absorption coefficient determined by a modified McDaniel method (1).	1. Xenon. Matheson Co., Inc. Technical grade. 2. Solvents. Source not given. Chemically pure grade except dimethylbenzene isomers which were technical grade.
	ESTIMATED ERROR: $\delta\beta/\beta = 0.05 - 0.10$
	REFERENCES: 1. Furman, N.H. "Scott's Standard Methods of Chemical Analysis" Van Nostrand Co., NY 1939, 5th ed., Vol. II, p. 2587.

COMPONENTS:	ORIGINAL MEASUREMENTS:
1. Xenon; Xe; 7440-63-3 2. Amsco 123-15	Steinberg, M.; Manowitz, B. Ind. Eng. Chem. 1959, 51, 47-50.

VARIABLES:	PREPARED BY:
T/K: 294.15 - 423.15 P/kPa: 101.325 (1 atm)	P.L. Long

EXPERIMENTAL VALUES:

	Xenon at 1 atm		Xenon at low conc in N_2	
T/K	Absorption Coefficient β	Henry's Constant K/atm	Xenon Initial ppm in N_2	Henry's Constant K/atm
294.15	3.39	-	-	-
297.15	3.30	-	-	-
297.15	3.42	31.0	662	33.4
333.15	1.60	67.0	655	63.0
383.15	1.18	90.0	659	97.5
423.15	0.76	139.0	1070	157.0

The authors define the Absorption Coefficient as the volume of gas, corrected to 288.15 K and 101.325 kPa, absorbed under a total system pressure of 101.325 kPa per unit volume of solvent at 288.15 K.

AUXILIARY INFORMATION

METHOD/APPARATUS/PROCEDURE:	SOURCE AND PURITY OF MATERIALS:
The absorption coefficient at one atm xenon was measured by modified McDaniel method (1). The Henry's constant $(K = (P/atm)/X_1)$ at low concentration of xenon was measured by static and dynamic tracer techniques. The authors state that log (Absorption Coefficient vs 1/T is linear and gives an enthalpy of solution of -3040 cal mol^{-1}.	1. Xenon. Matheson Co., Inc. Technical grade. 2. Amsco 123-15. American Mineral Spirits Co. No. 140. Paraffin 59.6 wt %, naphthene 27.3 wt %, aromatics 13.2 wt %.
	ESTIMATED ERROR: $\delta K/K = 0.05 - 0.10$ (McDaniel method) $\delta K/K = 0.18$ (Tracer technique)
	REFERENCES: 1. Furman, N.H. "Scott's Standard Methods of Chemical Analysis" Van Nostrand Co., NY 1939, 5th ed., Vol. II, p. 2587.

K.X.R.—N

COMPONENTS:	ORIGINAL MEASUREMENTS:
1. Xenon; Xe; 7440-63-3 2. 1-Propanol; C_3H_8O; 71-23-8	Komarenko, V.G.; Manzhelii, V.G. Ukr.Fiz.Zh.(Ukr.Ed.) 1968,13,387-391. Ukr. Phys. J. 1968, 13, 273-276.

VARIABLES:	PREPARED BY:
T/K: 193.15 - 243.15 P/kPa: 26.664 (200 mmHg)	T.D. Kittredge

EXPERIMENTAL VALUES:

T/K	Mol Fraction P/mmHg 200 $X_1 \times 10^2$	Mol Fraction P/mmHg 760 $X_1 \times 10^2$
193.15	2.20	8.36
203.15	1.648	6.262
213.15	1.183	4.495
223.15	0.862	3.28
233.15	0.663	2.52
243.15	0.515	1.96

Smoothed Data: $\ln X_1 = 188.924 - 262.126/(T/100) - 249.815 \ln (T/100)$
$+ 56.3087 \ (T/100)$

T/K	Mol Fraction $X_1 \times 10^2$	$\Delta G^{\circ}/kJmol^{-1}$	$\Delta H^{\circ}/kJmol^{-1}$	$\Delta S^{\circ}/JK^{-1}mol^{-1}$	$\Delta Cp^{\circ}/JK^{-1}mol^{-1}$
193.15	8.38	3.982	-8.58	-65.04	-268.5
203.15	6.215	4.692	-10.80	-76.25	-174.9
213.15	4.515	5.490	-12.08	-82.42	-81.24
223.15	3.31	6.325	-12.42	-84.02	+12.39
233.15	2.49	7.156	-11.83	-81.44	+106.0
243.15	1.97	7.942	-10.30	-75.03	+199.7

The mole fraction solubility at 101.325 kPa (760 mmHg) was calculated by Henry's law by the compiler.

AUXILIARY INFORMATION

METHOD/APPARATUS/PROCEDURE:	SOURCE AND PURITY OF MATERIALS:

The solvent was degassed by vacuum. A thin layer of alcohol, cooled to 125-175 K, was kept for 20 hours in a vacuum maintained at 10^{-3} mmHg.
     The degassed liquid was sealed under vacuum in an ampoule which was placed in the apparatus. The apparatus consisted of a manostat, a mercury compensator, and a solubility cell divided by a mercury seal. A gas pressure of 200 mmHg and the temperature were established. The foil ends of the ampoule were pierced. The gas dissolved as the liquid flowed through a series of small cups. The amount of gas dissolved was determined by the rise in mercury level in the compensator.
     Some measurements were made at 400 mmHg gas pressure. The results confirmed that Henry's law was obeyed.
	1. Xenon. Source not given. Purity by chromatographic method was 99.85 per cent.
	
	 2. 1-Propanol. Purified and analyzed in the All-Union Sci. Res. Inst. for Single Crystals & High-Purity Substances. Purity 99.97 weight per cent.

ESTIMATED ERROR:
$$\delta T/K = 0.05$$
$$\delta P/mmHg = 0.01$$
$$\delta X_1/X_1 = 0.005$$

REFERENCES:

COMPONENTS:	ORIGINAL MEASUREMENTS:
1. Xenon; Xe; 7440-63-3 2. Acetic Acid; $C_2H_4O_2$; 64-19-7	Steinberg, M.; Manowitz, B. Ind. Eng. Chem. 1959, 51, 47-50.

| VARIABLES:
 T/K: 301.15
 P/kPa: 101.325 (1 atm) | PREPARED BY:

 H.L. Clever |

EXPERIMENTAL VALUES:

T/K	Mol Fraction $X_1 \times 10^2$	Bunsen Coefficient α	Absorption Coefficient β
301.15	0.258	1.02	1.08

The authors define the Absorption Coefficient as the volume of gas, corrected to 288.15 K and 101.325 kPa, absorbed under a total system pressure of 101.325 kPa per unit volume of solvent at 288.15 K.

The mole fraction solubility and Bunsen coefficient were calculated by the compiler.

AUXILIARY INFORMATION

METHOD/APPARATUS/PROCEDURE:	SOURCE AND PURITY OF MATERIALS:
Absorption coefficient determined by a modified McDaniel method (1).	1. Xenon. Matheson Co., Inc. Technical grade. 2. Acetic acid. Source not given.

ESTIMATED ERROR:

$$\delta\beta/\beta = 0.05 - 0.10$$

REFERENCES:

1. Furman, N.H. "Scott's Standard Methods of Chemical Analysis" Van Nostrand Co., NY 1939, 5th ed., Vol. II, p. 2587.

COMPONENTS:	ORIGINAL MEASUREMENTS:
1. Xenon; Xe; 7440-63-3 2. Undecafluoro(trifluoromethyl)-cyclohexane (Perfluoromethyl-cyclohexane); C_7F_{14}; 355-02-2	Clever, H.L.; Saylor, J.H.; Gross, P.M. J. Phys. Chem. 1958, 62, 89-91.

VARIABLES:	PREPARED BY:
T/K: 289.15 - 316.25 Total P/kPa: 101.325 (1 atm)	P.L. Long

EXPERIMENTAL VALUES:

T/K	Mol Fraction $X_1 \times 10^2$	Bunsen Coefficient α	Ostwald Coefficient L
289.15	1.86	2.20	2.33
303.15	1.61	1.86	2.06
316.25	1.43	1.62	1.87

Smoothed Data: $\Delta G^O/\text{J mol}^{-1} = - RT \ln X_1 = -7375.1 + 58.643\,T$

Std. Dev. ΔG^O = 3.9, Coef. Corr. = 0.9999

$\Delta H^O/\text{J mol}^{-1} = -7375.1$, $\Delta S^O/\text{J K}^{-1}\text{mol}^{-1} = -58.643$

T/K	Mol Fraction $X_1 \times 10^2$	$\Delta G^O/\text{J mol}^{-1}$
288.15	1.88	9,522.8
293.15	1.78	9,816.0
298.15	1.69	10,109
303.15	1.61	10,402
308.15	1.54	10,696
313.15	1.47	10,989
318.15	1.40	11,282

The solubility values were adjusted to a partial pressure of xenon of 101.325 kPa (1 atm) by Henry's law.

The Bunsen coefficients were calculated by the compiler.

AUXILIARY INFORMATION

METHOD:
    Volumetric. The apparatus (1) is a modification of that used by Morrison and Billett (2). Modifications include the addition of a spiral solvent storage tubing, a manometer for constant reference pressure, and an extra gas buret for highly soluble gases.

SOURCE AND PURITY OF MATERIALS:
1. Xenon. Matheson Co., Inc. Both standard and research grades were used.

2. Perfluoromethylcyclohexane. du Pont FCS-326, shaken with concentrated H_2SO_4, washed, dried over Drierite and distilled. b.p. 75.95 to 76.05° at 753 mm., lit. b.p. 76.14 at 760 mm.

APPARATUS/PROCEDURE:
    (a) Degassing. 700 ml of solvent is shaken and evacuated while attached to a cold trap, until no bubbles are seen; solvent is then transferred through a 1 mm capillary tubing, released as a fine mist into a continuously evacuated flask. (b) Solvent is saturated with gas as it flows through 8 mm x 180 cm of tubing attached to a gas buret. Pressure is maintained at 1 atm as the gas is absorbed.

ESTIMATED ERROR:
$\delta T/K = 0.05$
$\delta P/\text{mmHg} = 3$
$\delta X_1/X_1 = 0.03$

REFERENCES:
1. Clever, H.L.; Battino, R.; Saylor, J.H.; Gross, P.M. J. Phys. Chem. 1957, 61, 1078.

2. Morrison, T.J.; Billett, F. J. Chem. Soc. 1948, 2033; ibid. 1952, 3819.

COMPONENTS:	ORIGINAL MEASUREMENTS:

1. Xenon; Xe; 7440-63-3
 	
2. Dichlorodifluoromethane
    (Freon-12); CCl_2F_2; 75-71-8
	Steinberg, M.; Manowitz, B.;
	    Pruzansky, J.
	
	US AEC BNL-542 (T-140).
	Chem. Abstr. 1959, 53, 21242g.

VARIABLES:	PREPARED BY:
T/K: 196.15 - 273.15	H. L. Clever

EXPERIMENTAL VALUES:

T/K	Absorption Coefficient	Henry's Constant K/atm	Mol Fraction $X_1 \times 10^2$	Bunsen Coefficient α	Ostwald Coefficient L
196.15	55.3	--	14.7	52.4	37.6
203.15	--	7.9	12.7	--	--
260.85	--	28.	3.6	--	--
273.15	7.4	37.	2.7	7.0	7.0

Smoothed Data: ΔG^O/J mol^{-1} = - RT ln X_1 = -9,778.5 + 65.508 T

Std. Dev. ΔG^O = 87, Coef. Corr. = 0.9994

ΔH^O/J mol^{-1} = -9,778.5, ΔS^O/J K^{-1} mol^{-1} = -65.508

T/K	Mol Fraction $X_1 \times 10^2$	ΔG^O/J mol^{-1}
193.15	16.7	2,874.4
203.15	12.4	3,529.5
213.15	9.43	4,184.6
223.15	7.36	4,839.6
233.15	5.87	5,494.7
243.15	4.77	6,149.8
253.15	3.94	6,804.9
263.15	3.31	7,460.0
273.15	2.81	8,115.0

AUXILIARY INFORMATION

METHOD /APPARATUS/PROCEDURE:	SOURCE AND PURITY OF MATERIALS:
Dynamic tracer technique (1). The Henry's constant is $K = (P/atm)/X_1$ The Henry's constants are probably from data smoothed by the authors. The report is discussed further in a later paper (2).	1. Xenon. 2. Dichlorodifluoromethane.

ESTIMATED ERROR:

$\delta X/X = 0.03 - 0.05$

(Compiler)

REFERENCES:

1. Steinberg, M.; Manowitz, B.
    Ind. Eng. Chem. 1959, 51, 47.

2. Steinberg, M.
    US AEC TID-7593, 1959, 217.
    Chem. Abstr. 1961, 55, 9083e.

COMPONENTS:	ORIGINAL MEASUREMENTS:

1. Xenon; Xe; 7440-63-3
 	
2. 1,1,2-Trichloro-1,2,2-trifluoro-
    ethane (Freon 113); $C_2Cl_3F_3$;
    76-13-1
	Linford, R.G.; Hildebrand, J.H.
	
	Trans. Faraday Soc. 1970, 66,
	577-581.

VARIABLES:	PREPARED BY:
T/K: 284.37 - 298.15 P/kPa: 101.325 (1 atm)	P.L. Long

EXPERIMENTAL VALUES:

T/K	Mol Fraction $X_1 \times 10^2$	Bunsen Coefficient α	Ostwald Coefficient L
284.37	2.568	5.03	5.24
287.75	2.447	4.76	5.01
290.40	2.362	4.36	4.64
295.09	2.230	4.28	4.62
298.15	2.149	4.11	4.49

Smoothed Data:

$$\Delta G^o/J\ mol^{-1} = -\ RT\ ln\ X_1 = -9060.7 + 62.327\ T$$

Std. Dev. ΔG^o = 3.9, Coef. Corr. = 0.9999

$$\Delta H^o/J\ mol^{-1} = -9060.7, \Delta S^o/J\ K^{-1}\ mol^{-1} = -62.327$$

T/K	Mol Fraction $X_1 \times 10^2$	$\Delta G^o/J\ mol^{-1}$
283.15	2.605	8587.1
288.15	2.437	8898.9
293.15	2.285	9210.4
298.15	2.147	9522.0

The Bunsen and Ostwald coefficients were calculated by the compiler.

AUXILIARY INFORMATION

METHOD:	SOURCE AND PURITY OF MATERIALS:
Saturation of liquid with gas at a partial pressure of gas equal to 1 atm.	1. Xenon. Source not given. Purest commercially obtainable, dried before use. 2. 1,1,2-Trichloro-1,2,2-trifluoro-ethane. Matheson, Coleman and Bell. Spectroquality.

APPARATUS/PROCEDURE:	ESTIMATED ERROR:
Dymond-Hildebrand apparatus (1) which uses an all-glass pumping system to spray slugs of degassed solvent into the gas. The amount of gas dissolved is calculated from initial and final pressures.	$\delta X_1/X_1 = 0.005$ (Evaluator)
	REFERENCES:
	1. Dymond, J.H.; Hildebrand, J.H. Ind. Eng. Chem. Fundam. 1967, 6, 130.

COMPONENTS:	ORIGINAL MEASUREMENTS:
1. Xenon; Xe; 7440-63-3 2. Trichloromethane (Chloroform); $CHCl_3$; 67-66-3	Yagi, M.; Kondo, K. Kakuriken Kenkyu Hokoku 1969, 2(2), 153-154. Chem. Abstr. 1971, 75, 54017c.

VARIABLES:	PREPARED BY:
T/K: room temperature P/kPa: 1.333 - 49.329 (10 - 370 mmHg)	H.L. Clever

EXPERIMENTAL VALUES:

T/K	Mol Fraction $X_1 \times 10^2$ at 101.325 kPa	Henry's Constant $K = P_1 (mmHg)/X_1$
298.15(?)	1.11	6.84×10^4

The mole fraction solubility at 101.325 kPa (1 atm) was calculated by the compiler.

Seven solubility measurements in degassed and three solubility measurements in non-degassed trichloromethane between 10 and 370 mmHg Xe pressure fell within experimental error of the Henry's law line. The ten solubilities were given on a graph along with the value of Henry's constant.

AUXILIARY INFORMATION

METHOD/APPARATUS/PROCEDURE:	SOURCE AND PURITY OF MATERIALS:
The solvent was degassed by vacuum at liquid N_2 temperature. 10 ml. gas and 5 ml. solvent were equilibrated for two hours at room temperature.	1. Xenon. Mixed tracer of ^{127}Xe, ^{129m}Xe, ^{133}Xe, ^{133m}Xe, ^{135}Xe, ^{135m}Xe, separated from parent ^{133}I and ^{135}I. 2. Tetrachloromethane. No information.

	ESTIMATED ERROR:
	$\delta T/K = 2$ $\delta K/K = 0.05$

	REFERENCES:

COMPONENTS:	ORIGINAL MEASUREMENTS:
1. Xenon; Xe; 7440-63-3 2. Tetrachloromethane (Carbon Tetrachloride); CCl_4; 56-23-5	Yagi, M.; Kondo, K. Kakuriken Kenkyu Hokoku 1969, 2(2), 153-154. Chem. Abstr. 1971, 75, 54017c.
VARIABLES: T/K: room temperature P/kPa: 1.333 - 49.329 (10 - 370 mmHg)	PREPARED BY: H.L. Clever

EXPERIMENTAL VALUES:

T/K	Mol Fraction $X_1 \times 10^2$ at 101.325 kPa	Henry's Constant $K = P_1 (mmHg)/X_1$
298.15(?)	1.32	5.75×10^4

The mole fraction solubility at 101.325 kPa (1 atm) was calculated by the compiler.

Nine solubility measurements in degassed and three solubility measurements in non-degassed tetrachloromethane between 10 and 370 mmHg Xe pressure fell within experimental error of the Henry's law line. The twelve solubilities were given on a graph along with the value of Henry's Constant.

W. L. Kay and R. A. Penneman [US AEC TID-5215 1955, 138 - 139; Chem. Abstr. 1956, 50, 16284e] used the McDaniel method to measure Ostwald coefficients of xenon in carbon tetrachloride at several temperatures. Their values are T/K/Ostwald: 301/2.4, 304/2.3 and 2.9, and 305/2.4. The values range from 0.94×10^{-2} to 1.13×10^{-2} mole fraction xenon at a pressure of 101.325 kPa. The value of Yagi and Kondo, which is higher by 15 to 30 percent, is probably a more reliable value.

AUXILIARY INFORMATION

METHOD/APPARATUS/PROCEDURE:	SOURCE AND PURITY OF MATERIALS:
The solvent was degassed by vacuum at liquid N_2 temperature. 10 ml. gas and 5 ml. solvent were equilibrated for two hours at room temperature.	1. Xenon. Mixed tracer of ^{127}Xe, ^{129m}Xe, ^{133}Xe, ^{133m}Xe, ^{135}Xe, ^{135m}Xe, separated from parent ^{133}I and ^{135}I. 2. Tetrachloromethane. No information.
	ESTIMATED ERROR: $\delta T/K = 2$ $\delta K/K = 0.05$
	REFERENCES:

COMPONENTS:	ORIGINAL MEASUREMENTS:
1. Xenon; Xe; 7440-63-3 2. Fluorobenzene; C_6H_5F; 462-06-6	Saylor, J.H.; Battino, R. J. Phys. Chem. 1958, 62, 1334-1337.

VARIABLES:
  T/K: 288.35 - 328.15
  P/kPa: 101.325 (1 atm)
	PREPARED BY:
	
	  H.L. Clever, A.L. Cramer

EXPERIMENTAL VALUES:

T/K	Mol Fraction $X_1 \times 10^2$	Bunsen Coefficient α	Ostwald Coefficient L
288.35	1.479	3.598	3.798
298.15	1.298	3.069	3.350
313.15	-	-	-
328.15	0.985	2.274	2.732

Smoothed Data:

$$\Delta G^O/J\ mol^{-1} = -RT\ \ln X_1 = -7897.3 + 62.504\ T$$

Std. Dev. ΔG^O = 28.0 , Coeff. Corr. =0.9998

$$\Delta H^O/J\ mol^{-1} = -7897.3,\ \Delta S^O/J\ K^{-1}\ mol^{-1} = -62.504$$

T/K	Mol Fraction $X_1 \times 10^2$	$\Delta G^O/J\ mol^{-1}$
288.15	1.47	10113
293.15	1.39	10426
298.15	1.31	10738
303.15	1.25	11051
308.15	1.18	11363
313.15	1.13	11676
318.15	1.08	11988
323.15	1.03	12301
328.15	0.982	12613

Solubility values were adjusted to a partial pressure of xenon of
101.325 kPa (1 atm) by Henry's law. Bunsen coefficients were
calculated by the compiler.

AUXILIARY INFORMATION

METHOD/APPARATUS/PROCEDURE:	SOURCE AND PURITY OF MATERIALS:
The solvent was degassed by evacuating the space above it, shaking, and then passing it as a fine mist into another evacuated container. The degassed liquid was saturated as it passed as a thin film inside a glass helix which contained the solute gas plus solvent vapor at a total pressure of 1 atm (1,2). The volume of liquid and the volume of gas absorbed are determined directly in a system of burets.	1. Xenon. Linde Air Products Co. 2. Fluorobenzene. Eastman Kodak Co., white label. Dried over P_4O_{10}, distilled, b.p. 84.28 - 84.68^{O}C.

	ESTIMATED ERROR:
	  $\delta T/K = 0.03$
	  $\delta P/mmHg = 1.0$
	  $\delta X_1/X_1 = 0.005$ (authors)

REFERENCES:

1. Morrison, T.J.; Billett, F.
   J. Chem. Soc. 1948, 2033.

2. Clever, H.L.; Battino, R.;
   Saylor, J.H.; Gross, P.M.
   J. Phys. Chem. 1957, 61, 1078.

COMPONENTS:	ORIGINAL MEASUREMENTS:
1. Xenon; Xe; 7440-63-3 2. Chlorobenzene; C_6H_5Cl; 108-90-7	Saylor, J.H.; Battino, R. J. Phys. Chem. 1958, 62, 1334-1337.

VARIABLES:	PREPARED BY:
T/K: 288.35 - 328.15 P/kPa: 101.325 (1 atm)	H.L. Clever, A.L. Cramer

EXPERIMENTAL VALUES:

T/K	Mol Fraction X_1 x 10^2	Bunsen Coefficient α	Ostwald Coefficient L
288.35	1.390	3.115	3.288
298.15	-	-	-
313.15	-	-	-
328.15	0.894	1.918	2.304

Smoothed Data: $\Delta G^O/J\ mol^{-1} = - RT\ ln\ X_1 = -8724.1 + 65.806\ T$

$\Delta H^O/J\ mol^{-1} = -8724.1$, $\Delta S^O/J\ K^{-1}\ mol^{-1} = -65.806$.

T/K	Mol Fraction X_1 x 10^2	$\Delta G^O/J\ mol^{-1}$
288.15	1.39	10238
293.15	1.31	10567
298.15	1.23	10896
303.15	1.16	11225
308.15	1.10	11554
313.15	1.04	11883
318.15	0.988	12212
323.15	0.939	12541
328.15	0.894	12870

Solubility values were adjusted to a partial pressure of xenon of
101.325 kPa (1 atm) by Henry's law. Bunsen coefficients were
calculated by the compiler.

AUXILIARY INFORMATION

METHOD/APPARATUS/PROCEDURE:	SOURCE AND PURITY OF MATERIALS:
The solvent was degassed by evacuating the space above it, shaking, and then passing it as a fine mist into another evacuated container. The degassed liquid was saturated as it passed as a thin film inside a glass helix which contained the solute gas plus solvent vapor at a total pressure of 1 atm (1,2). The volume of liquid and the volume of gas absorbed are determined in a system of burets.	1. Xenon. Linde Air Products Co. 2. Chlorobenzene. Eastman Kodak Co., white label. Dried over P_4O_{10}, distilled, b.p. 131.67 - 131.71°C.
	ESTIMATED ERROR:
The smoothed data above are based on only two experimental points. They should be used with caution.	$\delta T/K = 0.03$ $\delta P/mmHg = 1.0$ $\delta X_1/X_1 = 0.005$ (authors)
	REFERENCES: 1. Morrison, T.J.; Billett, F. J. Chem. Soc. 1948, 2033. 2. Clever, H.L.; Battino, R.; Saylor, J.H.; Gross, P.M. J. Phys. Chem. 1957, 61, 1078.

COMPONENTS:	ORIGINAL MEASUREMENTS:
1. Xenon; Xe; 7440-63-3 2. Bromobenzene; C_6H_5Br; 108-86-1	Saylor, J.H.; Battino, R. J. Phys. Chem. 1958, 62, 1334-1337.
VARIABLES: T/K: 288.35 - 328.15 P/kPa: 101.325 (1 atm)	PREPARED BY: H.L. Clever, A.L. Cramer

EXPERIMENTAL VALUES:

T/K	Mol Fraction X_1 x 10^2	Bunsen Coefficient α	Ostwald Coefficient L
288.35	1.222	2.651	2.798
298.15	1.057	2.270	2.478
313.15	-	-	-
328.15	0.824	1.716	2.062

Smoothed Data: $\Delta G^O/J\ mol^{-1} = - RT \ln X_1 = -7531.6 + 62.$

Std. Dev. ΔG^O = 52.544, Coeff. Corr. = 0.9992

$\Delta H^O/J\ mol^{-1} = -7531.6$, $\Delta S^O/J\ K^{-1}\ mol^{-1} = -62.$

T/K	Mol Fract X_1 x 10^2	$\Delta G^O/J\ mol^{-1}$
288.15	1.20	10591
293.15	1.14	10906
298.15	1.08	11220
303.15	1.03	11535
308.15	0.980	11849
313.15	0.936	12164
318.15	0.894	12478
323.15	0.855	12793
328.15	0.820	13107

Solubility values were adjusted to a partial pressure of xenon of 101.325 kPa (1 atm) by Henry's law. Bunsen coefficients were calculated by the compiler.

AUXILIARY INFORMATION

METHOD /APPARATUS/PROCEDURE:

The solvent was degassed by evacuating the space above it, shaking, and then passing it as a fine mist into another evacuated container. The degassed liquid was saturated as it passed as a thin film inside a glass helix which contained the solute gas plus solvent vapor at a total pressure of 1 atm (1,2). The volume of liquid and the volume of gas absorbed are determined in a system of burets.

SOURCE AND PURITY OF MATERIALS:

1. Xenon. Linde Air Products Co.

2. Bromobenzene. Eastman Kodak Co., white label. Dried over P_4O_{10}, distilled, b.p. 155.86 - 155.90°C.

ESTIMATED ERROR:

$\delta T/K = 0.03$
$\delta P/mmHg = 1.0$
$\delta X_1/X_1 = 0.005$ (authors)

REFERENCES:

1. Morrison, T.J.; Billett, F. J. Chem. Soc. 1948, 2033.

2. Clever, H.L.; Battino, R.; Saylor, J.H.; Gross, P.M. J. Phys. Chem. 1957, 61, 1078.

COMPONENTS:	ORIGINAL MEASUREMENTS:
1. Xenon; Xe; 7440-63-3 2. Iodobenzene; C_6H_5I; 591-50-4	Saylor, J.H.; Battino, R. J. Phys. Chem. 1958, 62, 1334-1337.

VARIABLES:
    T/K: 288.65 - 328.15
    P/kPa: 101.325 (1 atm)
	PREPARED BY:
	
	    H.L. Clever, A.L. Cramer

EXPERIMENTAL VALUES:

T/K	Mol Fraction $X_1 \times 10^2$	Bunsen Coefficient α	Ostwald Coefficient L
288.65	0.968	1.972	2.084
298.15	-	-	-
313.15	-	-	-
328.15	0.682	1.338	1.608

Smoothed Data: $\Delta G^O/\text{J mol}^{-1} = - RT \ln X_1 = -6982.2 + 62.748\ T$

$\Delta H^O/\text{J mol}^{-1} = -6982.2,\ \Delta S^O/\text{J K}^{-1}\text{mol}^{-1} = -62.748$

T/K	Mol Fraction $X_1 \times 10^2$	$\Delta G^O/\text{J mol}^{-1}$
288.15	0.973	11,099
293.15	0.926	11,412
298.15	0.882	11,726
303.15	0.842	12,040
308.15	0.805	12,354
313.15	0.771	12,667
318.15	0.739	12,981
323.15	0.709	13,295
328.15	0.682	13,609

Solubility values were adjusted to a partial pressure of xenon of 101.325 kPa (1 atm) by Henry's law. Bunsen coefficients were calculated by the compiler.

AUXILIARY INFORMATION

METHOD /APPARATUS/PROCEDURE:

   The solvent was degassed by evacuating the space above it, shaking, and then passing it as a fine mist into another evacuated container. The degassed liquid was saturated as it passed as a thin film inside a glass helix which contained the solute gas plus solvent vapor at a total pressure of 1 atm (1,2). The volume of liquid and the volume of gas absorbed are determined in a system of burets.

   The smoothed data above are based on only two experimental points. They should be used with caution.

SOURCE AND PURITY OF MATERIALS:

1. Xenon. Linde Air Products Co.

2. Iodobenzene. Eastman Kodak white label. Shaken with aq. $Na_2S_2O_3$, dried over P_4O_{10}, distilled, b.p. 77.40 - 77.60°C (20 mmHg).

ESTIMATED ERROR:
$$\delta T/K = 0.03$$
$$\delta P/\text{mmHg} = 1.0$$
$$\delta X_1/X_1 = 0.005 \text{ (authors)}$$

REFERENCES:

1. Morrison, T.J.; Billett, F. J. Chem. Soc. 1948, 2033.

2. Clever, H.L.; Battino, R.; Saylor, J.H.; Gross, P. M. J. Phys. Chem. 1957, 61, 1078.

COMPONENTS:	ORIGINAL MEASUREMENTS:
1. Xenon; Xe; 7440-63-3 ^{133}Xe; 14932-42-4 2. Carbon dioxide; CO_2; 124-38-9	Ackley, R.D.; Notz, K.J. Oak Ridge Natl. Lab. Oak Ridge, TN 37830 ORNL-5122, October 1976 Aval. Nat'l. Tech. Infor. Service Chem. Abstr. 1977, 86, 178229v.

VARIABLES:	PREPARED BY:
T/K: 218.35 - 303.65	A.L. Cramer H.L. Clever

EXPERIMENTAL VALUES:

T/K	Mol Fraction $X_1 \times 10^2$	Bunsen Coefficient α	Ostwald Coefficient L
223.15	2.22	13.33	10.89
233.15	1.95	11.26	9.61
243.15	1.73	9.62	8.56
253.15	1.53	8.15	7.55
263.15	1.37	6.94	6.69
273.15	1.23	5.88	5.88
283.15	1.15	5.11	5.30
293.15	1.12	4.43	4.75
298.15	1.14	4.12	4.50
303.15	1.17	3.69	4.10

The mole fraction solubility at a xenon partial pressure of 101.325 kPa (1 atm) was calculated by the compiler. The CO_2 density values were used from reference (1).

A smoothed data fit with thermodynamic values for the transfer of one mole of xenon from the gas at 101.325 kPa to the hypothetical unit mole fraction xenon liquid is on the next page.

AUXILIARY INFORMATION

METHOD:	SOURCE AND PURITY OF MATERIALS:
Tracer technique (2). Corrected for 5.27 day half-life attenuation. Collimated counter with equilibrated gas-liquid samples. Xenon gas was a mixture of Xenon-133 and research grade Xe. The total pressure of the system was the equilibrium pressure of liquid CO_2 + the Xe pressure.	1. Xenon. Cryogenic Rare Gas Labs. Ultra high purity grade. Xenon-133. Isotopes Div., ORNL. 2. Carbon dioxide. Matheson Co., Inc. Research grade, 99.5 mol percent.

APPARATUS/PROCEDURE:	ESTIMATED ERROR: $\delta T/K = 0.2$
	REFERENCES: 1. Notz, K.J.; Meservey, A.B. ORNL-5121, June 1976. 2. Notz, K.J.; Meservey, A.B.; Ackley, R.D. Trans. Am. Nucl. Soc. 1973, 17, 318.

COMPONENTS:	ORIGINAL MEASUREMENTS:
1. Xenon; Xe; 7440-63-3 ^{133}Xe; 14932-42-4 2. Carbon Dioxide; CO_2; 124-38-9	Ackley, R.D.; Notz, K.J. ORNL-5122, October 1976 Aval. Nat'l. Tech. Infor. Service
VARIABLES: T/K: 218.35 - 303.65	PREPARED BY: A.L. Cramer H.L. Clever

EXPERIMENTAL VALUES:

T/K	Mol Fraction $X_1 \times 10^2$	ΔG°/kJmol^{-1}	ΔH°/kJmol^{-1}	ΔS°/JK^{-1}mol^{-1}	ΔCp°/JK^{-1}mol^{-1}
223.15	2.20	7.077	-3.795	-48.72	-215
233.15	1.98	7.606	-5.556	-56.45	-138
243.15	1.74	8.194	-6.547	-60.63	-60.7
253.15	1.52	8.808	-6.770	-61.53	+16.2
263.15	1.35	9.415	-6.222	-59.43	93.2
273.15	1.23	9.987	-4.907	-54.53	170
283.15	1.16	10.497	-2.821	-47.04	247
293.15	1.13	10.920	+0.033	-37.14	324
303.15	1.16	11.232	3.657	-24.99	401

The smoothed data above was calculated from the linear regression of ln X_1 as a function of temperature according to the equation

$$\ln X_1 = 206.368 - 283.363/(T/100) - 232.255 \ln (T/100) + 46.2585 (T/100)$$

AUXILIARY INFORMATION

METHOD: See preceding page	SOURCE AND PURITY OF MATERIALS: See preceding page
APPARATUS/PROCEDURE:	ESTIMATED ERROR: See preceding page
	REFERENCES: See preceding page.

COMPONENTS:	ORIGINAL MEASUREMENTS:
1. Xenon; Xe; 7440-63-3 2. Carbon Disulfide; CS_2; 75-15-0	Powell, R.J. J. Chem. Eng. Data 1972, 17, 302-304.

VARIABLES:	PREPARED BY:
T/K: 298.15 P/kPa: 101.325 (1 atm)	P.L. Long

EXPERIMENTAL VALUES:

T/K	Mol Fraction $X_1 \times 10^2$	Bunsen Coefficient α	Ostwald Coefficient L	$R \dfrac{\Delta \log x_1}{\Delta \log T} = N$
298.15	1.042	3.89	4.25	-9.08

The author implies that solubility measurements were made between 273.15 and 308.15, but only the solubility at 298.15 was given in the paper. The slope $R (\Delta \log x_1 / \Delta \log T)$ was given. The smoothed data below were calculated by the compiler from the slope in the form:

$$\log x_1 = \log (1.042 \times 10^{-2}) + (-9.08/R) \log (T/298.15)$$

with $R = 1.9872$ cal K^{-1} mol^{-1}.

Smoothed data:

T/K	Mol Fraction $X_1 \times 10^2$
273.15	1.555
278.15	1.431
283.15	1.319
288.15	1.218
293.15	1.126
298.15	1.042
303.15	0.9657

The Bunsen and Ostwald Coefficients were calculated by the compiler.

AUXILIARY INFORMATION

METHOD /APPARATUS/PROCEDURE:	SOURCE AND PURITY OF MATERIALS:
Solvent is degassed by freezing and pumping, then boiling under reduced pressure. The Dymond and Hildebrand (1) apparatus, with all glass pumping system, is used to spray slugs of degassed solvent into the xenon. Amount of gas dissolved is calculated from the initial and final gas pressures.	1. Xenon. No source. Manufacturer's research grade, dried over $CaCl_2$ before use. 2. Carbon disulfide. No source. Manufacturer's spectrochemical grade.
	ESTIMATED ERROR: δ N /cal K^{-1} mol^{-1} = 0.1 $\delta X_1/X_1$ = 0.002
	REFERENCES: 1. Dymond, J.; Hildebrand, J.H. Ind. Eng. Chem. Fundam. 1967, 6, 130.

COMPONENTS:	ORIGINAL MEASUREMENTS:
1. Xenon; Xe; 7440-63-3 2. Sulfinylbismethane (Dimethyl Sulfoxide); C_2H_6OS (CH_3SOCH_3); 67-68-5	Dymond, J. H. J. Phys. Chem. 1967, 71, 1829-1831.
VARIABLES: T/K: 298.15 P/kPa: 101.325 (1 atm)	PREPARED BY: M. E. Derrick

EXPERIMENTAL VALUES:

T/K	Mol Fraction $X_1 \times 10^2$	Bunsen Coefficient α	Ostwald Coefficient L
298.15	0.170	0.535	0.584

The Bunsen and Ostwald coefficients were calculated by the compiler.

AUXILIARY INFORMATION

METHOD/APPARATUS/PROCEDURE:	SOURCE AND PURITY OF MATERIALS:
The liquid is saturated with the gas at a gas partial pressure of 1 atm. The apparatus is that described by Dymond and Hildebrand (1). The apparatus uses an all-glass pumping system to spray slugs of degassed solvent into the gas. The amount of gas dissolved is calculated from the initial and final gas pressure.	1. Xenon. Matheson Co., Dried. 2. Dimethyl Sulfoxide. Matheson, Coleman, and Bell Co. Spectroquality reagent, dried, and a fraction frozen out. Melting pt.: 18.37° C.
	ESTIMATED ERROR:
	REFERENCES: 1. Dymond, J.; Hildebrand, J. H. Ind. Eng. Chem. Fundam. 1967, 6, 130.

COMPONENTS:	ORIGINAL MEASUREMENTS:
1. Xenon; Xe; 7440-63-3 2. 1,1,2,2,3,3,4,4,4-nonafluoro-N,N-bis(nonafluorobutyl)-1-butanamine (Perfluorotributylamine); $(C_4F_9)_3N$; 311-89-7	Powell, R.J. J. Chem. Eng. Data 1972, 17, 302-304.

VARIABLES:	PREPARED BY:
T/K: 298.15 P/kPa: 101.325 (1 atm)	P.L. Long

EXPERIMENTAL VALUES:

T/K	Mol Fraction $X_1 \times 10^2$	Bunsen Coefficient α	Ostwald Coefficient L	$R \frac{\Delta \log x_1}{\Delta \log T} = N$
298.15	2.152	1.38	1.51	-5.01

The author implies that solubility measurements were made between 288.15 and 318.15, but only the solubility at 298.15 was given in the paper. The slope $R(\Delta \log x_1/\Delta \log T)$ was given. The smoothed data below were calculated by the compiler from the slope in the form:

$$\log x_1 = \log (2.152 \times 10^{-2}) + (-5.01/R) \log (T/298.15)$$

with $R = 1.9872$ cal K^{-1} mol^{-1}.

Smoothed data:

T/K	Mol Fraction $X_1 \times 10^2$
288.15	2.345
293.15	2.246
298.15	2.152
303.15	2.064
308.15	1.980
313.15	1.901
318.15	1.827

The Bunsen and Ostwald Coefficients were calculated by the compiler.

AUXILIARY INFORMATION

METHOD/APPARATUS/PROCEDURE:	SOURCE AND PURITY OF MATERIALS:
Solvent is degassed by freezing and pumping, then boiling under reduced pressure. The Dymond and Hildebrand (1) apparatus, with all glass pumping system, is used to spray slugs of degassed solvent into the xenon. Amount of gas dissolved is calculated from the initial and final gas pressures.	1. Xenon. No source. Manufacturer's research grade, dried over $CaCl_2$ before use. 2. Perfluorotributylamine. Minnesota Mining & Mfg. Co. Column distilled, used portion with b.p. = 447.85 - 448.64K, & single peak GC.

ESTIMATED ERROR:

$$\delta N /cal \ K^{-1} \ mol^{-1} = 0.1$$
$$\delta X_1/X_1 = 0.002$$

REFERENCES:

1. Dymond, J.; Hildebrand, J.H. Ind. Eng. Chem. Fundam. 1967, 6, 130.

COMPONENTS:	ORIGINAL MEASUREMENTS:

COMPONENTS:

1. Xenon; Xe; 7440-63-3

2. Nitrous Oxide; N_2O; 10024-97-2

ORIGINAL MEASUREMENTS:

Steinberg, M.; Manowitz, B.;
Pruzansky, J.

US AEC BNL-542 (T-140).
Chem. Abstr. 1959, 53, 21242g.

VARIABLES:

T/K: 193.15 - 243.65

PREPARED BY:

H. L. Clever

EXPERIMENTAL VALUES:

T/K	Absorption Coefficient	Henry's Constant K/atm	Mol Fraction $X_1 \times 10^2$	Bunsen Coefficient α	Ostwald Coefficient L
193.15	60.5	--	8.51	57.4	40.5$_5$
201.15	--	13.5	7.4	--	--
234.15	15.3	--	2.52	14.5	12.4
240.15	--	45	2.2	--	--
243.65	--	50	2.0	--	--

Smoothed Data: $\Delta G^O/J\ mol^{-1} = - RT\ ln\ X_1 = -11,702 + 80.501\ T$

Std. Dev. $\Delta G^O = 88.5$, Coef. Corr. $= 0.9989$

$\Delta H^O/J\ mol^{-1} = -11,702$, $\Delta S^O/J\ K^{-1}\ mol^{-1} = -80.501$

T/K	Mol Fraction $X_1 \times 10^2$	$\Delta G^O/J\ mol^{-1}$
193.15	9.12	3,845.9
203.15	6.37	4,650.9
213.15	4.60	5,455.9
223.15	3.42	6,260.9
233.15	2.61	7,065.9
243.15	2.04	7,870.9

AUXILIARY INFORMATION

METHOD/APPARATUS/PROCEDURE:

Dynamic tracer technique (1).

The Henry's constant is

$K = (P/atm)/X_1$

The Henry's constants are probably from data smoothed by the authors.

The report is discussed further in a later paper (2).

SOURCE AND PURITY OF MATERIALS:

1. Xenon.

2. Nitrous oxide.

ESTIMATED ERROR:

$\delta X/X = 0.03 - 0.05$

(Compiler)

REFERENCES:

1. Steinberg, M.; Manowitz, B.
Ind. Eng. Chem. 1959, 51, 47.

2. Steinberg, M.
US AEC TID-7593, 1959, 217.
Chem. Abstr. 1961, 55, 9083e.

187

COMPONENTS:	ORIGINAL MEASUREMENTS:

COMPONENTS:

1. Xenon; Xe; 7440-63-3

2. Nitromethane; CH_3NO_2; 75-52-5

ORIGINAL MEASUREMENTS:

Friedman, H.L.

J. Am. Chem. Soc. 1954, 76, 3294-3297.

VARIABLES:

T/K: 298.00
P/kPa: 101.325 (1 atm)

PREPARED BY:

P.L. Long

EXPERIMENTAL VALUES:

T/K	Mol Fraction $X_1 \times 10^3$	Bunsen Coefficient α	Ostwald Coefficient L
298.00			1.15
			1.11
			1.15
	0.201	1.04	1.14 av.

The author reports Ostwald coefficients measured at about 700 mmHg. The Bunsen coefficient and the mole fraction solubility at 101.325 kPa (1 atm) were calculated by the compiler with the assumptions that the gas is ideal, that Henry's law is obeyed and that the Ostwald coefficient is independent of pressure.

AUXILIARY INFORMATION

METHOD:

Gas absorption. The method was essentially that employed by Eucken and Herzberg (1). Modifications included a magnetic stirring device instead of shaking the saturation vessel, and balancing the gas pressure against a column of mercury with electrical contacts instead of balancing the gas pressure against the atmosphere.

SOURCE AND PURITY OF MATERIALS:

1. Xenon. Air Reduction Co. Reagent grade, 99.8 per cent pure by mass spectroscopy.

2. Nitromethane. Source not given. Distilled, dried by filtering at 253 K.

APPARATUS/PROCEDURE: The solvent was degassed by vacuum. The procedure, repeated 5-10 times, was to alternate 5-15 s evacuation and rapid stirring to produce cavitation. In the solubility measurement, gas, pre-saturated with solvent vapor, was brought into contact with about 80 ml of solvent in the saturation vessel. Initial conditions were established by a time extrapolation. Solubility equilibrium was approached from both under- and supersaturation by varying the rate.

ESTIMATED ERROR:

$\delta T/K = 0.05$
$\delta P/mmHg = 0.3$
$\delta L/L = 0.03$

REFERENCES:

1. Euken, A.; Herzberg, G. Z. Phys. Chem. 1950, 195, 1.

COMPONENTS:	ORIGINAL MEASUREMENTS:
1. Xenon; Xe; 7440-63-3 2. Benzenamine (Aniline); C_6H_7N; 62-53-3	Steinberg, M.; Manowitz, B. Ind. Eng. Chem. 1959, 51, 47-51.

VARIABLES:	PREPARED BY:
T/K: 303.15 P/kPa: 101.325 (1 atm)	H.L. Clever A.L. Cramer

EXPERIMENTAL VALUES:

T/K	Mol Fraction $X_1 \times 10^2$	Bunsen Coefficient α	Absorption Coefficient β
303.15	0.173	0.43	0.45

The authors define the Absorption coefficient as the volume of gas, corrected to 288.15 K and 101.325 kPa, absorbed under a total system pressure of 101.325 kPa per unit volume of solvent at 288.15 K.

The mole fraction solubilities and Bunsen coefficients were calculated by the compiler.

The solubility of xenon in benzenamine reported by von Antropoff at 303.15 K is about 9 per cent higher than the Steinberg and Manowitz value (see next page). There is not enough information to recommend one value over the other.

AUXILIARY INFORMATION

METHOD/APPARATUS/PROCEDURE:	SOURCE AND PURITY OF MATERIALS:
Absorption coefficient determined by a modified McDaniel method (1).	1. Xenon. Matheson Co., Inc. Technical grade. 2. Benzenamine. Source not given. Chemically pure grade.

ESTIMATED ERROR:

$$\delta\beta/\beta = 0.05 - 0.10$$

REFERENCES:

1. Furman, N.H. "Scott's Standard Methods of Chemical Analysis" Van Nostrand Co., NY, 1939, 5th ed., Vol. II, p. 2587.

COMPONENTS:	ORIGINAL MEASUREMENTS:
1. Xenon; Xe; 7440-63-3 2. Benzenamine (Aniline); C_6H_7N; 62-53-3	von Antropoff, A. Z. Electrochem. 1919, 25, 269-308.

VARIABLES:	PREPARED BY:
T/K: 283.15 - 323.15 P/kPa: 101.325 (1 atm)	

EXPERIMENTAL VALUES:

T/K	Kuenen Coefficient S	Mol Fraction X_1 x 10^2	Bunsen Coefficient α	Ostwald Coefficient L
283.15	0.6687	0.277	0.689	0.714
293.15	0.5006	0.208	0.512	0.549
298.15	0.4758			
298.15	0.4733	0.197	0.484	0.528
303.15	0.4546			
303.15	0.4426	0.189	0.462	0.513
313.15	0.4336			
313.15	0.4308	0.179	0.434	0.498
323.15	0.3892			
323.15	0.4113	0.171	0.411	0.486

The Kuenen coefficient was measured at pressures varying from 518 to 632 mmHg and corrected to 760 mmHg by Henry's law by the author.
The Bunsen, Ostwald and mole fraction solubilities were calculated by the compiler.
Smoothed Data: The solubility value at 283.15 K was not used.

$$\Delta G^O/J\ mol^{-1} = -RT\ ln\ X_1 = -4,957.1 + 68.381\ T$$
Std. Dev. $\Delta G^O = 28.9$, Coef. corr. = 0.9994
$$\Delta H^O/J\ mol^{-1} = -4,957.1,\ \Delta S^O/J\ K^{-1}\ mol^{-1} = -68.381$$

T/K	Mol Fraction X_1 x 10^2	$\Delta G^O/J\ mol^{-1}$	T/K	Mol Fraction X_1 x 10^2	$\Delta G^O/J\ mol^{-1}$
293.15	0.205	15,089	313.15	0.180	16,456
298.15	0.198	15,431	318.15	0.175	16,798
303.15	0.192	15,772	323.15	0.170	17,140
308.15	0.186	16,114			

AUXILIARY INFORMATION

METHOD /APPARATUS/PROCEDURE:	SOURCE AND PURITY OF MATERIALS:
Gas absorption. A modification of the apparatus of Estreicher. A calibrated buret connected to an absorption vessel by a flexible glass helix so the absorption vessel can be shaken. Solvent degassed by heat and vacuum (1).	1. Xenon. 2. Aniline was prepared from aniline sulfate. It was freshly distilled before use.
	ESTIMATED ERROR: $\delta S/S = 0.03$
	REFERENCES: 1. Estreicher, T. Z. Phys. Chem. 1899, 31, 176.

COMPONENTS:	ORIGINAL MEASUREMENTS:
1. Xenon; Xe; 7440-63-3 2. Nitrobenzene; $C_6H_5NO_2$; 98-95-3	Saylor, J.H.; Battino, R. J. Phys. Chem. 1958, 62, 1334-1337.

VARIABLES:	PREPARED BY:
T/K: 288.25 - 328.15 P/kPa: 101.325 (1 atm)	H.L. Clever, A.L. Cramer

EXPERIMENTAL VALUES:

T/K	Mol Fraction $X_1 \times 10^2$	Bunsen Coefficient α	Ostwald Coefficient L
288.25	0.627	1.388	1.465
298.15	-	-	-
313.15	-	-	-
328.15	0.483	1.032	1.240

Smoothed Data: $\Delta G^\circ/J\ mol^{-1} = -\ RT\ ln\ X_1 = -5143.0\ +\ 60.012\ T$

$\Delta H^\circ/J\ mol^{-1} = -5143.0$, $\Delta S^\circ/J\ K^{-1}\ mol^{-1} = -60.012$

T/K	Mol Fraction $X_1 \times 10^2$	$\Delta G^\circ/J\ mol^{-1}$
288.15	0.627	12,149
293.15	0.605	12,450
298.15	0.584	12,750
303.15	0.564	13,050
308.15	0.546	13,350
313.15	0.529	13,650
318.15	0.512	13,950
323.15	0.497	14,250
328.15	0.483	14,550

Solubility values were adjusted to a partial pressure of xenon of 101.325 kPa (1 atm) by Henry's law. Bunsen coefficients were calculated by the compiler.

AUXILIARY INFORMATION

METHOD/APPARATUS/PROCEDURE:

  The solvent was degassed by evacuating the space above it, shaking, and then passing it as a fine mist into another evacuated container. The degassed liquid was saturated as it passed as a thin film inside a glass helix which contained the solute gas plus solvent vapor at a total pressure of 1 atm (1,2). The volume of liquid and the volume of gas absorbed are determined in a system of burets.

  The smoothed data above are based on only two experimental points. They should be used with caution.

SOURCE AND PURITY OF MATERIALS:

1. Xenon. Linde Air Products Co.

2. Nitrobenzene. Eastman Kodak white label. Distilled from P_4O_{10}, b.p. 81.0 - 81.2°C (10 mmHg).

ESTIMATED ERROR:
$\delta T/K = 0.03$
$\delta P/mmHg = 1.0$
$\delta X_1/X_1 = 0.005$ (authors)

REFERENCES:

1. Morrison, T.J.; Billett, F. J. Chem. Soc. 1948, 2033.

2. Clever, H.L.; Battino, R.; Saylor, J.H.; Gross, P.M. J. Phys. Chem. 1957, 61, 1078.

COMPONENTS:	ORIGINAL MEASUREMENTS:
1. Xenon; Xe; 7440-63-3 2. Emulsions and Other Mixed Solvents	Steinberg, M.; Manowitz, B. Ind. Eng. Chem. 1959, 51, 47-50.

VARIABLES:
    T/K: 292.15 - 305.15
    P/kPa: 101.325 (1 atm)
	PREPARED BY:
	
	    P.L. Long

EXPERIMENTAL VALUES:

T/K	Absorption Coefficient β
Koppers Emulsion K-900 (50 wt % styrene - butadiene in water)	
305.15	0.95
Pine Oil	
292.15	2.25
Toluene; C_7H_8; 108-88-3	
297.15	3.17
Pine Oil (60 vol %) + Toluene (40 vol %)	
298.15	2.55
Ultrasene (80 wt % paraffin, 20 wt % naphthene)	
294.15	3.53

The authors define the Absorption Coefficient as the volume of gas, corrected to 288.15 K and 101.325 kPa, absorbed under a total system pressure of 101.325 kPa per unit volume of solvent at 288.15 K.

AUXILIARY INFORMATION

METHOD/APPARATUS/PROCEDURE:	SOURCE AND PURITY OF MATERIALS:
The absorption coefficients were determined by a modified McDaniel method (1). Dynamic tracer technique was used with the Ultrasene.	1. Xenon. Matheson Co., Inc. Technical pure grade. 2. Ultrasene. Atlantic Refining Co. No information given on other solvents.
	ESTIMATED ERROR: $\delta\beta/\beta = 0.05 - 0.10$
	REFERENCES: 1. Furman, N.H. "Scott's Standard Methods of Chemical Analysis" Van Nostrand Co., NY 1939, 5th ed., Vol. II, p. 2587.

COMPONENTS:	ORIGINAL MEASUREMENTS:
1. Xenon; Xe; 7440-63-3 2. Silicone Oils and Related Fluids	Steinberg, M.; Manowitz, B. Ind. Eng. Chem. 1959, 51, 47-50.

VARIABLES:
     T/K: 297.15 - 368.15
     P/kPa: 101.325 (1 atm)
	PREPARED BY:
	
	        P.L. Long

EXPERIMENTAL VALUES:

T/K	Absorption Coefficient β
Dow Corning Silicone Oil 200 (Dimethylsiloxane, 10 centistoke)	
297.15	1.50
368.15	0.17
Dow Corning Silicone Oil 200 (Dimethylsiloxane, 1 centistoke)	
303.15	2.45
Dow Corning Silicone Oil 702 (contains diphenyl groups)	
300.15	0.93
Dow Corning Anti-Foam A (60 wt % silicone oil in water)	
301.15	0.74
Dowtherm A (Diphenyl-Diphenyl Oxide)	
301.15	0.88

The authors define the Absorption Coefficient as the volume of gas, corrected to 288.15 K and 101.325 kPa, absorbed under a total system pressure of 101.325 kPa per unit volume of solvent at 288.15 K.

AUXILIARY INFORMATION

METHOD/APPARATUS/PROCEDURE:	SOURCE AND PURITY OF MATERIALS:
The absorption coefficient at one atmosphere of xenon was measured by a modified McDaniel method (1).	1. Xenon. Matheson Co., Inc. Technical grade. 2. Solvents. Dowtherm A from Dow Chemical Co. No statement about other solvents.
	ESTIMATED ERROR: $\delta\beta/\beta = 0.05 - 0.10$
	REFERENCES: 1. Furman, N.H. "Scott's Standard Methods of Chemical Analysis" Van Nostrand Co., NY 1939, 5th ed., Vol. II, p. 2587.

COMPONENTS:	EVALUATOR:
1. Xenon; Xe; 7440-63-3 2. Olive Oil	H. L. Clever Chemistry Department Emory University Atlanta, GA 30322 U.S.A. August 1978

CRITICAL EVALUATION:

The solubility of xenon in olive oil was measured by Lawrence, Loomis, Tobias and Turpin (1) at 295.15 and 310.15 K, by Yeh and Peterson (2) at 298.15, 303.15, 310.15 and 318.15 K and at 310.15 K by Ladefoged and Anderson (3) and by Kitani (4).

The data were converted to a mole fraction solubility at a partial pressure of xenon of 101.325 kPa (1 atm) assuming that olive oil is 1,2,3-propane-triyl ester of Z-9-octadecenoic acid, or triolein, of molecular weight 885.46. The data from the four laboratories shows considerable scatter. Yeh and Peterson made direct volumetric measurements at atmospheric pressure. The other workers used radiochemical techniques at low xenon partial pressure in the presence of a carrier gas, which can be subject to greater errors than the volumetric method. The data of Yeh and Peterson are internaly self-consistent. It was decided to accept the Yeh and Peterson data as tentative values of the xenon in olive oil solubility.

The Yeh and Peterson data were used in a linear regression of a Gibbs energy equation linear in temperature. The tentative values of the thermodynamic changes for the transfer of one mole of xenon from the gas at 101.325 kPa (1 atm) to the hypothetical unit mole fraction solution are

$$\Delta G^O/J\ mol^{-1} = -\ RT\ \ln X_1 = -8,733.3 + 50.649\ T$$

Std. Dev. = 4.9, Coef. Corr. = 0.9999

$$\Delta H^O/J\ mol^{-1} = -8,773.3, \quad \Delta S^O/J\ K^{-1}\ mol^{-1} = -50.649$$

A table of tentative mole fraction solubility and Gibbs energy values as a function of temperature appears below.

TABLE 1.   The solubility of xenon in olive oil. The tentative values of the mole fraction solubility at a xenon partial pressure of 101.325 kPa (1 atm) and the Gibbs energy change as a function of temperature.

T/K	Mol Fraction $X_1 \times 10^2$	$\Delta G^O/J\ mol^{-1}$
293.15	8.14	6,114.5
295.15	7.94	6,215.8
298.15	7.66	6,367.8
303.15	7.23	6,621.0
308.15	6.83$_5$	6,874.3
310.15	6.69	6,975.6
313.15	6.47	7,127.5
318.15	6.14	7,380.8

Continued on next page.

COMPONENTS:	EVALUATOR:
1. Xenon; Xe; 7440-63-3 2. Olive Oil	H.L. Clever Chemistry Department Emory University Atlanta, GA 30322 U.S.A. August 1978

CRITICAL EVALUATION:

Figure 1 shows the per cent deviation of all of the mole fraction solubility values from the smoothed data of Yeh and Peterson. Since olive oil is a natural product that may vary in composition and thus show variation in its solvent capacity it is not possible to classify any of the data as incorrect. Data sheets on all of the xenon in olive oil solubility reports are included.

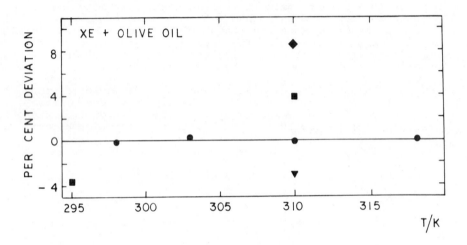

FIGURE 1.   Solubility of xenon in olive oil. Per cent deviation from
            regression line for Yeh and Peterson's data.
                ▼ Ladefoged and Anderson, ● Yeh and Peterson,
                ■ Lawrence, et al., and ◆ Kitani.

REFERENCES

1. Lawrence, J.H.; Loomis, W.F.; Tobias, C.A., Turpin, F.H.
    J. Physiol. 1946, 105, 197.

2. Yeh, S.-Y.; Peterson, R.E. J. Pharm. Sci. 1963, 52, 453.

3. Ladefoged, J.; Anderson, A.M. Phys. Med. Biol. 1967, 12, 353.

4. Kitani, K. Scand. J. Clin. Lab. Invest. 1972, 29, 167.

COMPONENTS:	ORIGINAL MEASUREMENTS:
1. Xenon; Xe; 7440-63-3 2. Olive oil	Lawrence, J. H.; Loomis, W. F.; Tobias, C. A.; Turpin, F. H. J. Physiol. 1946, 105, 197-204.

VARIABLES:	PREPARED BY:
T/K: 295.15 - 310.15	H. L. Clever A. L. Cramer

EXPERIMENTAL VALUES:

T/K	Mol Fraction $X_1 \times 10^2$	Bunsen Coefficient α	Ostwald Coefficient L
295.15	7.65	1.9	2.0_5
310.15	6.95	1.7	1.9_5

The mole fraction solubility and Ostwald coefficients were calculated by the compiler.

The molecular weight of olive oil was taken to be 885 and the density was calculated from $\rho = 0.9152 - 0.000468t/^{\circ}C$ (1) for the mole fraction calculation.

See the evaluation sheet for the solubility in olive oil for more information.

AUXILIARY INFORMATION

METHOD /APPARATUS/PROCEDURE:	SOURCE AND PURITY OF MATERIALS:
Radiochemical method. No details of the method given, but authors state they used an isotope of half life 34 days. Possibly the isotope was xenon-127.	No information given.

ESTIMATED ERROR:

$$\delta\alpha/\alpha = 0.05 \text{ (Compiler)}$$

REFERENCES:

1. Battino, R.; Evans, F.D.; Danforth, W. F. J. Am. Oil Chem. Soc. 1968, 45, 830.

COMPONENTS:	ORIGINAL MEASUREMENTS:
1. Xenon; Xe; 7440-63-3 2. Olive Oil	Yeh, S.Y.; Peterson, R.E. J. Pharm. Sci. 1963, 52, 453-458.

VARIABLES:
    T/K: 298.15 - 318.15
    P/kPa: 101.325 (1 atm)
	PREPARED BY:
	
	    H.L. Clever

EXPERIMENTAL VALUES:

T/K	Mol Fraction $X_1 \times 10^2$	Bunsen Coefficient $\alpha \pm$ Std. Dev.	Ostwald Coefficient L
298.15	7.65	1.8988 $\pm$ 0.0014	2.0725
303.15	7.25	1.7857 $\pm$ 0.0013	1.9749
310.15	6.68	1.6307 $\pm$ 0.0014	1.8532
318.15	6.14	1.4839 $\pm$ 0.0012	1.7248

The Bunsen coefficients are the average of three measurements. The Ostwald coefficients were fitted by the method of least squares to the equation log L = A/T + B by the authors. The same line fitted olive oil and the fats. From the slope and intercept they obtained

$$\Delta H^O = (-2273 \pm 80) \text{ cal mol}^{-1} \text{ and } \Delta S^O = (-6.4 \pm 0.2) \text{ cal K}^{-1} \text{ mol}^{-1}$$

The mole fractions were calculated by the compiler for a xenon partial pressure of 101.325 kPa (1 atm) assuming that olive oil has a molecular weight of 885.46 and a density of $\rho/g \text{ cm}^{-3} = 0.9152 - 4.68 \times 10^{-4} t/^oC$ (2).

See the evaluation of the xenon + olive oil system on pages 193 - 194 for the tentative thermodynamic changes of the solution process and the solubility values.

The values of the thermodynamic changes above, which are based on the temperature coefficient of the Ostwald coeifficient, are for the standard state change of transfering one mole of xenon from the gas phase at a concentration of one mole dm^{-3} to the solution at a concentration of one mole dm^{-3}. The values in the evaluation are for the standard state change of transfering one mole of xenon from the gas at 101.325 kPa (1 atm) to a hypothetical unit mole fraction solution.

AUXILIARY INFORMATION

METHOD /APPARATUS/PROCEDURE:	SOURCE AND PURITY OF MATERIALS:
Oil was dried and degassed by stirring under vacuum at 80°C for about 12 hr. A 50 ml. sample was placed in an absorption flask attached to a Geffken gas buret (1). The oil was constantly stirred and equilibrated with increments of gas until no change was observed in a differential oil manometer for ½ hr. Difference between initial and final buret readings indicated amount of gas absorbed. Absorption at successively lower temperatures was determined. The authors also measured the viscosity and surface tension of the liquid.	1. Xenon. Matheson Co. Research grade, maximum impurity 0.02 mol percent N_2 and 0.05 mol percent Kr. 2. Olive oil. Magnus, Mabee, and Raynard Co., U.S.P.

	ESTIMATED ERROR:
	    $\delta T/K = 0.05$
	    $\delta P/mmHg = 0.5$
	    $\delta\alpha/\alpha = 0.005$

	REFERENCES: 1. Geffken, G. Z. Physik Chem. 1904, 49, 257. 2. Battino, R.; Evans, F. D.; Danforth, W. F. J. Am. Oil Chem. Soc. 1968, 45, 830.

COMPONENTS:	ORIGINAL MEASUREMENTS:
1. Xenon-133; ^{133}Xe; 14932-42-4 2. Olive Oil	Ladefoged, J.; Andersen, A.M. Phys. Med. Biol. 1967, 12, 353-358.

VARIABLES:	PREPARED BY:
T/K: 310.15	A.L. Cramer

EXPERIMENTAL VALUES:

T/K	Mol Fraction X_1 x 10^2	Bunsen Coefficient α	Ostwald Coefficient L $\pm$ Std Dev	Number of Determinations
310.15	6.48	1.58	1.79 $\pm$ 0.04	8

The mole fraction solubility and Bunsen coefficient were calculated by the compiler.

The molecular weight of olive oil was taken to be 885 and the density was calculated from $\rho = 0.9152 - 0.000468t/^{\circ}C$ (1) for the mole fraction calculation.

See the evaluation sheet for the solubility in olive oil for more information.

AUXILIARY INFORMATION

METHOD /APPARATUS/PROCEDURE:

The olive oil in small vials was equilibrated with a mixture of air and xenon-133 for 24 h at 310.15 K with continuous stirring. After equilibration, 0.1 ml samples were transferred by syringes to a 2 ml rubber-capped vial and counted in a scintillation counter to a statistical error below 1 per cent.

SOURCE AND PURITY OF MATERIALS:

1. Xenon-133. Radiochemical Centre, Amersham, UK. Impurities less than 2 per cent.

2. Olive oil. Commercial Pharmacy. Met standards of Pharmacopia Nordeia and Pharmacopia Danica.

ESTIMATED ERROR:

REFERENCES:

1. Battino, R.; Evans, F.D. Danforth, W.F. J. Am. Oil Chem. Soc. 1968, 45, 830.

COMPONENTS:	ORIGINAL MEASUREMENTS:

COMPONENTS:

1. Xenon-133; $^{133}_{54}$Xe; 14932-42-4

2. Olive Oil

ORIGINAL MEASUREMENTS:

Kitani, K.

Scand. J. Clin. Lab. Invest. 1972, 29, 167-172.

VARIABLES:

T/K: 310.15

PREPARED BY:

P. L. Long
A. L. Cramer

EXPERIMENTAL VALUES:

T/K	Mol Fraction $X_1 \times 10^2$	Bunsen Coefficient α	Ostwald Coefficient $L \pm$ Std Dev	Number of Determinations
310.15	7.25	1.658	1.883 $\pm$ 0.036	9

The mole fraction solubility and Bunsen coefficient were calculated by the compiler.

The molecular weight of olive oil was taken to be 885, and the density (1) was calculated from $\rho = 0.9152 - 0.000468t/^{O}C$ for the mole fraction calculation.

See the evaluation sheet on olive oil for more information.

AUXILIARY INFORMATION

METHOD/APPARATUS/PROCEDURE:

Glass cuvettes were filled with the solvent. One-third was replaced by the radioactive gas and air. The sealed cuvette was rotated for two hours in a water bath at 37°C. The total pressure was adjusted to one atm by means of a thin needle. Samples from both the liquid and gas phase were counted by a scintillation detector and a Phillips pulse-height analyzer at the energy peak. Corrections for self absorption and scatter were made.

SOURCE AND PURITY OF MATERIALS:

1. Xenon-133. Radiochemical Centre, Amersham, UK.

2. Olive oil. Commercial sample.

ESTIMATED ERROR:

$\delta L/L = 0.02$

REFERENCES:

1. Battino, R.; Evans, E.D.; Danforth, W.F. J. Am. Oil Chem. Soc. 1968, 45, 830.

COMPONENTS:	EVALUATOR:
1. Xenon; Xe; 7440-63-3 2. Biological systems	H. L. Clever Chemistry Department Emory University Atlanta, GA 30322 U.S.A. September 1978

CRITICAL EVALUATION:

The solubility of xenon in biological fluids and tissues.

There are several factors that make it difficult to compare and evaluate the solubility of xenon in biological systems. First, the material from biological specimens may show a natural variation in properties which affects the solubility. Second, workers have used quite different experimental techniques to measure the solubility. Some have used classical volumetric methods with the xenon at a partial pressure near atmospheric pressure. Many have used radiochemical techniques either with natural xenon tagged with radioactive xenon at a total pressure near atmospheric or with a small unknown partial pressure of radioactive xenon. In neither of the radiochemical techniques is it necessary to know the total pressure to obtain an Ostwald coefficient. However, to compare the results of xenon solubility determinations by the volumetric method and by the radiochemical techniques one must assume the Ostwald coefficient is independent of pressure. This may not be true, especially if the gas associates with one or more components of the biological fluid. In these systems the solubility data are classed as tentative. Below are comments, which compare rather than evaluate, the solubility data in several types of biological systems.

Fat. Yeh and Peterson (1) found little difference between the solubility of xenon in olive oil and in human, rat, or dog fat. Conn (2) determined the distribution of xenon between water and dog fat. The distribution coefficient was converted to a gas-fat distribution coefficient on a weight basis, which if converted to a volume basis, assuming a density of 0.9 g cm^{-3} for fat, agrees with Yeh and Peterson's value.

Heme proteins. Hemoglobin and methemoglobin. Conn (2) showed that 5 to 12 weight percent solutions of human and dog hemoglobin and methemoglobin obey Henry's law at xenon partial pressures between 175 and 700 mmHg (23.33 - 93.33 kPa) at a temperature of 294.15 K. At a pressure of 700 mmHg (93.33 kPa) xenon, one mole of the protein associates with 1.9 mole of xenon, except the dog methemoglobin, which associates with 1.8 mole xenon. Schoenborn, Vogelhut and Featherstone (3) found 1.8 mole of xenon per mole of human hemoglobin at 293.15 K and a pressure of 760 mmHg (101.325 kPa). Yeh and Peterson (4) found 1.1 mole of xenon per mole of human hemoglobin at a temperature of 310.15 K and a xenon partial pressure of probably 720 mmHg (96.0 kPa). Schoenborn (5, 6) showed that Henry's law is obeyed between 0.5 and 1.5 atm (50.63 and 151.99 kPa) partial pressure xenon dissolved in a 5 per cent horse methemoglobin solution (solubility data not given) at temperatures between 273.15 and 313.15 K. Analysis of the solubility data showed about one mole of xenon bound per mole of protein at 313.15 K and nearly two moles bound at 273.15 K. The amount of xenon bound by reduced and oxyhemoglobin was the same within 5 percent as that bound by methemoglobin. X-ray diffraction analysis showed one xenon in the α sub-unit and the other in the β-sub-unit of the methemoglobin (two of each site per molecule, for a total of four sites). Wishnia (7) used a radiochemical method with xenon-133 and showed that Henry's law is not obeyed by solutions of hemoglobin in 1.8 molar $(NH_4)_2SO_4$ solution at 273.15 K (graph only). The curve was analyzed by a two step association of xenon and ferrihemoglobin.

$$Hb + Xe = HbXe$$

$$HbXe + Xe = HbXe_2.$$

He reported constants for the associations. Catchpool (8) reported on the solubility (graph) of xenon in a 0.5 per cent saline solution of human hemoglobin between 278.15 and 310.15 K at a xenon pressure of 1 atm (101.325 kPa). From his measurements he calculated 1.67 mole of xenon per mole of hemoglobin at 293.15 K and 1 atm (101.325 kPa) partial pressure

COMPONENTS:	EVALUATOR:
1. Xenon; Xe; 7440-63-3 2. Biological systems	H. L. Clever Chemistry Department Emory University Atlanta, GA 30322 U.S.A. September 1978

CRITICAL EVALUATION:

xenon gas. Meuhlbaecher, DeBon, and Featherstone (9) carried out experiments at 310.15 K in phosphate buffer solutions that showed that the solubility of xenon increases in the presence of bovine hemoglobin but not in the presence of either bovine gamma-globulin or bovine serum albumin.

The experiments on human, dog, bovine and horse hemoglobin are consistent with a two site association between hemoglobin and xenon which increases as the temperature decreases. The x-ray evidence indicates there are two sites for each of the two associations.

Myoglobin, metmyoglobin, cyanometuryoglobin. Schoenborn, Watson and Kendrew (10) equilibrated sperm whale metmyoglobin crystals with xenon at 2.5 atm (253.3 kPa) for 12 h. X-ray diffractions of the equilibrated crystal and normal material, analyzed by the difference Fourier method, showed that the xenon atom is bound to one specific site which is buried in the interior of the molecule. The xenon atom is nearly equidistant from the heme-linked histadine and a pyrrole ring of the heme group. Schoenborn and Nobbs (11) showed that the deoxymyoglobin crystal under similar conditions binds xenon to one specific site. Maestas and Ewing (12) showed that solutions of horse metmyoglobin do not obey Henry's law (graph) at 273.15 K and xenon partial pressures between 0.1 and 1.8 atm (10.133 - 182.385 kPa). The solubility curve was interpreted by a 1:1 association between metmyoglobin and xenon. However, Wishnia (7) interpreted his solubility measurements at 273.15 K in sperm whale myoglobin dissolved in either phosphate buffer (?) or 2 mol dm^{-3} $(NH_4)_2SO_4$ by a two step association. Ewing and Maestas (13) measured the solubility of xenon in 10 per cent horse heart myoglobin, metmyoglobin and cyanometmyoglobin between xenon partial pressures of 0.5 to 5 atm (50.63 - 506.6 kPa) at temperatures of 293.15, 298.15, and 303.15 K. The solubility (data not given) did not obey Henry's law. The solubility data were interpreted as by a single association for the xenon-cyanometmyoglobin and a two step association for the xenon-myoglobin and xenon-metmyoglobin associations. Keys and Lumry (14) report experiments that show the binding of oxygen and xenon by sperm whale myoglobin are independent up to pressures of 2 atm (202.65 kPa) xenon, but that CO and xenon binding are not independent. The evidence of one site in the crystaline material and two sites in the material in solution for xenon may be explained either by differences in configuration of the material or by a difference in accessibility of the site in the crystal and in aqueous solution.

Leghemoglobin. Leghemoglobin is a respiratory pigment found in the root nodules of most leguminous plants. Ewing and Ionescu (15) measured the solubility of xenon in 4-10 weight per cent solutions of the ferri- and ferro-leghemoglobin at 278.15, 288.15 and 298.15 K. The results (no solubility data presented) indicated a weak interaction between xenon and leghemoglobin solutions. However, the amount of xenon bound to the leghemoglobin is of the same order of magnitude as the experimental error and it was impossible to determine the stoichiometry of the interaction.

Brain Matter. Conn (2) reported water/tissue distribution coefficients for dog gray and white brain matter. Yeh and Peterson (16) reported xenon solubility in a total beef brain homogenate. Both Isbister , Schofield and Torrance (17) and Veall and Mallett (18) report xenon solubility in human gray and white matter homogenate. The solubility coefficient of xenon in human white brain matter from the two laboratories (17, 18) agrees within one per cent. The solubility of xenon in gray brain matter is approximately half the value in white matter and the agreement of the two laboratories is 20 per cent, which is outside the standard error range of the two measurements. Conn's values for dog gray and white matter, when converted to a xenon gas/tissue ratio are much higher than the values for human brain matter. Conn used the solid tissue rather than a homogenate, and he froze the tissue in liquid nitrogen before counting the absorbed radioactive xenon. The difference in the experimental techniques makes a comparison of the results of uncertain value.

COMPONENTS:	EVALUATOR:
1. Xenon; Xe; 7440-63-3 2. Biological systems	H. L. Clever Chemistry Department Emory University Atlanta, GA 30322 U.S.A. September 1978

CRITICAL EVALUATION:

Albumin. Conn (2) found 1.1 mole xenon per mole of human albumin at 294 K and 700 mmHg (93.33 kPa) xenon pressure. Yeh and Peterson (16) measured the solubility of xenon in three concentrations of albumin at 1 atm (101.325 kPa) and temperatures of 298.15, 303.15, and 310.15 K. These results give 0.78, 0.56 and 0.46 mole xenon per mole of albumin at 101.325 kPa and temperatures of 298.15, 303.15 and 310.15 K respectively. Ladefoged and Anderson (19) made one set of measurements at 310.15 K which appears to agree with the Yeh and Peterson value at 310.15 K.

Blood and blood components. The solubility of xenon in dog blood was studied by Conn (2). The solubility of xenon in human blood and blood components was studied by Yeh and Peterson (16), Anderson and Ladefoged (20), Ladefoged and Anderson (19), Isbister, Schofield and Torrance (17), Veall and Mallett (18) and Kitani (21). The data of Isbister et al., Veall and Mallett, and Kitani agree within a 3.5 per cent range for the distribution of xenon between the gas phase and red blood cells.

Tissues. Conn (2) studied dog tissues, Yeh and Peterson (16) studied rabbit leg muscle and Kitani and Winkler (21) studied human liver tissue of varying triglyceride content. Conn's results were for the solid tissue/water distribution, and they were converted to a tissue/gas distribution by us. Conn's 'solubility' values appear to be higher than the values determined by study of tissue homogenates.

REFERENCES

1. Yeh, S.Y.; Peterson, R.E. J. Pharm. Sci. 1963, 52, 453.
2. Conn, H.L. J. Appl. Physiol. 1961, 16, 1065.
3. Schoenborn, B.P.; Vogelhut, P.O.; Featherstone, R.M. Pharmacologist 1963, 5, 264.
4. Yeh, S.Y.; Peterson, R.E. J. Appl. Physiol. 1965, 20, 1041.
5. Schoenborn, B.P. Nature 1965, 208, 760.
6. Schoenborn, B.P. Fed. Proc. 1968, 27, 888.
7. Wishnia, A. Biochem. 1969, 8, 5064.
8. Catchpool, J.F. Fed. Proc. 1968, 27, 884.
9. Meuhlbaecher, C.; DeBon, F.L.; Featherstone, R.M. Inst. Anesth. Clinics 1963, 1, 937.
10. Schoenborn, B.P.; Watson, J.C.; Kendrew, J.C. Nature 1965, 207, 28.
11. Schoenborn, B.P.; Nobbs, C.L. Mol. Pharmacol. 1966, 2, 491.
12. Maestas, S.; Ewing, G.J. Curr. Mod. Biol. 1967, 1, 148.
13. Ewing, G.J.; Maestas, S. J. Phys. Chem. 1970, 74, 2341.
14. Keys, M.; Lumry, R. Fed. Proc. 1968, 27, 895.
15. Ewing, G.J.; Ionescu, L.G. J. Phys. Chem. 1972, 76, 591.
16. Yeh, S.-Y.; Peterson, R.E. J. Appl. Physiol. 1965, 20, 1041.
17. Isbister, W.J.; Schofield, P.F.; Torrance, H.B. Phys. Med. Biol. 1965, 10, 243.
18. Veall, N.; Mallett, B.L. Phys. Med. Biol. 1965, 10, 375.
19. Ladefoged, J.; Anderson, A.M. Phys. Med. Biol. 1967, 12, 353.
20. Kitani, K. Scand. J. Clin. Lab. Invest. 1972, 29, 167.
21. Kitani, K.; Winkler, K. Scand. J. Clin. Lab. Invest. 1972, 29, 173.

COMPONENTS:	ORIGINAL MEASUREMENTS:
1. Xenon; Xe; 7440-63-3 2. Human Fat	Yeh, S.Y.; Peterson, R.E. J. Pharm. Sci. 1963, 52, 453-458.

| VARIABLES:
 T/K: 298.15 - 318.15
 P/kPa: 101.325 (1 atm) | PREPARED BY:

 H.L. Clever |

EXPERIMENTAL VALUES:

Human fat 1			Human fat 2		
T/K	Bunsen Coefficient α + Std. Dev.	Ostwald Coefficient L	T/K	Bunsen Coefficient α + Std. Dev.	Ostwald Coefficient L
298.15	1.8878 + 0.0032	2.0606	298.15	1.8476 + 0.0026	2.0166
303.15	1.7688 + 0.0025	1.9630	303.15	1.7645 + 0.0025	1.9583
310.15	1.6143 + 0.0022	1.8345	310.15	1.6251 + 0.0025	1.8503
318.15	1.4753 + 0.0015	1.6903	318.15	1.4742 + 0.0028	1.7171

The Bunsen coefficients are the average of three measurements. The Ostwald coefficients were fitted by the method of least squares to the equation log L = A/T + B by the authors. The same line fitted olive oil and the fats. From the slope and intercept they obtained

$$\Delta H^O = (-2273 \pm 80) \text{ cal mol}^{-1} \text{ and } \Delta S^O = (-6.4 \pm 0.2) \text{ cal K}^{-1} \text{ mol}^{-1}$$

The thermodynamic changes are for the standard state transfer of one mole of xenon from the gas phase at a concentration of one mole dm^{-3} to the solution at a concetration of one mole dm^{-3}.

AUXILIARY INFORMATION

METHOD/APPARATUS/PROCEDURE:

Fat was dried and degassed by stirring under vacuum at 80°C for about 12 hr. A 50 ml. sample was placed in an absorption flask attached to a Geffken gas buret (1). The fat was constantly stirred and equilibrated with increments of gas until no change was observed in a differential oil manometer for ½ hr. Difference between initial and final buret readings indicated amount of gas absorbed. Absorption at successively lower temperatures was determined. The authors also measured the viscosity and surface tension of the liquid.

SOURCE AND PURITY OF MATERIALS:

1. Xenon. Matheson Co. Research grade, maximum impurity 0.02 mol percent N_2 and 0.05 mol per cent Kr.

2. Human omental fats obtained from two deceased patients (1 and 2). Extracted with petroleum ether (b.p. 309-338K). The ether was evaporated at 353 K under vacuum for several hours. Stored under refrigeration until use. Fat 2 appeared to have more stearine precipitate than fat 1 at 296 K.

ESTIMATED ERROR: $\delta T/K = 0.05$
 $\delta P/\text{mmHg} = 0.5$
 $\delta\alpha/\alpha = 0.005$

REFERENCES:

1. Geffken, G. Z. Physik Chem. 1904, 49, 257.

COMPONENTS:	ORIGINAL MEASUREMENTS:
1. Xenon; Xe; 7440-63-3 2. Rat-pooled Fat	Yeh, S.Y.; Peterson, R.E. J. Pharm. Sci. 1963, 52, 453-458.

VARIABLES:
     T/K: 298.15 - 318.15
     P/kPa: 101.325 (1 atm)
	PREPARED BY:
	
	     H.L. Clever

EXPERIMENTAL VALUES:

T/K	Bunsen Coefficient $\alpha \pm$ Std. Dev.	Ostwald Coefficient L
298.15	1.8376 ± 0.0065	2.0057
303.15	1.7197 ± 0.0055	1.9086
310.15	1.5712 ± 0.0050	1.7856
318.15	1.4276 ± 0.0045	1.6627

The Bunsen coefficients are the average of three measurements. The Ostwald coefficients were fitted by the method of least squares to the equation log L = A/T + B by the authors. The same line fitted olive oil and the fats. From the slope and intercept they obtained

$$\Delta H^{O} = (-2273 \pm 80) \text{ cal mol}^{-1} \text{ and } \Delta S^{O} = (-6.4 \pm 0.2) \text{ cal K}^{-1} \text{ mol}^{-1}$$

The thermodynamic changes are for the standard state transfer of one mole of xenon from the gas phase at a concentration of one mole dm^{-3} to the solution at a concentration of one mole dm^{-3}.

AUXILIARY INFORMATION

METHOD/APPARATUS/PROCEDURE:	SOURCE AND PURITY OF MATERIALS:
Fat was dried and degassed by stirring under vacuum at 80°C for about 12 hr. A 50 ml. sample was placed in an absorption flask attached to a Geffken gas buret (1). The fat was constantly stirred and equilibrated with increments of gas until no change was observed in a differential oil manometer for ½ hr. Difference between initial and final buret readings indicated amount of gas absorbed. Absorption at successively lower temperatures was determined. The authors also measured the viscosity and surface tension of the liquid.	1. Xenon. Matheson Co. Research grade, maximum impurity 0.02 mol percent N_2 and 0.05 mol percent Kr. 2. Rat retroperitoneal, mesenteric, omental, and hair clipped skin was cut into about 2.5 cm squares, dried at 353 K under vacuum, coarsely crushed, and then extracted with petroleum ether (b.p. 308-338 K) in a Soxhlet extractor. The ether was evaporated at 353 K under vacuum for several hours. Refrigerated until use.

	ESTIMATED ERROR: $\delta T/K = 0.05$
	            $\delta P/mmHg = 0.5$
	            $\delta\alpha/\alpha = 0.005$

	REFERENCES: 1. Geffken, G. Z. Physik Chem. 1904, 49, 257.

COMPONENTS:	ORIGINAL MEASUREMENTS:
1. Xenon; Xe; 7440-63-3 2. Dog Fat	Yeh, S.Y.; Peterson, R.E. J. Pharm. Sci. 1963, 52, 453-458.

VARIABLES:
      T/K: 298.15 - 318.15
      P/kPa: 101.325 (1 atm)
	PREPARED BY:
	
	        H.L. Clever

EXPERIMENTAL VALUES:

T/K	Bunsen Coefficient $\alpha \pm$ Std. Dev.	Ostwald Coefficient L
298.15	1.8393 ± 0.0025	2.0084
303.15	1.7557 ± 0.0020	1.9493
310.15	1.6113 ± 0.0021	1.8299
318.15	1.4589 ± 0.0015	1.6962

The Bunsen coefficients are the average of three measurements. The Ostwald coefficients were fitted by the method of least squares to the equation $\log L = A/T + B$ by the authors. The same line fitted olive oil and the fats. From the slope and intercept they obtained

$$\Delta H^{o} = (-2273 \pm 80) \text{ cal mol}^{-1} \text{ and } \Delta S^{o} = (-6.4 \pm 0.2) \text{ cal K}^{-1} \text{ mol}^{-1}$$

The thermodynamic changes are for the transfer of one mole of xenon from the gas phase at a concentration of one mole dm^{-3} to the solution at a concentration of one mole dm^{-3}.

AUXILIARY INFORMATION

METHOD/APPARATUS/PROCEDURE:

   Fat was dried and degassed by stirring under vacuum at 80°C for about 12 hr. A 50 ml. sample was placed in an absorption flask attached to a Geffken gas buret (1). The fat was constantly stirred and equilibrated with increments of gas until no change was observed in a differential oil manometer for ½ hr. Difference between initial and final buret readings indicated amount of gas absorbed. Absorption at successively lower temperatures was determined. The authors also measured the viscosity and surface tension of the liquid.

SOURCE AND PURITY OF MATERIALS:

1. Xenon. Matheson Co. Research grade. Maximum impurity 0.02 mol percent N_2 and 0.05 mol percent Kr.

2. Dog perineal, mesenteric, omental, and other adipose fats were extracted with petroleum ether (b.p. 309-338 K). The ether was evaporated at 353 K under vacuum for several hours. Stored under refrigeration until use.

ESTIMATED ERROR:
$$\delta T/K = 0.05$$
$$\delta P/mmHg = 0.5$$
$$\delta \alpha / \alpha = 0.005$$

REFERENCES:

1. Geffken, G. Z. Physik Chem. 1904, 49, 257.

205

COMPONENTS:	ORIGINAL MEASUREMENTS:
1. Xenon-133; $^{133}_{54}$Xe; 14932-42-4 2. Paraffin Oil	Ladefoged, J.; Anderson, A.M. Phys. Med. Biol. 1967, 12, 353-358.

VARIABLES:	PREPARED BY:
T/K: 310.15	A. L. Cramer

EXPERIMENTAL VALUES:

T/K	Ostwald Coefficient L $\pm$ Std Dev	Number of Determinations
310.15	1.96 $\pm$ 0.07	8

AUXILIARY INFORMATION

METHOD/APPARATUS/PROCEDURE:

The technique described by Anderson and Ladefoged (1) was used. The materials were placed in small test tubes together with a sample of water in an air tight chamber. The chamber contained air and 0.5 to 2 µCi of xenon-133. The samples were stirred continuously at 310.15 K for 24 hours. After equilibration the samples were transferred into syringes and samples of about 0.1 ml were counted in a scintillation counter. The Ostwald coefficients were calculated indirectly from the ratio of counts in water and the sample and the measured Ostwald coefficient in water. See the authors data sheet on Xe + H_2O on page 139.

SOURCE AND PURITY OF MATERIALS:

1. Xenon-133. Radiochemical Centre, Amersham, U.K. Two per cent impurity of ^{131}Xe and ^{85}Kr.

2. Paraffin Oil. Pharmacy. Met requirements of Pharmacopoea Nordica and Pharmacopoea Danica.

ESTIMATED ERROR:

See standard deviations above.

REFERENCES:

1. Anderson, A.M.; Ladefoged, J. J. Pharm. Sci. 1965, 54, 1684.

COMPONENTS:	ORIGINAL MEASUREMENTS:
1. Xenon; ^{133}Xe; 14932-42-4 2. Paraffin Oil 3. Lecithin	Kitani, K. Scand. J. Clin. Lab. Invest. 1972, 29, 167-172.

VARIABLES:
    T/K: 310.15
    P/kPa: 101.325 (1 atm)
	PREPARED BY:
	    P.L. Long
	    A.L. Cramer

EXPERIMENTAL VALUES:

T/K	Bunsen Coefficient α	Ostwald Coefficient L $\pm$ Std. Dev.	Replications
Paraffin Oil			
310.15	1.819	2.065 $\pm$ 0.043	16
	-	2.079*	-
Lecithin			
310.15	-	1.477*	-

*Extrapolated values. See equation below.

The Ostwald coefficient was independent of pressure.

The solubility value for lecithin is a value extrapolated from the solubility of the gas in mixtures of paraffin oil and lecithin. The coefficient of solubility of the gas is linear in lecithin per cent (graph in paper). The equation for the straight line in the paper appears to be in error. The equation below was estimated from the graph by the compiler.

$$L = 2.079 - (6.02 \times 10^{-3})(\text{Lecithin w/v per cent})$$

AUXILIARY INFORMATION

METHOD/APPARATUS/PROCEDURE:

A glass cuvette is filled with the liquid. One-third of the liquid is replaced with radioactive gas in air. The sealed cuvette is placed in a thermostated bath for 2 hours. The pressure is adjusted to 1 atm by means of a thin needle.

The radioactive assay, corrected for self absorption and scatter, is made by a scintillation detector and a Philips pulse-height analyzer.

SOURCE AND PURITY OF MATERIALS:

1. Xenon. Radiochemical Centre, Amersham, England.

2. Paraffin oil. Commercial quality.

3. Egg lecithin. Purified twice with ether and acetone.

ESTIMATED ERROR:

See standard deviation above.

REFERENCES:

COMPONENTS:	ORIGINAL MEASUREMENTS:

COMPONENTS:

1. Xenon; Xe; 7440-63-3
   Xenon-133; ^{133}Xe; 14932-42-4
   Xenon-135; ^{135}Xe; 14995-63-1

2. Dog Blood and Components

ORIGINAL MEASUREMENTS:

Conn, H. L.

J. Appl. Physiol. 1961, 16, 1065-1070.

VARIABLES:
   T/K: 294.15
   Xe P/kPa: 1.867 - 21.332
              (14-160 mmHg)

PREPARED BY:
   A. L. Cramer
   H. L. Clever

EXPERIMENTAL VALUES:

T/K	$\dfrac{Xe\ (counts\ m^{-1})\ g^{-1}\ Tissue}{Xe\ (counts\ m^{-1})\ g^{-1}\ Water}$	Solubility* Coefficient S
	Water	
294.15	1.00	0.118**
	Plasma	
294.15	1.45 ± 0.05	0.171
	Whole Blood (mean hemoglobin concentration 15 g 100 g^{-1} blood)	
294.15	2.49 ± 0.04	0.294
	Erythrocytes (mean hemoglobin conc. 35 g 100 g^{-1} red blood cells)	
294.15	3.75 ± 0.06	0.443
	Hemoglobin solution (mean hemoglobin conc. 35 g 100 g^{-1} solution)	
294.15	2.83 ± 0.05	0.334

*Calculated by the compiler. cm^3 xenon at 1 atm and 294.15 K g^{-1} tissue.

**cm^3 Xe 1 atm and 294 K/g water. Calculated from data in critical evaluation of xenon in water.

AUXILIARY INFORMATION

METHOD/APPARATUS/PROCEDURE:

A one ml aliquot of each solution was placed in a 100 ml glass tonometer and equilibrated with the gas mixture for 2 hours at 294 K. Three-tenths ml of each sample was withdrawn anaerobically into a 1-ml syringe and the radioactivity was determined in a scintillation counter. Water under the same partial pressure xenon was counted for the tissue/water ratio.

Some of the hemoglobin methemoglobin and albumin systems were determined by this method.

SOURCE AND PURITY OF MATERIALS:

1. Xenon. Air Reduction Sales. Specially purified. The gas was subjected to neutron bombardment in a nuclear reactor. Resultant xenon isotopes were mainly ^{133}Xe and ^{135}Xe.

2. Dog Blood and Components.

ESTIMATED ERROR:

See standard error of mean for 36 experiments above.

REFERENCES:

COMPONENTS:	ORIGINAL MEASUREMENTS:
1. Xenon; Xe; 7440-63-3 Xenon-133; ^{133}Xe; 14932-42-4 Xenon-135; ^{135}Xe; 14993-63-1 2. Albumin, Hemoglobin and Methemoglobin	Conn, H. L. J. Appl. Physiol. 1961, 16, 1065-1070.

VARIABLES:	PREPARED BY:
T/K: 294.15 Xe P/kPa: 23.33 - 93.325 (175 - 700 mmHg)	A. L. Cramer H. L. Clever

EXPERIMENTAL VALUES:

T/K	$\dfrac{cm^3\ Xe\ g^{-1}\ Protein}{cm^3\ Xe\ g^{-1}\ Water}$	$\dfrac{mol\ Xe*}{mol\ Protein}$	Solubility Coefficient S
	Human Albumin		
294.15	3.50 $\pm$ 0.3	1.1	0.413
	Human Hemoglobin		
294.15	6.43 $\pm$ 0.4	1.9	0.759
	Human Methemoglobin		
294.15	6.45 $\pm$ 0.4	1.9	0.761
	Dog Hemoglobin		
294.15	6.24 $\pm$ 0.1	1.9	0.736
	Dog Methemoglobin		
294.15	6.04 $\pm$ 0.4	1.8	0.713

*calculated from experiments done at a xenon partial
pressure of 700 mmHg.

The solubility coefficient, S, was calculated by the compiler by multiplying
the protein/water ratio by 0.118, which is the solubility coefficient of
xenon g^{-1} water at 294.15 K (See the critical evaluation of the solubility of
xenon in water.)

The hemoglobin and methemoglobin solutions were 5 to 12 weight percent.

The xenon gas uptake by human and dog hemoglobin and methemoglobin obeys
Henry's law.

AUXILIARY INFORMATION

METHOD /APPARATUS/PROCEDURE:

Blood was drawn and centrifuged. The
red blood cells were washed repeatedly
with saline to remove serum fats.
Hemoglobin was obtained by hemolysing
the red blood cells with distilled
water, and precipitating stroma   by
centrifugation. The mid-portion of
the supernatant fluid was withdrawn
for testing and use. Only one peak
was present in a starch block
electrophoresis.

Methemoglobin was obtained by oxi-
dizing hemoglobin with excess potassium ferricyanide or sodium nitrite.

One ml samples were drawn into a 5 ml
syringe and filled with gas with a
xenon partial pressure of 175 to 700
mmHg. The syringes were shaken 1-2
hours at 294 K. The gas phase was
completely expelled and the solution
radioactivity was determined in a
scintillation counter. Water samples
at the same xenon partial pressure
were used to calculate the water/
protein solubility ratio.

SOURCE AND PURITY OF MATERIALS:

1. Xenon. Air Reduction Sales.
   Specially purified. The gas was
   subjected to neutron bombardment
   in a nuclear reactor. Resultant
   xenon isotopes were mainly
   ^{133}Xe and ^{135}Xe.

2. Albumin. Pentex Corp. Crystaline.
   albumin, 2-4% g. Hemoglobin,
   methemoglobin. See procedure.

ESTIMATED ERROR:

See standard error of mean above.

REFERENCES:

COMPONENTS:	ORIGINAL MEASUREMENTS:
1. Xenon; Xe; 7440-63-3 2. Water; H_2O; 7732-18-5 3. Haemoglobin (human)	Schoenborn, B.P.; Vogelhut, P.O.; Featherstone, R.M. Pharmacologist 1963, 5, 264.

VARIABLES:
    T/K: 293.15
  Xe P/kPa: 101.325 (1 atm)
      pH: 7.0
	PREPARED BY:
	
	    H. L. Clever

EXPERIMENTAL VALUES:

T/K	Mol Xe per mol haemoglobin
293.15	1.8

AUXILIARY INFORMATION

METHOD / APPARATUS/PROCEDURE:	SOURCE AND PURITY OF MATERIALS:
Standard PVT measurement. No experimental details in either this abstract of a meeting paper or a later publication (1).	No information given on any of the components.
	ESTIMATED ERROR: Qualitative.

	REFERENCES:
	1. Schoenborn,B.P.;Featherstone,R.M.;
	    Vogelhut, P.O.; Süsskind, C.
	    Nature 1964, 202, 695.

COMPONENTS:	ORIGINAL MEASUREMENTS:
1. Xenon; Xe; 7440-63-3 2. Water; H_2O; 7732-18-5 3. Sodium Phosphate Buffer 4. Bovine Blood Components	Meuhlbaecher, C.; DeBon, F.L.; Featherstone, R.M. Inst. Anesth. Clinics 1963, 1, 937-952.
VARIABLES: T/K: 310.15	PREPARED BY: H.L. Clever

EXPERIMENTAL VALUES:
The Bunsen coefficients were presented on large scale graphs. The xenon Bunsen coefficient in sodium phosphate buffer was about 0.083. Comments about the individual systems follow:

T/K	Comments
	Bovine Gamma-Globulin + Phosphate Buffer (pH 6.3-6.5, Ionic Strength 0.16)
310.15	No apparent change in the xenon Bunsen coefficient as bovine gamma-globulin was increased from 0-8 per cent.
	Bovine Serum Albumin + Phosphate Buffer (pH 5.6-6.3, Ionic Strength 0.16)
310.15	No apparent change in the xenon Bunsen coefficient as the bovine serum albumin was increased from 0 to 10 per cent. At 20 per cent bovine serum albumin the Xenon Bunsen Coefficient appears to increase a little more than the 8 per cent uncertainty in the measurement.
	Bovine Hemoglobin + Phosphate Buffer (pH 6.3-6.6, Ionic Strength 0.16)
310.15	A linear increase in the xenon Bunsen coefficient as the bovine hemoglobin was increased from 0 to 20 per cent. The 20 per cent solution absorbed 58 per cent more xenon than did the buffer solution.

AUXILIARY INFORMATION

METHOD/APPARATUS/PROCEDURE: Gas chromatography	SOURCE AND PURITY OF MATERIALS: No information
	ESTIMATED ERROR: $\delta\alpha/\alpha = 0.08$
	REFERENCES:

COMPONENTS:	ORIGINAL MEASUREMENTS:
1. Xenon; Xe; 7440-63-3 2. Water; H_2O; 7732-18-5 3. Human Whole Blood and Blood Components	Yeh, S-Y.; Peterson, R.E. $\underline{J}$. $\underline{Appl}$. $\underline{Physiol}$. 1965, $\underline{20}$, 1041-1047.

VARIABLES:	PREPARED BY:
T/K: 310.15 Total P/kPa: 101.325 (1 atm)	A.L. Cramer

EXPERIMENTAL VALUES:

T/K	Absorption Coefficient mean β $\pm$ Std. Dev.	Number of Determinations	Absorption Coefficient g^{-1} Hemoglobin
Hemoglobin Solution (7.63% hemoglobin, 92.35% water, 1.57 mg cm^{-3} lipid)			
310.15	0.1046 $\pm$ 0.0057	3	0.3918
Hemoglobin solution (15.39 hemoglobin, 84.60% water, 2.60 mg cm^{-3} lipid)			
310.15	0.1241 $\pm$ 0.0032	4	0.3469
Whole Blood (425 ml. whole blood + 120 ml. 1.32% sodium citrate solution)			
310.15	0.1412 $\pm$ 0.0044	3	-

*The authors give a weighted average of 0.3661 cm^3 (STP) gas g^{-1} hemoglobin which is equivalent to 1.111 mole Xe $mole^{-1}$ hemoglobin (mol. wt. = 68,000). The values are for a partial pressure of xenon of (1-solution vapor pressure) atm. At one atm xenon the values would be about 6% higher.

AUXILIARY INFORMATION

METHOD/APPARATUS/PROCEDURE:	SOURCE AND PURITY OF MATERIALS:

In Yeh and Peterson (1) modification of Geffcken (2) apparatus, 45 ml. of liquid was frozen, evacuated, and melted repeatedly until no bubbles appeared in liquid under vacuum. Equilibration with gas and measurement of solubility followed (1).	1. Xenon. The Matheson Co. Research
	   grade, maximum impurity 0.02%
	   nitrogen and 0.05% krypton.
	2. Water. Distilled.
	3. Human Blood. 425 ml. from a
	   normal donor, mixed with 120 ml.
	   1.32% sodium citrate solution.
	   Frozen and thawed to hemolyze
	   red blood cells.
	4. Hemoblobin. From centrifuged
	   citrated human blood.

	ESTIMATED ERROR: $\delta T/K = 0.05$ $\delta P/mmHg = 0.2$ $\delta L/L = 0.0015$

	REFERENCES:
	1. Yeh, S-Y.; Peterson, R.E.
	    $\underline{J}$. $\underline{Pharm}$. $\underline{Sci}$. 1963, $\underline{52}$, 453.
	
	2. Geffcken, G.
	    $\underline{Z}$. $\underline{Physik}$. $\underline{Chem}$. 1904, $\underline{49}$, 257.

COMPONENTS:	ORIGINAL MEASUREMENTS:
1. Xenon-133; $^{133}_{54}$Xe; 14932-42-4 2. Human Blood	Anderson, A.M.; Ladefoged, J. J. Pharm. Sci. 1965, 54, 1684-1685.
VARIABLES: T/K: 310.15 Hematocrit/%: 0 - 88	PREPARED BY: A.L. Cramer H.L. Clever

EXPERIMENTAL VALUES:

A total of 64 determinations of the partition coefficient of xenon-133 between blood and water were carried out. Separate samples of water and of blood were equilibrated with the same air-radon-133 mixture in each experiment. The blood hematocrit varied from 0 to 88 weight per cent. The data was given in a graph and fitted by a linear regression to

Partition Coefficient = (1.02 ± 0.02) + (0.0112 ± 0.0004)(wt % hematocrit)

with an overall standard deviation of 0.08. The plasma/water partition coefficient is 1.02, and the erythorocyte/water partition coefficient is 2.14. A formula for the partition coefficient of tissue/blood as a function of hematocrit % was given.

A plasma/gas and erythorocyte/gas partition coefficient may be obtained from the equation value by mutiplying by 0.0834, the Ostwald solubility of xenon in water reported by the authors (see page 139).

The erythorocyte/xenon gas Ostwald coefficient of 0.178 calculated from the information above is 12 - 15 per cent lower than values reported on pages 213 and 214, and 6 per cent lower than the authors value in a later report (see page 215). See also the data on page 218.

AUXILIARY INFORMATION

METHOD/APPARATUS/PROCEDURE:	SOURCE AND PURITY OF MATERIALS:
The samples in small vials were equilibrated with a mixture of air and xenon-133 for 24 h at 310 K. The samples were stirred continuously. After equilibration 0.1 ml samples were transferred anaerobically by syringes to a 2 ml. rubber-capped vial and counted in a scintillation counter to a statistical error below 1 per cent. The hematocrit was varied by removal of plasma or erythrocytes from whole blood after centrifuging.	1. Xenon-133. The Radiochemical Centre, Amersham, England. Impurities (<2 per cent) were mainly xenon-133m with trace amounts of xenon-131 and krypton-85. 2. Blood. From voluntary donors.
	ESTIMATED ERROR:
	REFERENCES:

COMPONENTS:	ORIGINAL MEASUREMENTS:
1. Xenon-133; $^{133}_{54}$Xe; 14932-42-4 2. Saline and Blood Components	Isbister, W. H.; Schofield, P. F.; Torrance, H. B. Phys. Med. Biol. 1965, 10, 243-250.

VARIABLES:	PREPARED BY:
T/K: 310.15	A. L. Cramer H. L. Clever

EXPERIMENTAL VALUES:

T/K	Ostwald Coefficient L $\pm$ Std Error	Number of Determinations
	Saline (water)*	
310.15	0.0926 $\pm$ 0.0027	7
	Plasma	
310.15	0.1028 $\pm$ 0.0008	53
	Red Cells (Hematocrit 98%)	
310.15	0.2020 $\pm$ 0.0015	2

*The saline solution was not described. It may be the standard 0.9 weight percent solution of NaCl in water.

AUXILIARY INFORMATION

METHOD/APPARATUS/PROCEDURE:	SOURCE AND PURITY OF MATERIALS:
Two ml samples of blood were introduced into the sample tube, four ml of a concentrated xenon-133 air mixture were added to the tube which was then sealed. The tubes were shaken and equilibrated at 310 K for two hours. After equilibration the tubes were centrifuged at 310 K for 30 minutes at 3000 rpm. The three phases cells, plasma and gas were separately counted in a specially constructed lead collimator attached to a scintillation counter.	1. Xenon-133. No information given. 2. Blood components. Fresh human blood containing a suitable amount of Sequestrene.
	ESTIMATED ERROR: See standard error of mean above.
	REFERENCES:

214

COMPONENTS:	ORIGINAL MEASUREMENTS:
1. Xenon-133; $^{133}_{54}$Xe; 14932-42-4 2. Blood components	Veall, N.; Mallett, B. L. Phys. Med. Biol. 1965, 10, 375-380.
VARIABLES: T/K: 310.15	PREPARED BY: A. L. Cramer H. L. Clever

EXPERIMENTAL VALUES:

T/K	Solubility Coefficient S + Std Error	Number of Determinations
	Water	
310.15	0.0903 + 0.0005	9
	Plasma	
310.15	0.1025 + 0.0009	12
	Red Cells	
310.15	0.2100 + 0.0043	12

AUXILIARY INFORMATION

METHOD/APPARATUS/PROCEDURE:

Red cell measurements were made on blood samples of varying haematocrit. Plasma measurements were made on samples first haemolyzed by rapid freezing and thawing.

Approximately 1 ml samples were add to sample tube; about 0.1 ml of air containing about 10 µC of xenon-133 was added and the tube sealed. Sample tubes were equilibrated either by leaving over night lying horizontally at 310 K or inverted every 2-3 minutes for several hours. The samples were briefly centrifuged to remove air bubbles from liquid and then gas and liquid phases counted.

SOURCE AND PURITY OF MATERIALS:

1. Xenon-133. No information given.

2. Blood. Freshly drawn heparinized blood from normal people.

ESTIMATED ERROR:
See standard error of mean above.

δT/K = 0.5 (Compiler)

REFERENCES:

COMPONENTS:	ORIGINAL MEASUREMENTS:
1. Xenon-133; $^{133}_{54}$Xe; 14932-42-4 2. Plasma and Erythrocytes	Ladefoged, J.; Anderson, A.M. Phys. Med. Biol. 1967, 12, 353-358.
VARIABLES: 　　T/K:　310.15	PREPARED BY: 　　　A. L. Cramer 　　　H. L. Clever

EXPERIMENTAL VALUES:

T/K	Pre-saturating gas	Ostwald Coefficient L ± Std. Dev.	Number of Determinations
Water			
310.15	---	0.0834 ± 0.0002	107
Plasma*			
310.15	Air	0.091 ± 0.002	30
	Nitrogen	0.090 ± 0.002	36
	Oxygen	0.093 ± 0.002	27
Erythrocytes*			
310.15	Air	0.19 ± 0.008	30
	Nitrogen	0.20 ± 0.008	36
	Oxygen	0.17 ± 0.010	27

*A separate set of experiments in which the xenon partial
pressure was varied from 0.1 to 1 atm gave an Ostwald
coefficient of 0.089 ± 0.002 in Plasma and 0.189 ± 0.004
in erythrocytes.

AUXILIARY INFORMATION

METHOD/APPARATUS/PROCEDURE:

Blood from volunteer donors was
heparinized and centrifuged 30 minutes at 1500 G. Samples were placed
in small test tubes together with a
sample of water in an air tight box
containing 0.5 to 2 µCi xenon-133
in 100 ml of either air, nitrogen
or oxygen. The system was stirred
continuously at 310 K for 24 hours.
After equilibration the samples were
transferred into syringes, about 0.1
ml samples were counted in a scintillation counter and compared with
water value. Authors calculated the
Ostwald solubility coefficient in
blood of 30, 40 and 50 percent
hematocrit as 0.119, 0.129, and 0.139
respectively from the above values.

SOURCE AND PURITY OF MATERIALS:

1. Xenon-133. Radiochemical Centre,
    Amersham, U.K. Two per cent
    impurity of ^{131}Xe and ^{85}Kr.
    Xenon. Dansk Ilt og Briut.
    99.95 mol per cent.

2. Blood components. Volunteer
    donors.

ESTIMATED ERROR:

See standard deviations above.

REFERENCES:

1. Anderson, A.M.; Ladefoged, J.
    J. Pharm. Sci. 1965, 54, 1684.

COMPONENTS:	ORIGINAL MEASUREMENTS:
1. Xenon; Xe; 7440-63-3 2. Water; H_2O; 7732-18-5 3. Sodium Chloride; NaCl; 7647-14-5 4. Human Albumin; 9048-46-8	Yeh, S.-Y,; Peterson, R. E. J. Appl. Physiol. 1965, 20, 1041-1047

VARIABLES:	PREPARED BY:
T/K: 298.15 - 310.15 P/kPa: 101.325 (1 atm) Albumin/wt %: 5.12 - 27.93	A. L. Cramer H. L. Clever

EXPERIMENTAL VALUES:

T/K	Albumin wt %	Mol Fraction $X_1 \times 10^2$	Bunsen Coefficient α	Ostwald Coefficient $L \pm$ Std Dev	Number Determinations
298.15	5.12		0.0939	0.1025 $\pm$ 0.0028	4
303.15			0.0845	0.0938 $\pm$ 0.0025	3
310.15			0.0709	0.0805 $\pm$ 0.0008	2
298.15	15.34		0.1092	0.1192 $\pm$ 0.0028	3
303.15			0.0963	0.1069 $\pm$ 0.0047	3
310.15			0.0786	0.0892 $\pm$ 0.0020	3
298.15	27.93		0.1239	0.1352 $\pm$ 0.0070	4
303.15			0.1100	0.1221 $\pm$ 0.0046	4
310.15			0.0880	0.0999 $\pm$ 0.0087	4
298.15	100	42.26	0.2181	0.2382	
303.15		35.77	0.1761	0.1954	
310.15		31.51	0.1315	0.1493	

The Ostwald coefficients in 100 % human albumin were values extraploated from the values at lower concentration by the authors. The Bunsen coefficients and mole fraction solubilities were calculated by the compiler. The authors used an albumin molecular weight of 69,000 and a density of 1.0 to calculate mole ratios of xenon to albumin of 0.732, 0.557, and 0.460 at the temperatures 298.15, 303.15, and 310.15 K respectively.

AUXILIARY INFORMATION

METHOD	SOURCE AND PURITY OF MATERIALS:
A 45 cm^3 sample of albumin solution was frozen, evacuated, and melted repeatedly until no bubbles appeared in the liquid under vacuum in the Yeh and Peterson modification (1) of the Geffcken apparatus (2). Equilibration with gas and measurement of the solubility followed.	1. Xenon. The Matheson Co. Research grade, maximum impurity 0.02 % nitrogen and 0.05 % krypton. 2. Human serum albumin. Cutter Laboratories. A 25 % solution, stabilized with 0.02 M sodium caprylate and 0.02 M sodium acetyltryptophanate.
	ESTIMATED ERROR: $\delta T/K = 0.05$ $\delta P/mmHg = 0.2$ $\delta L/L = 0.0015$
	REFERENCES: 1. Yeh, S.-Y.; Peterson, R. E. J. Pharm. Sci. 1963, 52, 453. 2. Geffcken, G. Z. Phys. Chem. 1904, 49, 257.

COMPONENTS:	ORIGINAL MEASUREMENTS:

COMPONENTS:

1. Xenon-133; $^{133}_{54}$Xe; 14932-42-4

2. Water; H_2O; 7732-18-5

3. Sodium Chloride; NaCl; 7647-14-5, or Albumin

ORIGINAL MEASUREMENTS:

Ladefoged, J.; Anderson, A. M.

Phys. Med. Biol. 1967, 12, 353-358.

VARIABLES:

T/K: 310.15

PREPARED BY:

A. L. Cramer

EXPERIMENTAL VALUES:

T/K	Ostwald Coefficient $L \pm$ Std. Dev.	Number of Determinations
Water; H_2O; 7732-18-5		
310.15	0.0834 ± 0.0002	107
Standard Saline; 0.9% NaCl in water		
310.15	0.078 ± 0.007	5
Albumin, 200 g dm^{-3}		
310.15	0.099 ± 0.003	12

AUXILIARY INFORMATION

METHOD/APPARATUS/PROCEDURE:

The technique described by Anderson and Ladefoged (1) was used. The materials were placed in small test tubes together with a sample of water in an air tight chamber. The chamber contained air and 0.5 to 2 µCi of xenon-133. The samples were stirred continuously at 310.15 K for 24 hours. After equilibration the samples were transferred into syringes and samples of about 0.1 ml were counted in a scintillation counter. The Ostwald coefficients were calculated indirectly from the ratio of counts in water and the sample and the measured Ostwald coefficient in water. See the authors data sheet on Xe + H_2O.

SOURCE AND PURITY OF MATERIALS:

1. Xenon-133. Radiochemical Centre, Amersham, U.K. Two per cent impurity of ^{131}Xe and ^{85}Kr.

2. Albumin. Statens Serum Institut. Cohn's fraction V, 97% albumin and 3% α and β globulin. Heat treated 60°C for 10 hours.

ESTIMATED ERROR:

See standard deviations above.

REFERENCES:

1. Anderson, A. M.; Ladefoged, J. J. Pharm. Sci. 1965, 54, 1684.

K.X.R.—Q

COMPONENTS:	ORIGINAL MEASUREMENTS:
1. Xenon; ^{133}Xe; 14932-42-4 2. Water, Saline Solution, Plasma, and Human Red Blood Cells	Kitani, K. Scand. J. Clin. Lab. Invest. 1972, 29, 167-172.
VARIABLES: 　　T/K: 310.15 　　P/kPa: 101.325 (1 atm)	PREPARED BY: 　　P.L. Long 　　A.L. Cramer

EXPERIMENTAL VALUES:

T/K	Bunsen Coefficient α	Ostwald Coefficient L $\pm$ Std. Dev.	Replications
Water; H_2O; 7732-18-5			
310.15	0.0731	0.0830 $\pm$ 0.0017	17
Water + Sodium Chloride; NaCl; 7647-14-5 0.9 per cent Saline			
310.15	0.0687	0.0780 $\pm$ 0.0013	29
Plasma (3 Samples)			
310.15	0.0839	0.0953 $\pm$ 0.0006	6
	0.0821	0.0932 $\pm$ 0.0008	11
	0.0830	0.0942 $\pm$ 0.0007	7
Human Red Blood Cells			
310.15	-	0.1966	-

The Bunsen coefficients were calculated by the compiler assuming that the Ostwald coefficient was independent of pressure.

The solubility value for red blood cells is a value extrapolated from the solubility of the gas in mixtures of plasma and heparinized blood from healthy donors. The coefficient of solubility is linear in hematocrit per cent (graph in paper).

$$L = 0.09560 + (0.00101) (Hematocrit \%v/v) \quad (r = 0.997)$$

AUXILIARY INFORMATION

METHOD/APPARATUS/PROCEDURE:

A glass cuvette is filled with the liquid. One-third of the liquid is replaced with radioactive gas in air. The sealed cuvette is placed in a thermostated bath for 2 hours. The pressure is adjusted to 1 atm by means of a thin needle.

The radioactive assay, corrected for self absorption and scatter, is made by a scintillation detector and a Philips pulse-height analyzer.

SOURCE AND PURITY OF MATERIALS:

1. Xenon. Radiochemical Centre, Amersham, England.

2. Water and 0.9 per cent saline were prepared according to the criteria for purity in the Nordic Pharmacopeia. Heparinized blood from healthy donors was used.

ESTIMATED ERROR:

See standard deviations above.

REFERENCES:

COMPONENTS:	ORIGINAL MEASUREMENTS:
1. Xenon; Xe; 7440-63-3	Catchpool, J. F.
2. Water; H_2O; 7732-18-5	Fed. Proc. 1968, 27, 884 - 887.
3. Sodium Chloride; NaCl; 7647-14-5	
4. Methemoglobin	

VARIABLES:	PREPARED BY:
T/K: 278.15 - 308.15 P/kPa: 80.93 - 92.66 (607 - 695 mmHg)	H. L. Clever

EXPERIMENTAL VALUES:

T/K	Henry's Constant $K \times 10^{-7}$ $(P_1/mmHg)/(mol\ Xe/mol\ H_2O)$
278.15	0.28
283.15	0.34
288.15	0.40
293.15	0.45
298.15	0.52
303.15	0.60
308.15	0.66

The values of Henry's constant were read from a graph in the original paper. There is an increase in the temperature coefficient of Henry's constant between temperatures of 293 and 303 K.

The author calculated that at 293.15 K and a xenon partial pressure of 101.325 kPa (1 atm) 1.67 mole of xenon is combined with each mole of the methemoglobin. The result is based on a xenon Henry's constant of 0.88×10^7 in 0.5 per cent saline solution at 293.15 K.

AUXILIARY INFORMATION

METHOD:	SOURCE AND PURITY OF MATERIALS:
Volumetric apparatus with the gas pressure controlled by a quartz Bourdon tube, mirror-photocell, and a servomechanism. Measurements were made at all seven temperatures on one loading of the apparatus over a period of four days. Spectroscopic and density measurements before and after the experiment indicated no degradation of the sample. The sample was 0.00279 mole methemoglobin dm^{-3} of 0.5 per cent saline solution.	1. Xenon. } No information. 2. Water. } 3. Sodium Chloride. } 4. Hemoglobin. Prepared from human type 0 Rh negative blood. Packed cells washed six times with 1 per cent saline. Lipids and cellular debris were dissolved in toluene.
	ESTIMATED ERROR: $\delta T/K = 0.005$ $\delta P/mmHg = 0.005$
	REFERENCES:

COMPONENTS:	ORIGINAL MEASUREMENTS:
1. Xenon; Xe; 7440-63-3 2. Water; H_2O; 7732-18-5 3. Myoglobins	Ewing, G. J.; Maestas, S. J. Phys. Chem. 1970, 74, 2341-2344.

VARIABLES:	PREPARED BY:
T/K: 293.15 - 303.15 Xe P/kPa: 50.663 - 491.426 (0.5 - 4.85 atm)	H. L. Clever

EXPERIMENTAL VALUES:

T/K	Henry's Constant K_H/mmHg X_1^{-1}	K/m^{-1}	K'/m^{-1}
Water			
293.15	0.85×10^7	-	-
298.15	0.96×10^7	-	-
303.15	1.13×10^7	-	-
Cyanometmyoglobin			
293.15	-	186	-
298.15	-	145	-
303.15	-	115	-
Metmyoglobin			
293.15	-	200	∿8.2
298.15	-	146	∿7.2
303.15	-	130	∿2.3
Myoglobin			
293.15	-	109	2.3
298.15	-	94	2.6
303.15	-	85	0.5

The xenon solubility in the myoglobin solutions did not obey Henry's law.
The xenon solubility data are presented graphically in the original paper.

AUXILIARY INFORMATION

METHOD/APPARATUS/PROCEDURE:
Manometric method. The amount of
xenon absorbed by the solution was
determined by measuring the initial
pressure and the final pressure to
obtain the pressure change, ΔP, and
applying the ideal gas law

$$\Delta n = \frac{(\Delta P)V}{RT}$$

To determine the amount of xenon absorbed by the hemoprotein, the amount
of xenon calculated to dissolve in
water at the temperature and equilibrium pressure of the experiment was
subtracted from Δn. This quantity
divided by the number of moles of
myoglobin in the ∿10 weight per cent
solution (usually ∿10^{-4} mol) yielded
the amount of xenon dissolved per
mole of myoglobin.
The xenon adsorptions in the hemoprotein solutions were presented
graphically. No appreciable difference was noted in the results for
buffered (phosphate buffer pH 8.1)
and unbuffered (pH 6.95) metmyoglobin
solutions.

SOURCE AND PURITY OF MATERIALS:
1. Xenon. No information.

2. Horse heart myoglobin.
Calbiochem. The myoglobin was
prepared by reduction of metmyoglobin by sodium dithionite.
The cyanometmyoglobin was prepared from metmyoglobin by
reaction with KCN and $K_3Fe(CN)_6$
in a phosphate buffer solution.

ESTIMATED ERROR:
$$\delta T/K = 0.1$$
$$\delta K_H/\text{mmHg} = 0.05 \text{ (Compiler)}$$

The adsorptions were converted to
association constants for the
reactions
Hp + Xe = HpXe $K = m_{HpXe}/m_{Hp}m_{Xe}$

HpXe + Xe = HpXe$_2$

$K' = m_{HpXe_2}/m_{HpXe}m_{Xe}$ where m is
mol kg^{-1} H_2O.

COMPONENTS:	ORIGINAL MEASUREMENTS:
1. Xenon; Xe; 7440-63-3 Xenon-133; ^{133}Xe; 14932-42-4 Xenon-135; ^{135}Xe; 14995-63-1 2. Dog Tissues and Fat	Conn, H. L. J. Appl. Physiol. 1961, 16, 1065-1070.

VARIABLES:	PREPARED BY:
T/K: 294.15 Xe P/kPa: 0.933 - 8.666 (7-65 mmHg)	A. L. Cramer H. L. Clever

EXPERIMENTAL VALUES:

T/K	$\frac{\text{Xe (counts m}^{-1}) \text{ g}^{-1} \text{ Tissue}}{\text{Xe (counts m}^{-1}) \text{ g}^{-1} \text{ Water}}$	Solubility* Coefficient S
	Fat	
294.15	19.70 ± 0.40	2.32
	Skeletal Muscle	
294.15	1.62 ± 0.04	0.191
	Cardic Muscle	
294.15	1.79 ± 0.05	0.211
	Kidney	
294.15	1.62 ± 0.06	0.191
	Liver	
294.15	1.75 ± 0.04	0.207
	Brain Grey Matter	
294.15	1.84 ± 0.05	0.217
	Brain White Matter	
294.15	3.00 ± 0.06	0.354

*Calculated by compiler by multiplying tissue/water xenon ratio by 0.118 cm^3 Xe at 294.15 K and 101.325 kPa (1 atm) g^{-1} water. The multiplying factor was calculated from the data in the critical evaluation of xenon in water.

AUXILIARY INFORMATION

METHOD /APPARATUS/PROCEDURE:

Sample tissues weighing less than 100 mg and 1 mm or less in thickness were placed in individual cuvettes. They were equilibrated with a radioxenon-air mixture varying from 7 to 65 mmHg in several experiments. The tissues were allowed to equilibrate with the gas mixture for 1 to 8 hours. After equilibration the tissues were almost instantaneously removed from the container and plunged into a bath of liquid nitrogen. The Radioxenon concentration in the frozen samples was determined by counting the radioactivity of each in a well-type scintillation counter. The count was compared with the count in a liquid water sample equilibrated at the same xenon partial pressure.

SOURCE AND PURITY OF MATERIALS:

1. Xenon. Air Reduction Sales. Specially purified. The gas was subjected to neutron bombardment in a nuclear reactor. Resultant xenon isotopes were mainly ^{133}Xe and ^{135}Xe.

2. Tissues. Taken within 10 minutes after sacrifice of the animal.

ESTIMATED ERROR:

See standard error of mean from 36 determinations above.

REFERENCES:

COMPONENTS:	ORIGINAL MEASUREMENTS:
1. Xenon; Xe; 7440-63-3 2. Water; H$_2$O; 7732-18-5 3. Sodium chloride; NaCl; 7647-14-5 4. Rabbit leg muscle	Yeh, S-Y.; Peterson, R.E. J. Appl. Physiol. 1965, 20, 1041-1047.
VARIABLES: T/K: 298.15 - 310.15 P/kPa: 101.325 (1 atm)	PREPARED BY: A.L. Cramer

EXPERIMENTAL VALUES:

T/K	Absorption Coefficient mean β ± Std. Dev.	Number of Determinations	cm^3 Xe g^{-1} muscle*
298.15	0.1049 ± 0.0010	3	–
303.15	0.0944 ± 0.0013	3	–
310.15	0.0823 ± 0.0009	3	0.1049 - 0.1127

*Calculated by authors. They corrected for dilution with saline solution and assumed a muscle specific gravity of 1.07. The range of values reflects the range of lipid, water and protein content found by analysis of their samples.

AUXILIARY INFORMATION

METHOD /APPARATUS/PROCEDURE:

In Yeh and Peterson (1) modification of Geffcken (2) apparatus, 45 ml of liquid homogenate is frozen, evacuated, and melted repeatedly until no bubbles appear in liquid under vacuum. Equilibration with gas and measurement of solubility follow (1).

SOURCE AND PURITY OF MATERIALS:
1. Xenon. The Matheson Co. Research grade, maximum impurity 0.02% nitrogen and 0.05% krypton.
2. Rabbit leg muscle. Homogenized with 4x's its volume of saline (w/w), then 0.05% mercury chloride added as microbial poison.

ESTIMATED ERROR:
$$\delta T/K = 0.05$$
$$\delta P/mmHg = 0.2$$
$$\delta L/L = 0.05$$

REFERENCES:
1. Yeh, S-Y.; Peterson, R.E. J. Pharm. Sci. 1963, 52, 453.
2. Geffcken, G. Z. Physik Chem. 1904, 49, 257.

COMPONENTS:	ORIGINAL MEASUREMENTS:
1. Xenon-133; $^{133}_{54}$Xe; 14932-42-4 2. Human Liver Tissue	Kitani, K.; Winkler, K. Scand. J. Clin. Lab. Invest. 1972, 29, 173-176.
VARIABLES: T/K: 310.15	PREPARED BY: A.L. Cramer

EXPERIMENTAL VALUES:

Thirty three measurements of the solubility of xenon-133 were made in
liver tissue with triglyceride content varying between 1 and 20 weight
per cent. One measurement was made at 50 weight percent triglyceride.
The results were given in a graph. The data fitted the regression equation

$$L = 0.09524 + (0.016447)(\text{triglyceride wt \%})$$

with a regression coefficient of 0.97.

The triglyceride weight per cent was calculated as tripalmitin. The tissue
solubility was corrected for the water added to the sample. Below a liver
triglyceride content of 5 per cent the standard deviation was 0.7 per cent,
at higher triglyceride content the standard error increases, and reaches a
value of 5 per cent at 25 per cent triglyceride.

Tripalmitin is Hexadecanoic acid, 1,2,3-propanetriyl ester; $C_{51}H_{98}O_6$;
555-44-2.

AUXILIARY INFORMATION

METHOD /APPARATUS/PROCEDURE:

Five g of liver tissue was
homogenized at 277 K for 15 m after
addition of 5 cm^3 of water. A glass
cuvette (1 x 1 x 10 cm) was 2/3
filled with the homogenate, and
closed with a rubber stopper. The
space above the homogenate was
evacuated and filled with air containing xenon-133. The cuvette was
rotated for 2 h in a water bath at
310 K. The radioactivity in the
homogenate and in the air phase were
determined by a scintillation counter
placed in a thermostated box at 310 K
(1).

The triglyceride was determined
enzymatically in chloroform-
methanol extracts.

SOURCE AND PURITY OF MATERIALS:

1. Xenon-133. Radiochemical
   Centre. Amersham, U.K.

2. Liver tissue. Autopsy material.

ESTIMATED ERROR:

REFERENCES:

1. Kitani, K.
   Scand. J. Clin. Lab. Invest.
   1972, 29, 167.

COMPONENTS:	ORIGINAL MEASUREMENTS:
1. Xenon; Xe; 7440-63-3 2. Water; H_2O; 7732-18-5 3. Sodium chloride; NaCl; 7647-14-5 4. Beef brain	Yeh, S.-Y.; Peterson, R.E. J. Appl. Physiol. 1965, 20, 1041-1047.

VARIABLES:
    T/K: 298.15 - 310.15
    P/kPa: 101.325 (1 atm)
	PREPARED BY:
	    A.L. Cramer

EXPERIMENTAL VALUES:

T/K	Absorption Coefficient mean $\beta \pm$ Std. Dev.	Number of Determinations	cm^3 Xe g^{-1} muscle*
298.15	0.1174 ± 0.0001	3	–
303.15	0.1059 ± 0.0005	3	–
310.15	0.0928 ± 0.0010	3	0.164 - 0.186

*Calculated by the authors. They corrected for the dilution with saline solution and assumed a brain tissue specific gravity of 1.04. The range of values reflects the range of lipid, water and protein content found by analysis of their samples.

AUXILIARY INFORMATION

METHOD /APPARATUS/PROCEDURE:

In Yeh and Peterson (1) modification of Geffcken (2) apparatus, 45 ml of liquid homogenate is frozen, evacuated, and melted repeatedly until no bubbles appear in liquid under vacuum. Equilibration with gas and measurement of solubility follow (1).

SOURCE AND PURITY OF MATERIALS:
1. Xenon. The Matheson Co. Research grade, maximum impurity 0.02% nitrogen and 0.05% krypton.
2. Beef brain. Homogenized with 4 x volume normal saline (w/w); then mixed with 0.05% mercury chloride added as microbial poison.

ESTIMATED ERROR:
$$\delta T/K = 0.05$$
$$\delta P/mmHg = 0.2$$
$$\delta L/L = 0.05$$

REFERENCES:
1. Yeh, S-Y.; Peterson, R.E.
   J. Pharm. Sci. 1963, 52, 453.

2. Geffcken, G.
   Z. Physik Chem. 1904, 49, 257.

COMPONENTS:	ORIGINAL MEASUREMENTS:
1. Xenon-133; $^{133}_{54}$Xe; 14932-42-4 2. Human Brain Components	Veall, N.; Mallett, B. L. Phys. Med. Biol. 1965, 10, 375-380.

VARIABLES:	PREPARED BY:
T/K: 310.15	A. L. Cramer

EXPERIMENTAL VALUES:

T/K	Solubility Coefficient $S_T \pm$ Std Error	Number of Determinations
	Water	
310.15	0.0903 $\pm$ 0.0005	9
	Brain Grey Matter	
310.15	0.1196 $\pm$ 0.0032	14
	Brain White Matter	
310.15	0.2253 $\pm$ 0.0045	14
	Brain Homogenate*	
310.15	0.1616 $\pm$ 0.0035	12

*The fraction of grey matter in the brain was calculated to be 60.3 $\pm$ 6.3 per cent from the data above.

AUXILIARY INFORMATION

METHOD /APPARATUS/PROCEDURE:

Small samples of brain material were weighed, 1.0 ml of distilled water added and the mixture ground to a uniform fluid. The one ml tissue suspension was added to a sample tube, 0.1 ml of air containing about 10 µCi of Xenon-133 was added and the tube sealed. The tubes were inverted every few minutes for 2-3 hours and left overnight at 310 K. The tube was centrifuged and gas and liquid phases each counted. The solubility coefficient, S_T, is the gas tissue partition coefficient, referenced to 1 ml gas and 1 g tissue.

SOURCE AND PURITY OF MATERIALS:

1. Xenon-133. No information given.

2. Brain Matter. Fresh post-mortem brains with no evidence of abnormality.

ESTIMATED ERROR:

See standard error of mean above.

REFERENCES:

COMPONENTS:	ORIGINAL MEASUREMENTS:
1. Xenon-133; $^{133}_{54}$Xe; 14932-42-4 2. Brain material.	Isbister, W. H.; Schofield, P. F.; Torrance, H. B. Phys. Med. Biol. 1965, 10, 243-250.

VARIABLES:	PREPARED BY:
T/K: 310.15	A. L. Cramer

EXPERIMENTAL VALUES:

T/K	Partition Coefficient S ± Std Error	Number of Determinations
	Brain Grey Matter	
310.15	0.1466 ± 0.0052	18
	Brain white Matter	
310.15	0.2434 ± 0.0119	11

The authors calculated the partition coefficient of xenon-133 between blood and brain at varying haematocrit levels. See Table II of original paper.

AUXILIARY INFORMATION

METHOD/APPARATUS/PROCEDURE:	SOURCE AND PURITY OF MATERIALS:
Know weights of grey and white matter were collected in glass containers under a known volume of saline. Fine homogenates of the material were prepared. Two ml. samples of the homogenate were introduced into the special counting tube and xenon-133 air mixture was added. Equilibration at 310 K was complete in two hours. The gas and homogenate phases were counted in a specially constructed lead collimator attached to a scintillation counter. The partition coefficient between gas and tissue was calculated from the gas homogenate partion coefficient by the formula of Kety et al.	1. Xenon-133. No information given. 2. Brain material. Fresh adult human post-mortem material dissected and placed in glass container under saline solution.
	ESTIMATED ERROR: See standard error of mean above.

	REFERENCES:
	 1. Kety, S. S.; Harmel, M. H.;
	     Broomell, H. T.; Rhode, C. B.
	     J. Biol. Chem. 1948, 173, 487.

COMPONENTS:	EVALUATOR:
1. Radon-222; $^{222}_{86}Rn$; 14859-67-7 2. Water; H_2O; 7732-18-5	Rubin Battino Department of Chemistry Wright State University Dayton, Ohio, 45431 U.S.A. July 1978

CRITICAL EVALUATION:

Radon solubilities are special because we are dealing with a naturally radioactive isotope with a half-life of 3.8 days. The usual technique for determining the solubility of radon is to equilibrate it in a carrier gas like air or nitrogen with the liquid (or solution) it is to be dissolved in. The radon partial pressure is about 10^{-4} mm Hg (ca. 0.01 Pa). The solubility is then determined by measuring the radioactivity in the equilibrated gaseous phase and liquid phase. From this the Ostwald coefficient, L, is calculated at a given temperature according to

$$L = \frac{\text{concentration of gas per unit volume of liquid phase}}{\text{concentration of gas per unit volume of gas phase}} \qquad (1)$$

The Ostwald coefficient is normally defined as

$$L = \frac{\text{volume of gas at its partial pressure and T/K}}{\text{volume of solvent used to absorb the gas at T/K}} \qquad (2)$$

These two definitions are equal if there is no volume change on mixing or negligible volume change on mixing. For the radon partial gas pressure used in the experimental measurements, the latter is certainly the case. The mole fraction solubility at 101,325 Pa is then calculated via equation 3 with P_1 = 101,325 Pa.

$$x_1 = \left[\frac{R(T/K)}{V_1^o \, L \, P_1} + 1 \right]^{-1} \qquad (3)$$

where R is the gas constant and V_1^o is the molar volume of the solvent at T/K. The long extrapolation from 0.01 Pa to 101,325 Pa using Henry's law is questionable. However, by making this conversion we can treat the thermodynamic functions of all of the noble gases under the same standard state conditions. Boyle (1) did carry out a test of Henry's law by varying the Rn + Air pressure from 10^{-5} mm Hg to 10^{-4} mm Hg. The solubility (Ostwald Coefficient) at $14^{\circ}C$ varied irregularly from 0.299 to 0.307 which was considered to be within experimental error.

We used the data of three workers for smoothing. In fitting the equation those data points which differed from the smoothed curve by about two standard deviations or more were rejected and the data then re-fitted. The 40 points used for the final smoothing equation were obtained as follows (reference – number of data points used from that reference): 1-2; 2-6; 3-32. The fitting equation used was

$$\ln x_1 = A + B/(T/100K) + C \ln (T/100K) \qquad (4)$$

Using T/K as the variable rather than T/K gives coefficients of approximately equal magnitude. The best fit for the 40 data points was

$$\ln x_1 = -90.5481 + 130.026/(T/100K) + 35.0047 \ln (T/100K) \qquad (5)$$

where x_1 is the mole fraction solubility of radon at 101,325 Pa partial pressure of gas. The fit in $\ln x_1$ gave a standard deviation of 1.02% taken at the middle of the temperature range. Table 1 gives smoothed values of the mole fraction solubility at 101,325 Pa partial pressure of gas and the Ostwald coefficient at 5K intervals.

COMPONENTS:	EVALUATOR:
1. Radon-222; $^{222}_{86}$Rn; 14859-67-7 2. Water; H_2O; 7732-18-5	Rubin Battino Department of Chemistry Wright State University Dayton, Ohio, 45431 U.S.A. July 1978

CRITICAL EVALUATION:

TABLE 1. Smoothed values of radon solubility in water and thermodynamic functions[a] using equation 2. Mole fraction solubility at 101 325 Pa partial pressure of radon. Ostwald coefficient at equilibrium saturation pressure.

T/K	Mol Fraction X_1 x 10^4	Ostwald Coefficient L	$\Delta \bar{G}_1^\circ$/KJmol^{-1} [b]	$\Delta \bar{H}_1^\circ$/KJmol^{-1}	$\Delta \bar{S}_1^\circ$/JK^{-1}mol^{-1}
273.15	4.217	0.5249	17.65	-28.61	-169
278.15	3.382	0.4286	18.48	-27.15	-164
283.15	2.764	0.3565	19.29	-25.70	-159
288.15	2.299	0.3016	20.07	-24.24	-154
293.15	1.945	0.2593	20.83	-22.79	-149
298.15	1.671	0.2263	21.56	-21.33	-144
303.15	1.457	0.2003	22.27	-19.88	-139
308.15	1.288	0.1797	22.95	-18.42	-134
313.15	1.153	0.1632	23.61	-16.97	-130
318.15	1.046	0.1500	24.25	-15.51	-125
323.15	0.959	0.1395	24.86	-14.06	-120
328.15	0.889	0.1310	25.45	-12.60	-116
333.15	0.833	0.1243	26.02	-11.15	-112
338.15	0.788	0.1190	26.57	-9.69	-107
343.15	0.752	0.1149	27.09	-8.24	-103
348.15	0.724	0.1119	27.60	-6.78	-99
353.15	0.703	0.1099	28.08	-5.33	-95
358.15	0.688	0.1087	28.54	-3.87	-91
363.15	0.678	0.1082	28.98	-2.42	-86
368.15	0.673	0.1085	29.41	-0.96	-82
373.15	0.672	0.1095	29.81	0.49	-79

[a] $\Delta \bar{C}_{p_1}^\circ$ was independent of temperature and has a value of 291 J K^{-1} mol^{-1}.

[b] cal$_{th}$ = 4.184 joule

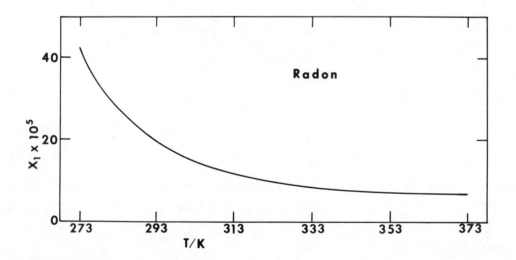

FIGURE 1. The mole fraction solubility of radon in water at a radon partial pressure of 101.325 kPa (1 atm).

COMPONENTS:	EVALUATOR:
1. Radon-222; $^{222}_{86}$Rn; 14859-67-7 2. Water; H_2O; 7732-18-5	Rubin Battino Department of Chemistry Wright State University Dayton, Ohio, 45431 U.S.A. July 1978

CRITICAL EVALUATION:

Table 1 also gives the thermodynamic functions $\Delta\bar{G}^\circ_1$, $\Delta\bar{H}^\circ_1$, $\Delta\bar{S}^\circ_1$, and $\Delta\bar{C}^\circ_{p_1}$ for the transfer of gas from the vapor phase at 101,325 Pa partial gas pressure to the (hypothetical) solution phase of unit mole fraction. These thermodynamic properties are calculated from the smoothing equation according to the following equations:

$$\Delta\bar{G}^\circ_1 = - RAT - 100RB - RCT \ln (T/100K) \qquad (6)$$

$$\Delta\bar{S}^\circ_1 = RA + RC \ln (T/100) + RC \qquad (7)$$

$$\Delta\bar{H}^\circ_1 = - 100RB + RCT \qquad (8)$$

$$\Delta\bar{C}^\circ_{p_1} = RC \qquad (9)$$

The results from three other workers was rejected for various reasons. The single point of Hofbauer's (4) was high by about 7%. Hofmann's seven values (5) were erratically high and low between 1 and 11%. Ramstedt's three measurements (6) were also high by about 2 to 4%. Although many references quote Valentiner (7), he used other people's data for his paper.

Figure 1 shows the temperature dependence of the solubility of radon in water. The curve was obtained from the smoothing equation. There appears to be a minimum at about 363K for the Ostwald coefficient. The mole fraction solubility minimum is at 371.45 according to the fitted equation although there is no minimum evident over the temperature range of the experimental measurements.

References

1. Boyle, R. W. Phil. Mag. 1911, 22, 840.

2. Kofler, M. Sitz. Akad. Wiss. Wien 1912, 121, 2169; Monatsh. 1913, 34, 389.

3. Szeparowicz, M. Sitz. Akad. Wiss. Wien 1920, 129, 437.

4. Hofbauer, G. Sitz. Akad. Wiss. Wien 1914, 123, 2001.

5. Hofmann, R. Phys. Z. 1905, 6, 337.

6. Ramstedt, E. J. Phys. Radium 1911, 8, 253.

7. Valentiner, S. Z. Physik. 1927, 42, 253.

COMPONENTS:	ORIGINAL MEASUREMENTS:
1. Radon-222; $^{222}_{86}$Rn; 14859-67-7 2. Water; H_2O; 7732-18-5	Hofmann, R. Phys. Z. 1905, 6, 337-340.
VARIABLES: T/K: 273.15 - 353.15	PREPARED BY: R. Battino

EXPERIMENTAL VALUES:

T/K	Mol Fraction $X_1 \times 10^4$	Ostwald Coefficient[a] L
273.15	4.178	0.52
276.15	1.947	0.245
293.15	1.725	0.23
313.15	1.201	0.17
333.15	0.905	0.135
343.15	0.785	0.12
353.15	0.768	0.12

[a] Ostwald coefficient. Since the original measurements were made in terms of radon concentrations in the gas and liquid phase, it is assumed that there is a negligible volume change on mixing.

The mole fraction solubility at 101.325 Pa partial pressure of radon was calculated by the compiler.

AUXILIARY INFORMATION

METHOD/APPARATUS/PROCEDURE:	SOURCE AND PURITY OF MATERIALS:
Radon in a carrier gas such as air is equilibrated with water and counts taken of each phase to determine the radon concentrations and hence the Ostwald coefficient. The Rn partial pressure was of the order of 10^{-4} mmHg.	1. Radon 2. Water
	ESTIMATED ERROR: $\delta X_1/X_1 = 0.05$ (compiler)
	REFERENCES:

COMPONENTS:	ORIGINAL MEASUREMENTS:
1. Radon-222; $^{222}_{86}$Rn; 14859-67-7 2. Water; H_2O; 7732-18-5	Ramstedt, E. J. Phys. Radium 1911, 8, 253-256.
VARIABLES: T/K: 273.15 - 291.15	PREPARED BY: R. Battino

EXPERIMENTAL VALUES:

T/K	Mol Fraction X_1 x 10^4	Ostwald Coefficient[a] L
273.15	4.178	0.52
282.15	3.034	0.39
291.15	2.152	0.285

[a]Ostwald coefficient. Since the original measurements were made in terms of radon concentrations in the gas and liquid phase, it is assumed that there is a negligible volume change on mixing.

The mole fraction solubility at 101.325 Pa partial pressure of radon was calculated by the compiler.

AUXILIARY INFORMATION

METHOD/APPARATUS/PROCEDURE:	SOURCE AND PURITY OF MATERIALS:
Radon in a carrier gas such as air is equilibrated with water and counts taken to determine the radon concentration in each phase and hence the Ostwald coefficient. The radon partial pressure is of the order of 10^{-4} mm Hg.	1. Radon - 2. Water -
	ESTIMATED ERROR:
	REFERENCES:

COMPONENTS:	ORIGINAL MEASUREMENTS:
1. Radon-222; $^{222}_{86}$Rn; 14859-67-7 2. Water; H_2O; 7732-18-5	Boyle, R. W. Phil. Mag. 1911, 22, 840-854.
VARIABLES: T/K: 273.15 - 312.25	PREPARED BY: R. Battino

EXPERIMENTAL VALUES:

T/K	Mol Fraction X_1 x 10^4	Ostwald Coefficient[a] L
273.15	4.066	0.506
277.45	3.354	0.424
278.85	3.133	0.398
283.15	2.636	0.340
287.15	2.318	0.303
290.75	2.117*	0.280
293.15	1.838	0.245
299.95	1.513	0.206
304.75	1.397*	0.193
307.95	1.262	0.176
308.35	1.218	0.170
312.25	1.133	0.160

[a] Ostwald coefficient. Since original measurements were made in terms of concentrations this assumes that the volume of the solvent and the solution are identical within experimental error.

* solubility values which were used in the final smoothing equation for the recommended values given in the critical evaluation.

The mole fraction solubility at 101.325 Pa partial pressure of radon was calculated by the compiler.

AUXILIARY INFORMATION

METHOD /APPARATUS/PROCEDURE:	SOURCE AND PURITY OF MATERIALS:
Radon in a carrier gas such as air is equilibrated with the degassed water. Counts are made of the saturated solution and the equilibrated gas phase to determine the Ostwald coefficient. The Rn partial pressure ranged from 0.00001 to 0.0001 mm Hg.	1. Radon - 2. Water -
	ESTIMATED ERROR:
	REFERENCES:

| COMPONENTS: | ORIGINAL MEASUREMENTS: |

COMPONENTS:

1. Radon; $^{222}_{86}$Rn; 14859-67-7

2. Water; H_2O; 7732-18-5

3. Ferrous Sulfate; $FeSO_4$; 7720-78-7

ORIGINAL MEASUREMENTS:

Kofler, M.

Physik. Z. 1908, 9, 6-8.

VARIABLES:

T/K: 291.15

Specific Gravity: 1.226

PREPARED BY:

W. Gerrard

EXPERIMENTAL VALUES:

T/K	Specific Gravity	Ferrous Sulfate mol kg^{-1} H_2O	Ostwald Coefficient L
291.15	–	0.0	0.280
	1.226	1.58	0.081

The solubility in water was interpolated from the author's data from reference (1).

The ferrous sulfate molality was calculated by the Editor from data taken from the International Critical Tables, Vol. III.

The author reported his solubility values as an Absorption Coefficient = (Concentration of Rn in the liquid phase)/(Concentration in the gas phase). We have labelled it as an Ostwald coefficient. The solubility was measured at a radon partial pressure of less than 0.1 kPa at equilibrium. The liquid was also saturated with air as a carrier gas at a pressure of about 100 kPa.

AUXILIARY INFORMATION

METHOD:

Measurement of radioactivity in the liquid and gaseous phases. Described by Kofler (1,2), who gave a diagram.

SOURCE AND PURITY OF MATERIALS:

Not specified

APPARATUS/PROCEDURE:

The apparatus consisted of two glass bulbs connected by a tap. Each bulb had a separate tap. The assembly is shaken. Radioactivity is measured by an aluminum leaf electroscope.

ESTIMATED ERROR:

REFERENCES:

1. Kofler, M.
    Sitzungsber. Akad. Wiss. Wien, Abt. 2A 1912, 121, 2169.

2. Kofler, M.
    Monatsh. 1913, 34, 389.

COMPONENTS:	ORIGINAL MEASUREMENTS:
1. Radon; $^{222}_{86}$Rn; 14859-67-7	Kofler, M.
2. Water; H_2O; 7732-18-5	
3. Barium Nitrate; $Ba(NO_3)_2$; 10022-31-8	Sitzungsber. Akad. Wiss. Wien, Math.-Naturwiss. Kl. Abt. 2A 1913, 122, 1473 - 1479.

VARIABLES:	PREPARED BY:
T/K: 273.45 - 314.55 Specific Gravity: 1.006 - 1.0624	W. Gerrard

EXPERIMENTAL VALUES:

$Ba(NO_3)_2$

T/K	Specific* Gravity	Barium Nitrate mol kg^{-1} H_2O	Ostwald Coefficient L
273.45	–	0.0	0.530
279.35			0.418
286.55			0.324
303.45			0.202
314.55			0.161
274.75	1.006	0.0346	0.452
288.15			0.274
300.65			0.198
273.45	1.012	0.0630	0.474
279.35			0.368
286.55			0.285
303.45			0.182
278.25	1.036	0.179	0.366
286.95			0.254
314.55			0.142
281.55	1.055	0.271	0.310
287.75	1.0624	0.298	0.242
300.65			0.177

*All specific gravities were measured at 291.15 K.

The comments on water solubility, salt molality and absorption coefficient on the other Kofler data sheets apply.

AUXILIARY INFORMATION

METHOD:	SOURCE AND PURITY OF MATERIALS:
Measurement of radioactivity in the liquid and gaseous phases. Described by Kofler (1,2), who gave a diagram.	Not specified

APPARATUS/PROCEDURE:	ESTIMATED ERROR:
The apparatus consisted of two glass bulbs connected by a tap. Each bulb had a separate tap. The assembly is shaken. Radioactivity is measured by an aluminum leaf electroscope.	
	REFERENCES:
	1. Kofler, M. Sitzungsber. Akad. Wiss. Wien, Abt. 2A 1912, 121, 2169. 2. Kofler, M. Monatsh. 1913, 34, 389.

COMPONENTS:	ORIGINAL MEASUREMENTS:
1. Radon; $^{222}_{86}$Rn; 14859-67-7 2. Water; H_2O; 7732-18-5 3. Sodium Chloride; NaCl; 7647-14-5	Kofler, M. Sitzungsber. Akad. Wiss. Wien, <u>Math.-Naturwiss. Kl. Abt. 2A</u> 1913, <u>122</u>, 1473 - 1479.

VARIABLES:	PREPARED BY:
T/K: 273.35 - 290.15 Specific Gravity: 1.008 - 1.215	W. Gerrard

EXPERIMENTAL VALUES:

NaCl

T/K	Specific* Gravity	Sodium Chloride mol kg^{-1} H_2O	Ostwald Coefficient L
273.35	–	0.0	0.532
273.65			0.526
278.95			0.423
280.35			0.403
282.95			0.366
284.65			0.345
291.15			0.280
298.35			0.229
305.35			0.194
323.65			0.140
291.15	1.008	0.220	0.239
273.65	1.009	0.245	0.417
291.15	1.021	0.542	0.202
291.15	1.039	1.00	0.163
291.15	1.0392	1.01	0.163
273.95	1.071	1.87	0.220
279.65			0.176
285.95			0.154
291.15	1.096	2.59	0.096

*All specific gravity measurements were at 291.15 K except the value 1.145 which was at 289.15 K.

AUXILIARY INFORMATION

METHOD:	SOURCE AND PURITY OF MATERIALS:
Measurement of radioactivity in the liquid and gaseous phases. Described by Kofler (1,2), who gave a diagram.	Not specified

APPARATUS/PROCEDURE:	ESTIMATED ERROR:
The apparatus consisted of two glass bulbs connected by a tap. Each bulb had a separate tap. The assembly is shaken. Radioactivity is measured by an aluminum leaf electroscope. The comments on water solubility, salt molality, and absorption coefficients on the other Kofler data sheets apply.	

REFERENCES:

1. Kofler, M. <u>Sitzungsber. Akad. Wiss. Wien,</u> <u>Abt. 2A</u> 1912, <u>121</u>, 2169.

2. Kofler, M. <u>Monatsh.</u> 1913, <u>34</u>, 389.

COMPONENTS:	ORIGINAL MEASUREMENTS:
1. Radon; $^{222}_{86}$Rn; 14859-67-7 2. Water; H_2O; 7732-18-5 3. Sodium Chloride; NaCl; 7647-14-5	Kofler, M. Sitzungsber. Akad. Wiss. Wien, Math.-Naturwiss. Kl. Abt. 2A 1913, 122, 1473 - 1479.

VARIABLES:
    T/K: 273.35 - 290.15
 	
Specific Gravity: 1.008 - 1.215
	PREPARED BY:
	        W. Gerrard

EXPERIMENTAL VALUES:
NaCl

T/K	Specific* Gravity	Sodium Chloride mol kg^{-1} H_2O	Ostwald Coefficient L
274.95	1.103	2.80	0.157
277.15			0.145
282.65			0.122
286.95			0.110
288.65			0.108
297.35			0.084
298.85			0.079
291.15	1.121	3.35	0.077
273.35	1.145	4.05	0.100
273.65			0.100
278.95			0.090
280.35			0.089
282.95			0.086
284.65			0.084
298.35			0.062
305.35			0.0575
323.65			0.049
291.15	1.203	6.20**	0.042
280.35	1.215	6.65**	0.048
283.15			0.043
290.15			0.042

*All specific gravity measurements were at 291.15 K except the value 1.145
which was at 289.15 K.
**greater than saturation (?).

AUXILIARY INFORMATION

METHOD:
 	
        See preceding page.
	SOURCE AND PURITY OF MATERIALS:
	
	        See preceding page

APPARATUS/PROCEDURE:
 	
        See preceding page.
	ESTIMATED ERROR:
	
	REFERENCES:
	
	        See preceding page.

COMPONENTS:	ORIGINAL MEASUREMENTS:
1. Radon; $^{222}_{86}$Rn; 14859-67-7 2. Water; H_2O; 7732-18-5 3. Tetrapotassium, (OC-6-11)-hexa-kis (cyano-C)-Ferrate (4-) (Potassium Ferrocyanide), $K_4Fe(CN)_6$; 13943-58-3	Kofler, M. Physik. Z. 1908, 9, 6-8.

VARIABLES:	PREPARED BY:
T/K: 291.15 Specific Gravity: 1.107	W. Gerrard

EXPERIMENTAL VALUES:

T/K	Specific Gravity	Potassium Ferrocyanide mol kg^{-1} H_2O	Ostwald Coefficient L
291.15	–	0.0	0.280
	1.107	0.490	0.102

The solubility in water was interpolated from the author's data from reference (1).

The potassium ferrocyanide molality was calculated by the Editor from data taken from the International Critical Tables, Vol. III.

The author reported his solubility values as an Absorption Coefficient = (Concentration of Rn in the liquid phase)/(Concentration in the gas phase). We have labelled it as an Ostwald coefficient. The solubility was measured at a radon partial pressure of less than 0.1 kPa at equilibrium. The liquid was also saturated with air as a carrier gas at a pressure of about 100 kPa.

AUXILIARY INFORMATION

METHOD:	SOURCE AND PURITY OF MATERIALS:
Measurement of radioactivity in the liquid and gaseous phases. Described by Kofler (1,2), who gave a diagram.	Not specified

APPARATUS/PROCEDURE:	ESTIMATED ERROR:
The apparatus consisted of two glass bulbs connected by a tap. Each bulb had a separate tap. The assembly is shaken. Radioactivity is measured by an aluminum leaf electroscope.	
	REFERENCES: 1. Kofler, M. Sitzungsber. Akad. Wiss. Wien, Abt. 2A 1912, 121, 2169. 2. Kofler, M. Monatsh. 1913, 34, 389.

COMPONENTS:	ORIGINAL MEASUREMENTS:
1. Radon; $^{222}_{86}$Rn; 14859-67-7 2. Water; H_2O; 7732-18-5 3. Potassium Permanganate; $KMnO_4$; 7722-64-7	Kofler, M. Physik. Z. 1908, 9, 6-8.

VARIABLES:
    T/K: 291.15
 Specific Gravity: 1.038
	PREPARED BY:
	    W. Gerrard

EXPERIMENTAL VALUES:

T/K	Specific Gravity	Potassium permanganate mol kg^{-1} H_2O	Ostwald Coefficient L
291.15	–	0.0	0.280
	1.038	0.380	0.194

The solubility in water was interpolated from the author's data from reference (1).

The potassium permanganate molality was calculated by the Editor from data taken from the International Critical Tables, Vol. III.

The author reported his solubility values as an Absorption Coefficient = (Concentration of Rn in the liquid phase)/(Concentration in the gas phase). We have labelled it as an Ostwald coefficient. The solubility was measured at a radon partial pressure of less than 0.1 kPa at equilibrium. The liquid was also saturated with air as a carrier gas at a pressure of about 100 kPa.

AUXILIARY INFORMATION

METHOD:	SOURCE AND PURITY OF MATERIALS:
Measurement of radioactivity in the liquid and gaseous phases. Described by Kofler (1,2), who gave a diagram.	Not specified

APPARATUS/PROCEDURE:	ESTIMATED ERROR:
The apparatus consisted of two glass bulbs connected by a tap. Each bulb had a separate tap. The assembly is shaken. Radioactivity is measured by an aluminum leaf electroscope.	

	REFERENCES:
	 1. Kofler, M.
	      Sitzungsber. Akad. Wiss. Wien,
	      Abt. 2A 1912, 121, 2169.
	
	 2. Kofler, M.
	      Monatsh. 1913, 34, 389.

| COMPONENTS: | ORIGINAL MEASUREMENTS: |

COMPONENTS:

1. Radon; $^{222}_{86}$Rn; 14859-67-7

2. Water; H_2O; 7732-18-5

3. Potassium Chloride; KCl; 7447-40-7

ORIGINAL MEASUREMENTS:

Kofler, M.

<u>Physik</u>. <u>Z</u>. 1908, <u>9</u>, 6 - 8.

VARIABLES:

T/K: 291.15

Specific Gravity: 1.000 - 1.175

PREPARED BY:

W. Gerrard

EXPERIMENTAL VALUES:

T/K	Specific Gravity	Potassium Chloride mol kg^{-1} H_2O	Ostwald Coefficient L
291.15	1.000	0.0	0.280
	1.028	0.641	0.187
	1.0449	1.032	0.163
	1.069	1.62	0.133
	1.100	2.42	0.114
	1.175	4.63	0.061

The solubility in water was interpolated from the author's data from reference (1).

The potassium chloride molalities were calculated by the Editor from data taken from the International Critical Tables, Vol. III.

The author reported his solubility values as an Absorption Coefficient = (Concentration of Rn in the liquid phase)/(Concentration in the gas phase). We have labelled it as an Ostwald coefficient. The solubility was measured at a radon partial pressure of less than 0.1 kPa at equilibrium. The liquid was also saturated with air as a carrier gas at a pressure of about 100 kPa.

AUXILIARY INFORMATION

METHOD:

Measurement of radioactivity in the liquid and gaseous phases. Described by Kofler (1,2), who gave a diagram.

SOURCE AND PURITY OF MATERIALS:

Not specified

APPARATUS/PROCEDURE:

The apparatus consisted of two glass bulbs connected by a tap. Each bulb had a separate tap. The assembly is shaken. Radioactivity is measured by an aluminum leaf electroscope.

ESTIMATED ERROR:

REFERENCES:

1. Kofler, M.
   <u>Sitzungsber</u>. <u>Akad</u>. <u>Wiss</u>. <u>Wien</u>, <u>Abt</u>. 2A 1912, <u>121</u>, 2169.

2. Kofler, M.
   <u>Monatsh</u>. 1913, <u>34</u>, 389.

COMPONENTS:	ORIGINAL MEASUREMENTS:

COMPONENTS:

1. Radon; $^{222}_{86}$Rn; 14859-67-7

2. Water; H_2O; 7732-18-5

3. Ethanol; C_2H_5OH; 64-17-5

ORIGINAL MEASUREMENTS:

Kofler, M.

Physik. Z. 1908, 9, 6-8.

VARIABLES:

T/K: 291.15

Specific Gravity: 0.800 - 0.977

PREPARED BY:

W. Gerrard

EXPERIMENTAL VALUES:

T/K	Specific Gravity	Ethanol Mol Fraction X_3	Ostwald Coefficient L
291.15	-	0.0	0.280
	0.977	0.061	0.300
	0.944	0.182	0.436
	0.919	0.268	0.765
	0.885	0.403	1.301
	0.800	0.928	5.606

The solubility in water was interpolated from the author' data from reference (1).

The ethanol mole fractions were calculated by the Editor from data taken from the International Critical Tables, Vol. III.

The author reported his solubility values as an Absorption Coefficient = (Concentration of Rn in the liquid phase)/(Concentration in the gas phase). We have labelled it as an Ostwald coefficient. The solubility was measured at a radon partial pressure of less than 0.1 kPa at equilibrium. The liquid was also saturated with air as a carrier gas at a pressure of about 100 kPa.

AUXILIARY INFORMATION

METHOD:

Measurement of radioactivity in the liquid and gaseous phases. Described by Kofler (1,2), who gave a diagram.

SOURCE AND PURITY OF MATERIALS:

Not specified.

APPARATUS/PROCEDURE:

The apparatus consisted of two glass bulbs connected by a tap. Each bulb had a separate tap. The assembly is shaken. Radioactivity is measured by an aluminum leaf electroscope.

ESTIMATED ERROR:

REFERENCES:

1. Kofler, M.
Sitzungsber. Akad. Wiss. Wien, Abt. 2A 1912, 121, 2169.

2. Kofler, M.
Monatsh, 1913, 34, 389.

COMPONENTS:	ORIGINAL MEASUREMENTS:
1. Radon; $^{222}_{86}$Rn; 14859-67-7 2. Water; H_2O; 7732-18-5 3. Sucrose; $C_{12}H_{22}O_{11}$; 57-50-1	Kofler, M. Physik. Z. 1908, 9, 6-8.

VARIABLES:
      T/K: 291.15
 Specific Gravity: 1.082 - 1.214
	PREPARED BY:
	        W. Gerrard

EXPERIMENTAL VALUES:

T/K	Specific Gravity	Sucrose mol kg^{-1} H_2O	Ostwald Coefficient L
291.15	-	0.0	0.280
	1.082	0.744	0.188
	1.214	2.61	0.114

The solubility in water was interpolated from the author's data from reference (1).

The sucrose molalities were calculated by the Editor from data taken from the International Critical Tables, Vol. II, for 293.15 K.

The author reported his solubility values as an Absorption Coefficient = (Concentration of Rn in the liquid phase)/(Concentration in the gas phase). We have labelled it as an Ostwald coefficient. The solubility was measured at a radon partial pressure of less than 0.1 kPa at equilibrium. The liquid was also saturated with air as a carrier gas at a pressure of about 100 kPa.

AUXILIARY INFORMATION

METHOD:	SOURCE AND PURITY OF MATERIALS:
Measurement of radioactivity in the liquid and gaseous phases. Described by Kofler (1,2), who gave a diagram.	Not specified

APPARATUS/PROCEDURE:	ESTIMATED ERROR:
The apparatus consisted of two glass bulbs connected by a tap. Each bulb had a separate tap. The assembly is shaken. Radioactivity is measured by an aluminum leaf electroscope.	

	REFERENCES:
	 1. Kofler, M.
	     Sitzungsber. Akad. Wiss. Wien,
	     Abt. 2A 1912, 121, 2169.
	
	 2. Kofler, M.
	     Monatsh. 1913, 34, 389.

COMPONENTS:	ORIGINAL MEASUREMENTS:
1. Radon; $^{222}_{86}$Rn; 14859-67-7	Kofler, M.
2. Water; H_2O; 7732-18-5	
3. Urea; $(NH_2)_2CO$; 57-13-6	Sitzungsber. Akad. Wiss. Wien, Math.-Naturwiss. Kl. Abt. 2A 1913, 122, 1473 - 1479.

VARIABLES:	PREPARED BY:
T/K: 273.95 - 303.65 Specific Gravity: 1.004 - 1.070	W. Gerrard

EXPERIMENTAL VALUES:
Urea

T/K	Specific* Gravity	Urea mol kg^{-1} H$_2$O	Ostwald Coefficient L
273.95	−	0.0	0.519
278.65			0.429
286.75			0.322
298.75			0.227
306.65			0.188
289.55	1.004	0.336	0.270
307.15			0.179
276.75	1.012	0.862	0.412
288.95			0.273
306.65			0.181
273.95	1.016	1.125	0.439
288.35			0.273
305.35			0.187
273.95	1.029	2.095	0.418
278.65			0.354
286.75			0.278
298.75			0.210
306.65			0.174
273.75	1.070	5.850	0.370
287.95			0.255
303.65			0.187

*All of the specific gravities were measured
at 291.15 K except the value 1.016 which
was measured at 290.15 K.

AUXILIARY INFORMATION

METHOD:	SOURCE AND PURITY OF MATERIALS:
Measurement of radioactivity in the liquid and gaseous phases. Described by Kofler (1,2), who gave a diagram.	Not specified
	DATA CLASS:

APPARATUS/PROCEDURE:	ESTIMATED ERROR:
The apparatus consisted of two glass bulbs connected by a tap. Each bulb had a separate tap. The assembly is shaken. Radioactivity is measured by an aluminum leaf electroscope. The comments on water solubility, molality, and absorption coefficient on the other Kofler data sheets apply.	REFERENCES: 1. Kofler, M. Sitzungsber. Akad. Wiss. Wien, Abt. 2A 1912, 121, 2169. 2. Kofler, M. Monatsh. 1913, 34, 389.

COMPONENTS:	EVALUATOR:
1. Radon; $^{222}_{86}$Rn; 14859-67-7 2. Non-aqueous Liquids	William Gerrard Department of Chemistry The Polytechnic of North London Holloway, London N7 8DB U. K. May 1977

CRITICAL EVALUATION: PREAMBLE

The determination of the solubility of radon in liquids is based on a technique so different from those used for other gases that a special explanation is required. In the gaseous phase, the concentration of radon is absolutely small, the partial pressure of radon being less than 0.1 kPa at equilibrium. Furthermore, it is present in a carrier gas (nitrogen or air), the initial pressure of the carrier gas being about 100 kPa. The liquid is therefore saturated with the carrier gas at the equilibrium partial pressure of that gas. The concentration of radon in the gaseous phase, and that in the liquid phase are separately determined by the measurement of gamma-ray emission by means of an electroscope, a condenser, or a sodium iodide scintillation counter. Up to 1910, the electroscope technique was primitive, and apparently not reliable. In 1911, Boyle (1) described the procedure of holding the vessel containing one of the phases near enough to the electroscope to cause an effective ionization which could be compared with that produced by a standard amount of radium. Ramstedt (2) put the vessel into a condenser. Nussbaum and Hursh (3) however, used a sodium iodide scintillation counter.

The primary results were presented as the equivalent of a ratio of concentrations: concentration in the liquid phase/concentration in the gaseous phase; and this ratio was called the "absorption coefficient." It was actually stated, or tacitly assumed, that Henry's law was "obeyed," and that this ratio was independent of pressure. The ratio of concentrations was assumed to be equal to the volume of radon absorbed by an equal volume of the liquid phase, i.e. in effect the Ostwald absorption coefficient; but with the implication that the partial pressure of radon was 101.325 kPa. Nussbaum and Hursh (3), for example, used nitrogen containing a concentration of radon equivalent to 1μc per dm^3 of nitrogen. The radioactivity was assayed by counting the gamma-rays emitted by the short-lived decay products in radioactive equilibrium with radon. The ratio: (net counts per min./cm^3 of the original liquid)/(net counts per min./cm^3 of gas) was called the Ostwald solubility coefficient, and given the symbol α'.It must be emphasized that this ratio is pinned to a very small partial pressure of radon.

For the comparison of solubilities on a molecular basis, the absorption coefficients in their primary form may not be used; they must be converted into mole ratios and mole fractions. The main purpose in doing this conversion is to check the primary data by means of the reference line procedure described by Gerrard (4), and to make easier the comparison of radon solubility values with the solubility values of other gases. The following statements indicate the basis of the conversion.

        1.   The Ostwald coefficient is independent of pressure. In principle this is not valid; although deviations may be neglected in certain specific examples, they may not be in others.

        2.   The gram-mole volume of radon is taken to be 22,290 cm^3 at 273.15/K and a pressure, p_{Rn}, of 101.325 kPa. The gram-mole volume for other temperatures is given by 22,290 x T/273.15

        3.   The density, ρ_S, of the liquid, S, is that for the temperature of observation of the primary data.

        4.   The molecular weight of the liquid, S, is taken to be the simplest formula weight, e.g. C_2H_5OH (46.07) and CH_3CO_2H (60.05).

        5.   The mole ratio is given by: (absorption coefficient/gram-mole volume) x (M_S/ρ_S) where M_S is the molecular weight, and ρ_S is the density of S. The mole fraction X_{gas} obtained by (mole ratio)/(1 + mole ratio).

COMPONENTS:	EVALUATOR:
1. Radon; $^{222}_{86}$Rn; 14859-67-7 2. Non-aqueous Liquids	William Gerrard Department of Chemistry The Polytechnic of North London Holloway, London N7 8DB U. K. May 1977

CRITICAL EVALUATION: EVALUATION

There are two independent operations which have to be considered in assessing the reliability of the reported data.

(1) The determination of concentration of radon by the measurement of
gamma-ray emission is for a system in equilibrium at a partial pressure,
p_{Rn}, of radon less than 0.1 kPa, and with carrier gas, initially at a
pressure of about 100 kPa.

(2) The invocation of Henry's Law is a basis for accepting the concentration ratio for conditions under (1) as equal to the absorption coefficient
deemed independent of pressure, and therefore the same value for p_{Rn} =
101.325 kPa.

Evaluation of the primary data expressed as absorption coefficients for the
very small p_{Rn} values is a matter of assessing the accuracy of the measurement of radioactivity. The measurement of temperature is accurate enough
when this is specifically carried out. The pressure, p_{Rn}, is not stated,
and the assumption that the absorption coefficient is independent of pressure entails a decision identified as operation (2). The evaluator has
used his own reference line procedure (4) in evaluating the radon systems.

The assessment of reliability of the primarily observed absorption coefficient (for the very small p_{Rn}) cannot lead to the same grading as that
resulting from the assumption that the primary coefficient is the same at
p_{Rn} = 101.325 kPa. In the evaluation, the assessment is based on the
primary absorption coefficients for the small prevailing p_{Rn} values.
Nevertheless, the calculated mole fraction, x_{gas}, checked by the reference
line procedure of Gerrard (4), has been used as a guide to the reliability
of the primary data. It must be emphasised that none of the primary
measurements of radioactivity was carried out at a measured pressure, p_{Rn}.

All the primarily observed absorption coefficients of Boyle (1), Ramstedt
(2), and of Nussbaum and Hursh (3) are acceptable. The values by Hofbauer
(5) are much less reliable. Lurie's (6) values tend to be much too large,
and Traubenberg's (7) values are too small.

The publications (8, 9, 10, 11) have been cited as sources of data; but
they do not contain original data.

The values given by Gabrilova (12) for petroleum oils are not admissible;
because the systems were not in equilibrium. Onuma (13) reviewed the
extraction of radon by organic solvents. Krestov and Nedel'ko (14)
estimated solubility data for radon in 51 organic liquids from the basis
of data relating to the other noble gases. They also gave estimates for
solutions of the dihalogenobenzenes in xylene.

261

COMPONENTS:	EVALUATOR:
1. Radon; $^{222}_{86}$Rn; 14859-67-7 2. Non-aqueous Liquids	William Gerrard Department of Chemistry The Polytechnic of North London Holloway, London N7 8DB U. K.

CRITICAL EVALUATION: REFERENCES

1. Boyle, R.W. Phil. Mag. 1911, 22, 840.

2. Ramstedt, E. Le Radium 1911, 8, 253.

3. Nussbaum, E.; Hursh, J.B. J. Phys. Chem. 1958, 62, 81.

4. Gerrard, W. Solubility of Gases and Liquids - A Graphic Approach, Plenum, New York, 1976.

5. Hofbauer, G. Sitzungsber. Akad. Wiss. Wien, Abt. 2A, 1914, 123 2001.

6. Lurie, A. Thesis, University of Grenoble, 1910.

7. Traubenberg, H.F.R. Phys. Z. 1904, 5, 130.

8. Schulze, A. Z. Phys. Chem., 1920, 95, 257.

9. Swinne, R. Z. Phys. Chem., 1913, 84, 348.

10. Semenchenko, V.K. and Shakhparanov, M.I. Zh. Fiz. Khim., 1948, 22, 243.

11. Szeparowicz, M. Sitzungsber. Akad. Wiss. Wien Abt IIa, 1920, 129, 437.

12. Gavrilova, E.N. Trudy Vses. Nauch. Issles. Inst. Metro, 1939, 26, 39.

13. Onum, N. Bunseki Kagaku, 1967, 16, 274.

14. Krestov, G.A. and Nedel'ko, B.E. Tr. Ivanov. Khim. Tekhnol. Inst., 1972, 32, 102.

COMPONENTS:	ORIGINAL MEASUREMENTS:
1. Radon; $^{222}_{86}$Rn; 14859-67-7 2. Hexane; C_6H_{14}; 110-54-3	Ramstedt, E. Le Radium 1911, 8, 253 - 256.
VARIABLES: T/K: 255.15 - 291.15	PREPARED BY: W. Gerrard May 1977

EXPERIMENTAL VALUES:

T/K	Mol Fraction X_1 x 10^2	Ostwald Coefficient
255.15	17.4	35.2
273.15	11.8	23.4
291.15	8.32	16.56

The Ostwald coefficient was measured at a radon partial pressure of less than 0.1 kPa at equilibrium. The radon was present in a carrier gas (air or nitrogen) at an initial pressure of about 101 kPa.

The mole fraction solubility was calculated by the compiler. It was assumed that the Ostwald coefficient was independent of pressure and that the gram-mole volume of radon is 22,290 cm^3 at 273.15 K and 101.325 kPa.

Smoothed Data: ΔG^0/J mol^{-1} = -RT ln X_1 = -12,662 + 64.151 T

Std. Dev. ΔG = 6, Coef. Corr. = 0.9999

ΔH^0/J mol^{-1} = -12,662, ΔS^0/J $K^{-1}mol^{-1}$ = -64.151

T/K	Mol Fraction X x 10^2	ΔG^0/J mol^{-1}
258.15	16.3	3,898.5
268.15	13.1	4,540.0
278.15	10.6	5,181.5
288.15	8.80	5,823.0
298.15	7.37	6,463.0

The smoothed data fit was added by the Volume Editor.

AUXILIARY INFORMATION

METHOD: Measurement of radioactivity in the liquid and in the gaseous phase.	SOURCE AND PURITY OF MATERIALS: 1. Radon. 2. Hexane. Dried and distilled, b.p. 66 - 70 ^{o}C.
APPARATUS/PROCEDURE: Two tube connected by a wide tap. To determine concentration of radon, each tube is placed in a condenser.	ESTIMATED ERROR: REFERENCES:

COMPONENTS:	ORIGINAL MEASUREMENTS:
1. Radon; $^{222}_{86}$Rn; 14859-67-7 2. Cyclohexane; C_6H_{12}; 110-82-7	Ramstedt, E. Le Radium 1911, 8, 253-256.
VARIABLES: T/K: 291.15	PREPARED BY: W. Gerrard

EXPERIMENTAL VALUES:

T/K	Mol Fraction X_1 x 10^2	Ostwald Coefficient L
291.15	7.59	18.04

The Ostwald coefficient was measured at a radon partial pressure of less than 0.1 kPa at equilibrium. The radon was present in a carrier gas (air or nitrogen) at an initial pressure of about 101 kPa.

The mole fraction solubility was calculated by the compiler. It was assumed that the Ostwald coefficient was independent of pressure and that the gram-mole volume of radon is 22,290 cm^3 at 273.15 K and 101.325 kPa.

AUXILIARY INFORMATION

METHOD: Measurement of radioactivity in the liquid and in the gaseous phase.	SOURCE AND PURITY OF MATERIALS: 1. Radon. 2. Cyclohexane. Dried and distilled, b.p. 80°C.
APPARATUS/PROCEDURE: Two tubes connected by a wide tap. To determine concentration of radon, each tube is placed in a condenser.	ESTIMATED ERROR: REFERENCES:

COMPONENTS:	ORIGINAL MEASUREMENTS:
1. Radon; $^{222}_{86}$Rn; 14859-67-7 2. Benzene; C_6H_6; 71-43-2	Ramstedt, E. Le Radium 1911, 8, 253-256.
VARIABLES: T/K: 276.15 - 291.15	PREPARED BY: W. Gerrard

EXPERIMENTAL VALUES:

T/K	Mol Fraction X_1 x 10^2	Ostwald Coefficient L
276.15	6.16	16.54
291.15	4.57	12.82

The Ostwald coefficient was measured at a radon partial pressure of less
than 0.1 kPa at equilibrium. The radon was present in a carrier gas (air or
nitrogen) at an initial pressure of about 101 kPa.
The mole fraction solubility was calculated by the compiler. It was assumed
that the Ostwald coefficient was independent of pressure and that the gram-
mole volume of radon is 22,290 cm^3 at 273.15 K and 101.325 kPa.

Smoothed data: ΔG^O/J mol^{-1} = -RT ln X_1 = -13,306 + 71.355 T

ΔH^O/J mol^{-1} =-13,306, ΔS^O/J K^{-1} mol^{-1} = -71.355

T/K	Mol Fraction X_1 x 10^2	ΔG^O/J mol^{-1}
278.15	5.91	6,541.9
283.15	5.34	6,898.6
288.15	4.84	7,255.4
293.15	4.40	7,612.2
298.15	4.02	7,968.9

The smoothed data fit was added by the Volume Editor.

AUXILIARY INFORMATION

METHOD: Measurement of radioactivity in the liquid and in the gaseous phase.	SOURCE AND PURITY OF MATERIALS: 1. Radon. 2. Benzene. Dried and distilled.
APPARATUS/PROCEDURE: Two tubes connected by a wide tap. To determine concentration of radon, each tube is placed in a condenser.	ESTIMATED ERROR: REFERENCES:

COMPONENTS:	EVALUATOR:
1. Radon-222; $^{222}_{86}$Rn; 14859-67-7 2. Methylbenzene (Toluene); C_7H_8; 108-88-3	William Gerrard Department of Chemistry The Polytechnic of North London Holloway, London N7 8DB U.K. May 1977

CRITICAL EVALUATION:

The solubility of radon-222 in methylbenzene is reported by four laboratories. Hofman (1) reports radon solubilities at 194.15 and 293.15 K,
Lurie (2) at 293.15 K, Boyle (3) at 287.15 K and Ramstedt (4) at 255.15,
273.15 and 291.15 K.

The Hofman value at 194.15 K should be used with caution. If the Ostwald
coefficient at 194.15 K and an approximate radon pressure of 0.1 kPa has
the same value at 101.325 kPa the mole fraction solubility is 0.258 at
101.325 kPa radon pressure. However, the normal boiling point of radon is
211.4 K and thus the mole fraction solubility must, in effect, be unity
at 194.15 K and 101.325 kPa.

The Hofman and the Lurie Ostwald coefficients at 293.15 K agree well but
appear to be too small.

The Ostwald coefficient values of Ramstedt are probably the best available,
but they are classed as tentative. The Ramstedt data, smoothed by a linear
regression of a Gibbs energy equation linear in temperature, are 3.3 per
cent higher than the Boyle value at 287.15 K and about 8 per cent higher
than the Hofman and the Lurie values at 293.15 K.

The tentative values of the thermodynamic properties of solution for the
transfer of one mole of radon from the gas at 101.325 kPa to the hypothetical unit mole fraction solution are

$$\Delta G^{O}/J\ mol^{-1} = -RT\ ln\ X_1 = -12{,}673 + 67.505\ T$$

$$Std.\ Dev.\ \Delta G^{O} = 2.8,\ Coef.\ corr. = 0.9999$$

$$\Delta H^{O}/J\ mol^{-1} = -12{,}673,\ \Delta S^{O}/J\ K^{-1}\ mol^{-1} = -67.505$$

A table of tentative mole fraction solubility and Gibbs energy values as a
function of temperature appears on the Ramstedt data sheet, and below.

REFERENCES

1. Hofman, R. Phys. Z. 1905, 6, 339.

2. Lurie, A. Thesis 1910, University of Grenoble.

3. Boyle, R. W. Phil. Mag. 1911, 22, 840.

4. Ramstedt, E. Le Radium 1911, 8, 253.

TABLE 1. The solubility of radon-222 in methylbenzene. The tentative mole
fraction solubility of radon at a partial pressure of 101.325 kPa
(1 atm), and the Gibbs energy change as a function of temperature.

T/K	Mol Fraction $X_1 \times 10^2$	$\Delta G^{O}/J\ mol^{-1}$
258.15	10.92	4,752.5
263.15	9.76	5,090.0
268.15	8.76	5,427.5
273.15	7.90	5,765.1
278.15	7.14	6,102.6
283.15	6.49	6,440.1
288.15	5.91	6,777.6
293.15	5.40	7,116.1
298.15	4.95	7,453.6

COMPONENTS:	ORIGINAL MEASUREMENTS:
1. Radon-222; $^{222}_{86}$Rn; 14859-67-7 2. Methylbenzene (Toluene); C_7H_8;	Hofmann, R. Phys. Z. 1905, 6, 339-340.
VARIABLES: T/K: 194.15 - 293.15	PREPARED BY: W. Gerrard

EXPERIMENTAL VALUES:

T/K	Ostwald Coefficient L
194.15	66.69
293.15	11.79

The author reported an Absorption coefficient which appears equivalent to the Ostwald coefficient. The coefficient was measured at a radon-222 partial pressure of less than 0.1 kPa at equilibrium. The radon was present in a carrier gas (air or nitrogen) at an initial pressure of about 101 kPa.

AUXILIARY INFORMATION

METHOD: Determination of concentration of radon in the gas and liquid phases by means of radioactivity.	SOURCE AND PURITY OF MATERIALS: No specific information.
APPARATUS/PROCEDURE: Electroscope, and glass containers for measurement of radioactivity.	ESTIMATED ERROR: Of doubtful accuracy because of the primitive electroscope technique.
	REFERENCES: 1. Traubenberg, H.F.R. Phys. Z. 1904, 5, 130.

COMPONENTS:	ORIGINAL MEASUREMENTS:
1. Radon-222; $^{222}_{86}$Rn; 14859-67-7 2. Methylbenzene (Toluene); C_7H_8; 108-88-3	Lurie, A. Thesis University of Grenoble, 1910 Microfilm available. See also Tables annuelles de con- stantes et donnees numeriques de chemie, de physique et de technologie 1913 (for 1911), 2, 401.
VARIABLES: T/K: 293.15	PREPARED BY: W. Gerrard

EXPERIMENTAL VALUES:

T/K	Mol Fraction $X_1 \times 10^2$	Ostwald Coefficient L
293.15	4.97	11.76

The author reported a coefficient of absorption, α, which appears to have the same meaning as the ratio of concentrations: concentration of radon in the liquid/concentration of radon in the gas. The coefficient of absorption has been labelled as the Ostwald coefficient.

The coefficient of absorption was measured at a radon partial pressure of less then 0.1 kPa at equilibrium. The radon was present in a carrier gas (air or nitrogen) at an initial pressure of about 101 kPa.

The mole fraction solubility at 101.325 kPa was calculated by the compiler. It was assumed that the Ostwald coefficient was independent of pressure and that the gram-mole volume of radon is 22,290 cm^3 at 273.15 K and 101.325 kPa.

AUXILIARY INFORMATION

METHOD /APPARATUS/PROCEDURE:	SOURCE AND PURITY OF MATERIALS:
The concentration of radon in the gas and liquid phases determined by measurements of radioactivity. An aluminum foil electroscope was used to measure the radioactivity. Diagrams given by Lurie.	1. Radon. 2. Methylbenzene (Toluene). Purity of liquid not specified.
	ESTIMATED ERROR:
	REFERENCES:

COMPONENTS:	ORIGINAL MEASUREMENTS:
1. Radon-222; $^{222}_{86}$Rn; 14859-67-7 2. Methylbenzene (Toluene); C_7H_8; 108-88-3	Boyle, R.W. Phil. Mag. 1911, 22, 840-854.
VARIABLES: T/K: 287.15	PREPARED BY: W. Gerrard

EXPERIMENTAL VALUES:

T/K	Mol Fraction X_1 x 10^2	Ostwald Coefficient L
287.15	5.80	13.7

Boyle defined the coefficient of solubility, s, as $(e_1/V_1)/(e_2/V_2)$ where e_1 is emanation in volume V_1 of liquid, and e_2 is emanation in volume V_2 of gas. The partial pressure of radon is less than 0.1 kPa. The liquid is also saturated with carrier gas (air or another gas) at the prevailing pressure, originally about 101 kPa.

The mole fraction solubility at 101.325 kPa (1 atm) was calculated by the compiler. It was assumed that the Ostwald coefficient was independent of pressure and that the gram-mole volume of radon is 22,290 cm^3 at 273.15 K and 101.325 kPa.

AUXILIARY INFORMATION

METHOD:	SOURCE AND PURITY OF MATERIALS:
Radon-222 in a carrier gas (air, or other gas not specified) is shaken with a measured volume of liquid. Radioactivity is determined in the gas and liquid phases by gamma-ray electroscope.	1. Radon-222 2. Methylbenzene (Toluene).

APPARATUS/PROCEDURE:	ESTIMATED ERROR: $\delta T/K = 0.2$ $\delta L/L = 0.03$
The sampling bulb and mixing bulb are connected by a three-way tap. For the measurement of radioactivity, the bulbs are separately placed near to the electroscope. A diagram is given by Boyle.	REFERENCES:

COMPONENTS:	ORIGINAL MEASUREMENTS:
1. Radon; $^{222}_{86}$Rn; 14859-67-7 2. Methylbenzene (Toluene); C_7H_8; 108-88-3	Ramstedt, E. Le Radium 1911, <u>8</u>, 253-256.
VARIABLES: T/K: 255.15 - 291.15,	PREPARED BY: W. Gerrard

EXPERIMENTAL VALUES:

T/K	Mol Fraction X_1 x 10^2	Ostwald Coefficient L
255.15	11.7	27.0
273.15	7.91	18.4
291.15	5.59	13.24

The Ostwald coefficient was measured at a radon partial pressure of less than 0.1 kPa at equilibrium. The radon was present in a carrier gas (air or nitrogen) at an initial pressure of about 101 kPa.

The mole fraction solubility was calculated by the compiler. It was assumed that the Ostwald coefficient was independent of pressure and that the gram-mole volume of radon is 22,290 cm^3 at 273.15 K and 101.325 kPa.

Smoothed data: ΔG^o/J mol^{-1} = -RT ln X_1 = -12,673 + 67.505 T

Std. Dev. ΔG = 2.9, Coef. Corr. = 0.9999

ΔH^o/J mol^{-1} = -12,673, ΔS^o/J K^{-1} mol^{-1} = -67.505

T/K	Mol Fraction X_1 x 10^2	ΔG^o/J mol^{-1}
258.15	10.92	4,752.5
263.15	9.76	5,090.0
268.15	8.76	5,427.5
273.15	7.90	5,765.1
278.15	7.14	6,102.6
283.15	6.49	6,440.1
288.15	5.91	6,777.6
293.15	5.40	7,115.1
298.15	4.95	7,453.6

The smoothed data fit was added by the Volume Editor.

AUXILIARY INFORMATION

METHOD:	SOURCE AND PURITY OF MATERIALS:
Measurement of radioactivity in the liquid and in the gaseous phase.	1. Radon. 2. Methylbenzene (Toluene). Dried and distilled.
APPARATUS/PROCEDURE: Two tubes connected by a wide tap. To determine concentration of radon, each tube is placed in a condenser.	ESTIMATED ERROR:
	REFERENCES:

COMPONENTS:	ORIGINAL MEASUREMENTS:
1. Radon; $^{222}_{86}$Rn; 2. Dimethylbenzene (xylene); C_8H_{10}; 1330-20-7	Ramstedt, E. Le Radium 1911, 8, 253-256.
VARIABLES: T/K: 291.15	PREPARED BY: W. Gerrard

EXPERIMENTAL VALUES:

T/K	Mol Fraction $X_1 \times 10^2$	Ostwald Coefficient L
291.15	6.15	12.75

The Ostwald coefficient was measured at a radon partial pressure of less than 0.1 kPa at equilibrium. The radon was present in a carrier gas (air or nitrogen) at an initial pressure of about 101 kPa.

The mole fraction solubility was calculated by the compiler. It was assumed that the Ostwald coefficient was independent of pressure and that the gram-mole volume of radon is 22,290 cm^3 at 273.15 K and 101.325 kPa.

AUXILIARY INFORMATION

METHOD: Measurement of radioactivity in the liquid and in the gaseous phase.	SOURCE AND PURITY OF MATERIALS: 1. Radon. 2. Dimethylbenzene (xylene). Dried and distilled. Mixture of variable composition. 1,3-Dimethylbenzene 40 - 48 %, 1,2-dimethylbenzene 20 - 24 %, 1,4-dimethylbenzene 16 - 20 %, and ethylbenzene 10 - 15 %.
APPARATUS/PROCEDURE: Two tubes connected by a wide tap. To determine concentration of radon, each tube is placed in a condenser.	ESTIMATED ERROR:
	REFERENCES:

COMPONENTS:	ORIGINAL MEASUREMENTS:
1. Radon-222; $^{222}_{86}$Rn; 14859-67-7 2. Dimethylbenzene; C_8H_{10}; 1330-20-7	Lurie, A. Thesis University of Grenoble, 1910 Microfilm available. See also Tables annuelles de constantes et donnees numeriques de chemie, de physique et de technologie 1913 (for 1911), 2, 401.

VARIABLES:	PREPARED BY:
T/K: 253.15 - 343.15	W. Gerrard

EXPERIMENTAL VALUES:

T/K	Mol Fraction $X_1 \times 10^2$	Ostwald Coefficient L
253.15	-	27.55
273.15	-	19.56
291.15	7.32	15.4
293.15	-	15.1
323.15	-	7.2
343.15	-	4.8

The author reported a coefficient of absorption, α, which appears to have the same meaning as the ratio of concentrations: concentration of radon in the liquid/concentration of radon in the gas. The coefficient of absorption has been labelled as the Ostwald coefficient.

The coefficient of absorption was measured at a radon partial pressure of less than 0.1 kPa at equilibrium. The radon was present in a carrier gas (air or nitrogen) at an initial pressure of about 101 kPa.

The mole fraction solubility at 101.325 kPa was calculated by the compiler. It was assumed that the Ostwald coefficient was independent of pressure and that the gram-mole volume of radon is 22,290 cm^3 at 273.15 K and 101.325 kPa.

AUXILIARY INFORMATION

METHOD /APPARATUS/PROCEDURE:	SOURCE AND PURITY OF MATERIALS:
The concentration of radon in the gas and liquid phases determined by measurements of radioactivity. An aluminum foil electroscope was used to measure the radioactivity. Diagrams given by Lurie.	1. Radon 2. Dimethylbenzene. Purity of liquid not specified. Mixture of isomers. 1,3-Dimethylbenzene 40 - 48%, 1,2-dimethylbenzene 20 - 24%, 1,4-dimethylbenzene 16 - 20%, and ethylbenzene 10 - 15%.
	ESTIMATED ERROR:
	REFERENCES:

COMPONENTS:	ORIGINAL MEASUREMENTS:
1. Radon; $^{222}_{86}$Rn; 14859-67-7 2. Methanol; CH_4O; 67-56-1	Hofbauer, G. Sitzungsber. Akad. Wiss. Wien, <u>Math. Naturwiss. Kl. Abt. 2A</u> 1914, <u>123,</u> 2001 - 2009.
VARIABLES: T/K: 276.95 - 302.75	PREPARED BY: W. Gerrard

EXPERIMENTAL VALUES:

T/K	Mol Fraction $X_1 \cdot x\ 10^2$	Ostwald Coefficient L
276.95	1.32	7.43
289.35	0.980	5.71
302.75		(5.37)*

*Operational defect.

The author reported a coefficient of solubility, α', based on the radon concentration ratio in the liquid and gaseous phases at equilibrium by radioactivity. The coefficient of solubility is labelled an Ostwald coefficient above. It was measured at a radon partial pressure of probably less than 0.1 kPa at equilibrium. The liquid was also saturated with air as a carrier gas at an initial pressure of about 100 kPa.

The mole fraction solubility was calculated by the compiler. It was assumed that the Ostwald coefficient was independent of pressure and that the gram-mole volume of radon is 22,290 cm^3 at 273.15 K and 101.325 kPa.

Smoothed Data: $\Delta G^O/J\ mol^{-1} = -RT\ ln\ X_1 = -16,003 + 93.764\ T$

$\Delta H^O/J\ mol^{-1} = -16,003,\ \Delta S^O/J\ K^{-1}\ mol^{-1} = -93.764$

T/K	Mol Fraction $X_1\ x\ 10^2$	$\Delta G^O/J\ mol^{-1}$
273.15	1.45	9,608
278.15	1.28	10,077
283.15	1.13	10,546
288.15	1.01	11,014
293.15	0.90	11,484

The smoothed data fit was added by the Volume Editor.

AUXILIARY INFORMATION

METHOD/APPARATUS/PROCEDURE:

The method was based on the technique of Kofler (1,2). The radon radio-activity was measured in both the liquid and gaseous phases.

The apparatus consists of two glass bulbs connected by a tap. Each bulb carries two separate taps. Both Hofbauer and Kofler give a diagram.

SOURCE AND PURITY OF MATERIALS:

1. Radon.

2. Methanol. Purified by distillation. The density was given.

ESTIMATED ERROR:

REFERENCES:

1. Kofler, M.
   <u>Sitzungsber. Akad. Wiss. Wien</u>
   1912, <u>121,</u> 2169.

2. Kofler, M.
   <u>Monatsh.</u> 1913, <u>34,</u> 389.

COMPONENTS:	EVALUATOR:
1. Radon-222; $^{222}_{86}$Rn; 14859-67-7 2. Ethanol; C_2H_6O; 64-17-5	H. L. Clever Chemistry Department Emory University Atlanta, GA 30322 U.S.A. August 1978

CRITICAL EVALUATION:

Four laboratories report the solubility of radon-222 in ethanol. Traubenberg (1) reports a solubility relative to water at room temperature, Boyle (2) reports an Ostwald coefficient at 287.15 K, Ramstedt (3) reports Ostwald coefficients at 255.15, 273.15 and 291.15 K, and Hofbauer (4) reports four Ostwald coefficients between 275.15 and 310.95 K.

In the experience of the evaluator both Boyle's and Ramstedt's data are usually more reliable than Hofbauer's data. For the radon + ethanol system Hofbauer's solubility values are a consistent 15.0 to 16.5 percent higher than Ramstedt's values over the temperature range of common measurement of 273.15 to 298.15 K. For this system Boyle's single value agrees better with the Hofbauer data than the Ramstedt data. Although the evidence is conflicting, the evaluator prefers the Ramstedt data.

The tentative values of the thermodynamic changes for the transfer of one mole of radon from the gas at 101.325 kPa (1 atm) to the hypothetical unit mole fraction solution are the values from the Ramstedt data:

$$\Delta G^{o}/J\ mol^{-1} = -RT\ ln\ X_1 = -11,900 + 75.819\ T$$

Std. Dev. ΔG^{o} = 12.9, Coef. Corr. = 0.9998

$$\Delta H^{o}/J\ mol^{-1} = -11,900,\ \Delta S^{o}/J\ K^{-1}\ mol^{-1} = -75.819$$

A table of tentative mole fraction solubility and Gibbs energy values as a function of temperature at 5 degree interval between 258.15 and 298.15 K appears on the radon + ethanol data sheet of Ramstedt, and below.

REFERENCES

1. Traubenberg, H. F. R. Phys. Z. 1904, 5, 130.

2. Boyle, R. W. Phil. Mag. 1911, 22, 840.

3. Ramstedt, E. Le Radium 1911, 8, 253.

4. Hofbauer, G. Sitzungsber. Akad. Wiss. Wien. Math. Naturwiss. Kl. Abt. 2A 1914, 123, 2001.

TABLE 1. The solubility of radon-222 in ethanol.The tentative mole fraction solubility of radon at a partial pressure of 101.325 kPa (1 atm), and the Gibbs energy change on solution as a function of temperature.

T/K	Mol Fraction $X_1 \times 10^2$	$\Delta G^{o}/J\ mol^{-1}$
258.15	2.80	7,673.0
263.15	2.52	8,052.1
268.15	2.28	8,431.2
273.15	2.07	8,810.3
278.15	1.88	9,189.4
283.15	1.72	9,568.4
288.15	1.57	9,947.5
293.15	1.44	10,326
298.15	1.33	10,705

COMPONENTS:	ORIGINAL MEASUREMENTS:
1. Radon-222; $^{222}_{86}$Rn; 14859-67-7 2. Ethanol; C_2H_6O; 64-17-5	Traubenberg, H.F.R. Phys. Z. 1904, 5, 130 - 134.
VARIABLES:	PREPARED BY: W. Gerrard

EXPERIMENTAL VALUES:

T/K	Ratio compared with water*	Ostwald Coefficient L
"Room temperature"	16.1	4.83

*water was taken as 0.3

The author reported his solubility value as an absorption coefficient based
on the concentration ratio of radon in the gas and in the liquid phases.
It has been labelled as an Ostwald coefficient above. The solubility was
measured at a radon partial pressure of less than 0.1 kPa. The liquid was
also saturated with air as a carrier gas at a pressure of about 100 kPa.

AUXILIARY INFORMATION

METHOD: Measurement of radioactivity.	SOURCE AND PURITY OF MATERIALS: Not stated.
APPARATUS/PROCEDURE: Electroscope for measurement of radioactivity.	ESTIMATED ERROR: Traubenberg's values appear to be low when compared with values reported by other workers.
	REFERENCES:

COMPONENTS:	ORIGINAL MEASUREMENTS:
1. Radon-222; $^{222}_{86}Rn$; 14859-67-7 2. Ethanol; C_2H_6O; 64-17-5	Boyle, R. W. Phil. Mag. 1911, 22, 840- 854.
VARIABLES: T/K: 287.15	PREPARED BY: W. Gerrard

EXPERIMENTAL VALUES:

T/K	Mol Fraction X_1 x 10^2	Ostwald Coefficient L
287.15	1.81	7.34

Boyle defined the coefficient of solubility, s, as $(e_1/V_1)/(e_2/V_2)$ where e_1 is emanation in volume V_1 of liquid, and e_2 is emanation in volume V_2 of gas. The partial pressure of radon is less than 0.1 kPa. The liquid is also saturated with carrier gas (air or another gas) at the prevailing pressure, originally about 101 kPa.

The mole fraction solubility at 101.325 kPa (1 atm) was calculated by the compiler. It was assumed that the Ostwald coefficient was independent of pressure and that the gram-mole volume of radon is 22,290 cm^3 at 273.15 K and 101.325 kPa.

AUXILIARY INFORMATION

METHOD:	SOURCE AND PURITY OF MATERIALS:
Radon-222 in a carrier gas (air, or other gas not specified) is shaken with a measured volume of liquid. Radioactivity is determined in the gas and liquid phases by gamma-ray electroscope.	1. Radon-222. 2. Ethanol. Density at 287.15 K is 0.796 g cm^{-3}.

APPARATUS/PROCEDURE:	ESTIMATED ERROR:
The sampling bulb and mixing bulb are connected by a three-way tap. For the measurement of radioactivity, the bulbs are separately placed near to the electroscope. A diagram is given by Boyle.	$\delta T/K = 0.2$ $\delta L/L = 0.03$
	REFERENCES:

COMPONENTS:	ORIGINAL MEASUREMENTS:
1. Radon; $^{222}_{86}$Rn; 14859-67-7 2. Ethanol (Ethyl Alcohol); C_2H_6O; 64-17-5	Ramstedt, E. Le Radium 1911, 8, 253-256.
VARIABLES: T/K: 255.15 - 291.15	PREPARED BY: W. Gerrard

EXPERIMENTAL VALUES:

T/K	Mol Fraction X_1 x 10^2	Ostwald Coefficient L
255.15	2.98	11.4
273.15	2.08	8.28
291.15	1.49	6.17

The Ostwald coefficient was measured at a radon partial pressure of less than 0.1 kPa at equilibrium. The radon was present in a carrier gas (air or nitrogen) at an initial pressure of about 101 kPa.
The mole fraction solubility was calculated by the compiler. It was assumed that the Ostwald coefficient was independent of pressure and that the gram-mole volume of radon is 22,290 cm^3 at 273.15 K and 101.325 kPa.

Smoothed data: ΔG^O/J mol^{-1} = -RT ln X_1 = -11,900 + 75.819 T

Std. Dev. ΔG = 12.9, Coef. Corr. = 0.9999

ΔH^O/J mol^{-1} = -11,900, ΔS^O/J K^{-1} mol^{-1} = -75.819

T/K	Mol Fraction X_1 x 10^2	ΔG^O/J mol^{-1}
258.15	2.80	7,673.0
263.15	2.52	8,052.1
268.15	2.28	8,431.2
273.15	2.07	8,810.3
278.15	1.88	9,189.4
283.15	1.72	9,568.4
288.15	1.57	9,947.5
293.15	1.44	10,326
298.15	1.33	10,705

The smoothed data fit was added by the Volume Editor.

AUXILIARY INFORMATION

METHOD: Measurement of radioactivity in the liquid and in the gaseous phase.	SOURCE AND PURITY OF MATERIALS: 1. Radon. 2. Ethanol (Ethyl Alcohol). Dried and distilled.
APPARATUS/PROCEDURE: Two tubes connected by a wide tap. To determine concentration of radon, each tube is placed in a condenser.	ESTIMATED ERROR:
	REFERENCES:

COMPONENTS:	ORIGINAL MEASUREMENTS:

COMPONENTS:

1. Radon; $^{222}_{86}$Rn; 14859-67-7

2. Ethanol (Ethyl Alcohol); C_2H_6O; 64-17-5

ORIGINAL MEASUREMENTS:

Hofbauer, G.

Stizungsber. Akad. Wiss. Wien, Math. Naturwiss. Kl. Abt. 2A 1914, 123, 2001 - 2009.

VARIABLES:

T/K: 275.15 - 310.95

PREPARED BY:
W. Gerrard

EXPERIMENTAL VALUES:

T/K	Mol Fraction $X_1 \cdot x\ 10^2$	Ostwald Coefficient L
275.15	2.37	9.48
287.75	--	(7.42)*
288.65	1.76	7.23
288.85	--	(7.43)*
302.35	1.39	5.91
310.95	1.28	5.52

*Operational defect.

The mole fraction solubility was calculated by the compiler. It was assumed that the Ostwald coefficient was independent of pressure and that the gram-mole volume of radon is 22,290 cm^3 at 273.15 K and 101.325 kPa.

Smoothed data: ΔG^O/J mol^{-1} = -RT ln X$_1$ = -12,344 + 76.160 T

Std. Dev. ΔG^O = 70.6, Coef. Corr. = 0.9983

ΔH^O/J mol^{-1} = -12,344, ΔS^O/J K^{-1} mol^{-1} = -76.160

T/K	Mol Fraction X_1 x 10^2	ΔG^O/J mol^{-1}
273.15	2.41	8,459.2
278.15	2.19	8,840.1
283.15	1.99	9,220.9
288.15	1.82	9,601.7
293.15	1.66	9,982.4
298.15	1.53	10,363
303.15	1.41	10,744
308.15	1.30	11,125
313.15	1.20	11,506

AUXILIARY INFORMATION

METHOD/APPARATUS/PROCEDURE:
   The method was based on the technique of Kofler (1,2). The radon radioactivity was measured in both the liquid and gaseous phases.
   The apparatus consists of two glass bulbs connected by a tap. Each bulb carries two separate taps. Both Hofbauer and Kofler give a diagram.

The author reported a coefficient of solubility, α', based on the radon concentration ratio in the liquid and gaseous phases at equilibrium by radioactivity. The coefficient of solubility is labelled an Ostwald co-efficient above. It was measured at a radon partial pressure of probably less than 0.1 kPa at equilibrium. The liquid was also saturated with air as a carrier gas at an initial pressure of about 100 kPa.

SOURCE AND PURITY OF MATERIALS:

1.   Radon.

2.   Ethanol (Ethyl Alcohol). Purified by distillation. The density was given.

ESTIMATED ERROR:

REFERENCES:

1.   Kofler, M.
      Sitzungsber. Akad. Wiss. Wien
      1912, 121, 2169.

2.   Kofler, M.
      Monatsh. 1913, 34, 389.

COMPONENTS:	ORIGINAL MEASUREMENTS:
1. Radon; $^{222}_{86}$Rn; 14859-67-7 2. 1-Propanol; C_3H_8O; 71-23-8	Hofbauer, G. Sitzungsber. Akad. Wiss. Wien, Math. Naturwiss. Kl. Abt. 2A 1914, 123, 2001.
VARIABLES: T/K: 286.75 - 303.35	PREPARED BY: W. Gerrard

EXPERIMENTAL VALUES:

T/K	Mol Fraction $X_1 \times 10^2$	Ostwald Coefficient L
286.75	2.84	9.12
303.35	--	(7.69)*

*Operational defect.

The author reported a coefficient of solubility, α', based on the radon concentration ratio in the liquid and gaseous phases at equilibrium by radioactivity. The coefficient of solubility is labelled an Ostwald coefficient above. It was measured at a radon partial pressure of probably less that 0.1 kPa at equilibrium. The liquid was also saturated with air as a carrier gas at an initial pressure of about 100 kPa.

The mole fraction solubility was calculated by the compiler. It was assumed that the Ostwald coefficient was independent of pressure and that the gram-mole volume of radon is 22,290 cm^3 at 273.15 K and 101.325 kPa.

AUXILIARY INFORMATION

METHOD/APPARATUS/PROCEDURE:	SOURCE AND PURITY OF MATERIALS:
The method was based on the technique of Kofler (1,2). The radon radioactivity was measured in both the liquid and gaseous phases. The apparatus consists of two glass bulbs connected by a tap. Each bulb carries two separate taps. Both Hofbauer and Kofler give a diagram.	1. Radon. 2. 1-Propanol. Purified by distillation. The density was given.
	ESTIMATED ERROR:

	REFERENCES:
	
	1. Kofler, M.
	   Sitzungsber. Akad. Wiss. Wien
	   1912, 121, 2169.
	
	2. Kofler, M.
	   Monatsh. 1913, 34, 389.

COMPONENTS:	ORIGINAL MEASUREMENTS:
1. Radon; $^{222}_{86}$Rn; 14859-67-7 2. 2-Propanol; C_3H_8O; 67-63-0	Hofbauer, G. Sitzungsber. Akad. Wiss. Wien, Math. Naturwiss. Kl. Abt. 2A 1914, 123, 2001 - 2009.
VARIABLES: T/K: 287.45 - 300.35	PREPARED BY: W. Gerrard

EXPERIMENTAL VALUES:

T/K	Mol Fraction X_1 x 10^2	Ostwald Coefficient L
287.45	2.29	7.24
300.35	1.90	6.16

The author reported a coefficient of solubility, α', based on the radon concentration ratio in the liquid and gaseous phases at equilibrium by radioactivity. The coefficient of solubility is labelled an Ostwald Coefficient above. It was measured at a radon partial pressure of probably less than 0.1 kPa at equilibrium. The liquid was also saturated with air as a carrier gas at an initial pressure of about 100 kPa.

The mole fraction solubility was calculated by the compiler. It was assumed that the Ostwald coefficient was independent of pressure and that tha gram-mole volume of radon is 22,290 cm^3 at 273.15 K and 101.325 kPa.

Smoothed Data: $\Delta G^O/J\ mol^{-1} = -RT\ ln\ X_1 = -10,389 + 67.541\ T$

$\Delta H^O/J\ mol^{-1} = -10,389, \Delta S^O/J\ K^{-1}\ mol^{-1} = -67.541$

T/K	Mol Fraction X_1 x 10^2	$\Delta G^O/J\ mol^{-1}$
283.15	2.45	8,735
288.15	2.27	9,073
293.15	2.10	9,411
298.15	1.96	9,749
303.15	1.83	10,086

The smoothed data fit was added by the Volume Editor.

AUXILIARY INFORMATION

METHOD/APPARATUS/PROCEDURE:

The method was based on the technique of Kofler (1,2). The radon radioactivity was measured in both the liquid and gaseous phases.

The apparatus consists of two glass bulbs connected by a tap. Each bulb carries two separate taps. Both Hofbauer and Kofler give a diagram.

SOURCE AND PURITY OF MATERIALS:

1. Radon.

2. 2-Propanol. Purified by distillation. The density was given.

ESTIMATED ERROR:

REFERENCES:
1. Kofler, M. Sitzungsber. Akad. Wiss. Wien 1912, 121, 2169.

2. Kofler, M. Monatsh. 1913, 34, 389.

COMPONENTS:	ORIGINAL MEASUREMENTS:
1. Radon; $^{222}_{86}$Rn; 14859-67-7 2. 1-Butanol; $C_4H_{10}O$; 71-36-3	Hofbauer, G. Sitzungsber. Akad. Wiss. Wien, Math. Naturwiss. Kl. Abt. 2A 1914, 123, 2001 - 2009.

VARIABLES:
 	
    T/K: 284.35
	PREPARED BY:
	
	    W. Gerrard

EXPERIMENTAL VALUES:

T/K	Mol Fraction $X_1 \times 10^2$	Ostwald Coefficient L
284.35	3.72	9.79

The author reported a coefficient of solubility, α', based on the radon concentration ratio in the liquid and gaseous phases at equilibrium by radioactivity. The coefficient of solubility is labelled an Ostwald coefficient above. It was measured at a radon partial pressure of probably less than 0.1 kPa at equilibrium. The liquid was also saturated with air as a carrier gas at an initial pressure of about 100 kPa.

The mole fraction solubility was calculated by the compiler. It was assumed that the Ostwald coefficient was independent of pressure and that the gram-mole volume of radon is 22,290 cm^3 at 273.15 K and 101.325 kPa. The mole fraction solubility was calculated for a radon partial pressure of 101.325 kPa.

AUXILIARY INFORMATION

METHOD /APPARATUS/PROCEDURE:	SOURCE AND PURITY OF MATERIALS:
The method was based on the technique of Kofler (1,2). The radon radio-activity was measured in both the liquid and gaseous phases. The apparatus consists of two glass bulbs connected by a tap. Each bulb carries two separate taps. Both Hofbauer and Kofler give a diagram.	1. Radon. 2. 1-Butanol. Purified by distillation. The density was given.
	ESTIMATED ERROR:
	REFERENCES: 1. Kofler, M. Sitzungsber. Akad. Wiss. Wien 1912, 121, 2169. 2. Kofler, M. Monatsh. 1913, 34, 389.

COMPONENTS:	ORIGINAL MEASUREMENTS:

COMPONENTS:

1. Radon; $^{222}_{86}$Rn; 14859-67-7

2. 2-Butanol; $C_4H_{10}O$; 78-92-2

ORIGINAL MEASUREMENTS:

Hofbauer, G.

Sitzungsber. Akad. Wiss. Wien, Math. Naturwiss. Kl. Abt. 2A 1914, 123, 2001 - 2009.

VARIABLES:

T/K; 289.85

PREPARED BY:

W. Gerrard

EXPERIMENTAL VALUES:

T/K	Mol Fraction $X_1 \times 10^2$	Ostwald Coefficient L
289.85	2.86	7.58

The author reported a coefficient of solubility, α', based on the radon concentration ratio in the liquid and gaseous phases at equilibrium by radioactivity. The coefficient of solubility is labelled an Ostwald coefficient above. It was measured at a radon partial pressure of probably less than 0.1 kPa at equilibrium. The liquid was also saturated with air as a carrier gas at an initial pressure of about 100 kPa.

The mole fraction solubility was calculated by the compiler. It was assumed that the Ostwald coefficient was independent of pressure and that the gram-mole volume of radon is 22,290 cm^3 at 273.15 K and 101.325 kPa. The mole fraction solubility is for a radon partial pressure of 101.325 kPa.

AUXILIARY INFORMATION

METHOD/APPARATUS/PROCEDURE:

The method was based on the technique of Kofler (1,2). The radon radioactivity was measured in both the liquid and gaseous phases.

The apparatus consists of two glass bulbs connected by a tap. Each bulb carries two separate taps. Both Hofbauer and Kofler give a diagram.

SOURCE AND PURITY OF MATERIALS:

1. Radon.

2. 2-Butanol. Purified by distillation. The density was given.

ESTIMATED ERROR:

REFERENCES:

1. Kofler, M.
    Sitzungsber. Akad. Wiss. Wien 1912, 121, 2169.

2. Kofler, M.
    Monatsh. 1913, 34, 389.

K.X.R.—U

COMPONENTS:	ORIGINAL MEASUREMENTS:
1. Radon; $^{222}_{86}$Rn; 14859-67-7 2. 2-Methyl-1-Propanol; $C_4H_{10}O$; 78-83-1	Hofbauer, G. Sitzungsber. Akad. Wiss. Wien. Math. Naturwiss. Kl. Abt. 2A 1914, 123, 2001 - 2009.

VARIABLES:	PREPARED BY:
T/K: 276.55 - 305.15	W. Gerrard

EXPERIMENTAL VALUES:

T/K	Mol Fraction X_1 x 10^2	Ostwald Coefficient L
276.55	4.12	10.42
289.45	3.30	8.41
305.15		(6.63)*

*Operational defect.

The author reported a coefficient of solubility, α', based on the radon concentration ratio in the liquid and gaseous phases at equilibrium by radioactivity. The coefficient of solubility is labelled an Ostwald coefficient above. It was measured at a radon partial pressure of probably less than 0.1 kPa at equilibrium. The liquid was also saturated with air as a carrier gas at an initial pressure of about 100 kPa.

The mole fraction solubility was calculated by the compiler. It was assumed that the Ostwald coefficient was independent of pressure and that the gram-mole volume of radon is 22,290 cm^3 at 273.15 K and 101.325 kPa.

Smoothed Data: $\Delta G^O/J\ mol^{-1} = -RT\ \ln X_1 = -11.450 + 67.919\ T$

$\Delta H^O/J\ mol^{-1} = -11,450$, $\Delta S^O/J\ K^{-1}\ mol^{-1} = -67.919$

T/K	Mol Fraction X_1 x 10^2	$\Delta G^O/J\ mol^{-1}$	
273.15	4.38	7,102	
278.15	4.00	7,442	The smoothed data
283.15	3.67	7,782	fit was added by
288.15	3.37	8,121	the Volume Editor.
293.15	3.11	8,461	

AUXILIARY INFORMATION

METHOD/APPARATUS/PROCEDURE:	SOURCE AND PURITY OF MATERIALS:
The method was based on the technique of Kofler (1,2). The radon radioactivity was measured in both the liquid and gaseous phases. The apparatus consists of two glass bulbs connected by a tap. Each bulb carries two separate taps. Both Hofbauer and Kofler give a diagram. The mole fraction solubility values above are for a radon partial pressure of 101.325 kPa.	1. Radon. 2. 2-Methyl-1-propanol. Purified by distillation. The density was given.
	ESTIMATED ERROR:
	REFERENCES: 1. Kofler, M. Sitzungsber. Akad. Wiss. Wien 1912, 121, 2169. 2. Kofler, M. Monatsh. 1913, 34, 389.

283

COMPONENTS:	ORIGINAL MEASUREMENTS:
1. Radon; $^{222}_{86}$Rn; 14859-67-7 2. 3-Methyl-1-Butanol; $C_5H_{12}O$; 123-51-3	Hofbauer, G. Sitzungsber. Akad. Wiss. Wien, Math. Naturwiss. Kl. Abt. 2A 1914, 123, 2001 - 2009.

VARIABLES:	PREPARED BY:
T/K: 273.25 - 298.25	W. Gerrard

EXPERIMENTAL VALUES:

T/K	Mol Fraction $X_1 \times 10^2$	Ostwald Coefficient L
273.25	5.24	11.33
280.05	4.58	10.03
288.15	4.02	9.02
298.25	3.41	7.86

The author reported a coefficient of solubility, α', based on the radon concentration ratio in the liquid and gaseous phases at equilibrium by radioactivity. The coefficient of solubility is labelled an Ostwald coefficient above. It was measured at a radon partial pressure of probably less than 0.1 kPa at equilibrium. The liquid was also saturated with air as a carrier gas at an initial pressure of about 100 kPa.

The mole fraction solubility was calculated by the compiler. It was assumed that the Ostwald coefficient was independent of pressure and that the gram-mole volume of radon is 22,290 cm^3 at 273.15 K and 101.325 kPa.

Smoothed Data: $\Delta G^O/J\ mol^{-1} = -RT \ln X_1 = -11,544 + 66.799\ T$

Std. Dev. ΔG^O = 11.3, Coef. Corr. = 0.9999

$\Delta H^O/J\ mol^{-1} = -11,546$, $\Delta S^O/J\ K^{-1}\ mol^{-1} = -66.799$

T/K	Mol Fraction $X_1 \times 10^2$	$\Delta G^O/J\ mol^{-1}$
273.15	5.23	6,702.6
278.15	4.77	7,036.6
283.15	4.37	7,370.6
288.15	4.01	7,704.5
293.15	3.70	8,038.5
298.15	3.41	8,372.5

AUXILIARY INFORMATION

METHOD /APPARATUS/PROCEDURE:

The method was based on the technique of Kofler (1,2). The radon radioactivity was measured in both the liquid and gaseous phases.

The apparatus consists of two glass bulbs connected by a tap. Each bulb carries two separate taps. Both Hofbauer and Kofler give a diagram.

The mole fraction solubility values above were calculated for a radon partial pressure of 101.325 kPa.

SOURCE AND PURITY OF MATERIALS:

1. Radon.

2. 3-Methyl-1-Butanol. Purified by distillation. The density was given.

ESTIMATED ERROR:

REFERENCES:
1. Kofler, M.
   Sitzungsber. Akad. Wiss. Wien 1912, 121, 2169.

2. Kofler, M.
   Monatsh. 1913, 34, 389.

COMPONENTS:	ORIGINAL MEASUREMENTS:
1. Radon-222; $^{222}_{86}$Rn; 14859-67-7 2. 1-Pentanol (Amyl Alcohol); $C_5H_{12}O$; 71-41-0	Lurie, A. Thesis University of Grenoble, 1910 See also Tables annuelles de constantes et donnees numeriques de chemie, de physique et de technologie 1913 (for 1911), 2, 401.
VARIABLES: T/K: 291.15	PREPARED BY: W. Gerrard

EXPERIMENTAL VALUES:

T/K	Mol Fraction $X_1 \times 10^2$	Ostwald Coefficient L
291.15	4.72	10.6

The author reported a coefficient of absorption, α, which appears to have the same meaning as the ratio of concentrations: concentration of radon in the liquid/concentration of radon in the gas. The coefficient of absorption has been labelled as the Ostwald coefficient.

The coefficient of absorption was measured at a radon partial pressure of less than 0.1 kPa at equilibrium. The radon was present in a carrier gas (air or nitrogen) at an initial pressure of about 101 kPa.

The mole fraction solubility at 101.325 kPa was calculated by the compiler. It was assumed that the Ostwald coefficient was independent of pressure and that the gram-mole volume of radon is 22,290 cm^3 at 273.15 K and 101.325 kPa.

AUXILIARY INFORMATION

METHOD/APPARATUS/PROCEDURE:	SOURCE AND PURITY OF MATERIALS:
The concentration of radon in the gas and liquid phases determined by measurements of radioactivity. An aluminum foil electroscope was used to measure the radioactivity. Diagrams given by Lurie.	1. Radon. 2. 1-Pentanol (Amyl Alcohol). Purity of liquid not specified.
	ESTIMATED ERROR:
	REFERENCES:

285

COMPONENTS:	ORIGINAL MEASUREMENTS:

COMPONENTS:

1. Radon-222; $^{222}_{86}$Rn; 14859-67-7

2. 1-Pentanol (Amyl Alcohol); $C_6H_{12}O$; 71-41-0

ORIGINAL MEASUREMENTS:

Boyle, R.W.

Phil. Mag. 1911, 22, 840-854.

VARIABLES:

T/K: 287.15

PREPARED BY:

W. Gerrard

EXPERIMENTAL VALUES:

T/K	Mol Fraction $X_1 \times 10^2$	Ostwald Coefficient L
287.15	4.11	9.31

Boyle defined the coefficient of solubility, s, as $(e_1/V_1)/(e_2/V_2)$ where e_1 is emanation in volume V_1 of liquid, and e_2 is emanation in volume V_2 of gas. The partial pressure of radon is less than 0.1 kPa. The liquid is also saturated with carrier gas (air or another gas) at the prevailing pressure, originally about 101 kPa.

The mole fraction solubility at 101.325 kPa (1 atm) was calculated by the compiler. It was assumed that the Ostwald coefficient was independent of pressure and that the gram-mole volume of radon is 22,290 cm^3 at 273.15 K and 101.325 kPa.

AUXILIARY INFORMATION

METHOD:

Radon-222 in a carrier gas (air, or other gas not specified) is shaken with a measured volume of liquid.

Radioactivity is determined in the gas and liquid phases by gamma-ray electroscope.

SOURCE AND PURITY OF MATERIALS:

1. Radon-222.

2. 1-Pentanol (Amyl Alcohol).

APPARATUS/PROCEDURE:

The sampling bulb and mixing bulb are connected by a three-way tap. For the measurement of radioactivity, the bulbs are separately placed near to the electroscope. A diagram is given by Boyle.

ESTIMATED ERROR:

$\delta T/K = 0.2$
$\delta L/L = 0.03$

REFERENCES:

COMPONENTS:	ORIGINAL MEASUREMENTS:
1. Radon; $^{222}_{86}$Rn; 14859-67-7 2. 1,2,3-Propanetriol (Glycerol); $C_3H_8O_3$; 56-81-5	Ramstedt, E. Le Radium 1911, 8, 253-256.

VARIABLES: T/K: 291.15	PREPARED BY: W. Gerrard

EXPERIMENTAL VALUES:

T/K	Mol Fraction X_1 x 10^2	Ostwald Coefficient L
291.15	0.0645	0.21

The Ostwald coefficient was measured at a radon partial pressure of less than 0.1 kPa at equilibrium. The radon was present in a carrier gas (air or nitrogen) at an initial pressure of about 101 kPa.

The mole fraction solubility was calculated by the compiler. It was assumed that the Ostwald coefficient was independent of pressure and that the gram-mole volume of radon is 22,290 cm^3 at 273.15 K and 101.325 kPa. The mole fraction solubility is calculated for a radon partial pressure of 101.325 kPa.

AUXILIARY INFORMATION

METHOD: Measurement of radioactivity in the liquid and in the gaseous phase.	SOURCE AND PURITY OF MATERIALS: 1. Radon-222. 2. 1,2,3-Propanetriol (Glycerol). Dried and distilled.
APPARATUS/PROCEDURE: Two tubes connected by a wide tap. To determine concentration of radon, each tube is placed in a condenser.	ESTIMATED ERROR: REFERENCES:

COMPONENTS:	ORIGINAL MEASUREMENTS:
1. Radon-222; $^{222}_{86}$Rn; 14859-67-7 2. 1,2,3-Propanetriol (Glycerol); $C_3H_8O_3$; 56-81-5	Lurie, A. Thesis University of Grenoble, 1910 Microfilm available. See also Tables annuelles de constantes et donnees numeriques de chemie, de physique et de technologie 1913 (for 1911), <u>2</u>, 401.

VARIABLES:	PREPARED BY:
T/K: 276.15 - 323.15	W. Gerrard May 1977

EXPERIMENTAL VALUES:

T/K	Ostwald Coefficient L
276.15	2.87
291.15	1.7
303.15	0.6
323.15	0.09

The author reported a coefficient of absorption, α, which appears to have the same meaning as the ratio of concentrations: concentration of radon in the liquid/concentration of radon in the gas. The coefficient of absorption has been labelled as the Ostwald coefficient.

The coefficient of absorption was measured at a radon partial pressure of less than 0.1 kPa at equilibrium. The radon was present in a carrier gas (air or nitrogen) at an initial pressure of about 101 kPa.

AUXILIARY INFORMATION

METHOD/APPARATUS/PROCEDURE:	SOURCE AND PURITY OF MATERIALS:
The concentration of radon in the gas and liquid phases determined by measurements of radioactivity. An aluminum foil electroscope was used to measure the radioactivity. Diagrams given by Lurie.	1. Radon. 2. 1,2,3-Propanetriol (Glycerol). Purity of liquid not specified.
	DATA CLASS:
	ESTIMATED ERROR:
	REFERENCES:

288

COMPONENTS:	ORIGINAL MEASUREMENTS:
1. Radon-222; $^{222}_{86}$Rn; 14859-67-7 2. 1,1-Oxybisethane (Diethyl Ether); $C_4H_{10}O$; 60-29-7	Ramstedt, E. Le Radium 1911, 8, 253 - 256.
VARIABLES: T/K: 255.15 - 291.15	PREPARED BY: W. Gerrard

EXPERIMENTAL VALUES:

T/K	Mol Fraction $X_1 \times 10^2$	Ostwald Coefficient L
255.15	12.1	29.1
273.15	8.62	20.9
291.15	6.16	15.08

The Ostwald coefficient was measured at a radon partial pressure of less than 0.1 kPa at equilibrium. The radon was present in a carrier gas (air or nitrogen) at an initial pressure of about 101 kPa.

The mole fraction solubility of radon at a partial pressure of 101.325 kPa was calculated by the compiler. It was assumed that the Ostwald coefficient was independent of pressure and that the gram-mole volume of radon is 22,290 cm^3 at 273.15 K and 101.325 kPa.

Smoothed Data: $\Delta G^0/J\ mol^{-1} = -RT\ ln\ X_1 = -11,599 + 62.956\ T$

Std. Dev. $\Delta G^0 = 27.1$, Coef. corr. = 0.9997

$\Delta H^0/J\ mol^{-1} = -11,599$, $\Delta S^0/J\ K^{-1}\ mol^{-1} = -62.956$

T/K	Mol Fraction $X_1 \times 10^2$	$\Delta G^0/J\ mol^{-1}$
258.15	11.4	4,653.5
263.15	10.3	4,968.3
268.15	9.35	5,283.1
273.15	8.50	5,597.9
278.15	7.76	5,912.6
283.15	7.10	6,227.4
288.15	6.52	6,542.2
293.15	6.00	6,857.0
298.15	5.54	7,171.7

The smoothed data was added by the Volume Editor.

AUXILIARY INFORMATION

METHOD: Measurement of radioactivity in the liquid and in the gaseous phase.	SOURCE AND PURITY OF MATERIALS: 1. Radon-222. 2. 1,1-Oxybisethane. Dried and distilled.
APPARATUS/PROCEDURE: Two tubes connected by a wide tap. To determine concentration of radon, each tube is placed in a condenser.	ESTIMATED ERROR: REFERENCES:

COMPONENTS:	ORIGINAL MEASUREMENTS:
1. Radon; $^{222}_{86}$Rn; 14859-67-7 2. 2-Propanone (Acetone); C_3H_6O; 67-64-1	Ramstedt, E. Le Radium 1911, 8, 253-256.

VARIABLES:	PREPARED BY:
T/K: 255.15 - 291.15	W. Gerrard

EXPERIMENTAL VALUES:

T/K	Mol Fraction $X_1 \times 10^2$	Ostwald Coefficient L
255.15	3.49	10.8
273.15	2.50	7.99
291.15	1.89	6.30

The Ostwald coefficient was measured at a radon partial pressure of less than 0.1 kPa at equilibrium. The radon was present in a carrier gas (air or nitrogen) at an initial pressure of about 101 kPa.

The mole fraction solubility was calculated by the compiler. It was assumed that the Ostwald coefficient was independent of pressure and that the gram-mole volume of radon is 22,290 cm^3 at 273.15 K and 101.325 kPa.

Smoothed Data: ΔG^O/J mol^{-1} = -RT ln X_1 = -10,518 + 69.138 T

           Std. Dev. ΔG = 8.8, Coef. Corr. 1.000

      ΔH^O/J mol^{-1} = 10,518, ΔS^O/J K^{-1} mol^{-1} = -69.138

T/K	Mol Fraction $X_1 \times 10^2$	ΔG^O/J mol^{-1}
258.15	3.29	7,330.4
263.15	2.99	7,676.1
268.15	2.74	8,021.7
273.15	2.51	8,367.4
278.15	2.31	8,713.1
283.15	2.13	9,058.8
288.15	1.97	9,404.5
293.15	1.83	9,750.1
298.15	1.70	10,096

The smoothed data fit was added by the Volume Editor.

AUXILIARY INFORMATION

METHOD:	SOURCE AND PURITY OF MATERIALS:

Measurement of radioactivity in the liquid and in the gaseous phase.
 	
  The mole fraction solubility values were calculated for a radon partial pressure of 101.325 kPa.
	1. Radon-222.
	
	2. 2-Propanone (Acetone). Dried
	    and distilled.

APPARATUS/PROCEDURE:	ESTIMATED ERROR:
Two tubes connected by a wide tap. To determine concentration of radon, each tube is placed in a condenser.	
	REFERENCES:

COMPONENTS:	ORIGINAL MEASUREMENTS:
1. Radon; $^{222}_{86}$Rn; 14859-67-7 2. Formic Acid; CH_2O_2; 64-18-6	Nussbaum, E.; Hursh, J. B. J. Phys. Chem. 1958, 62, 81 - 84.

VARIABLES:	PREPARED BY:
T/K: 298.15 -,323.15	W. Gerrard

EXPERIMENTAL VALUES:

T/K	Mol Fraction X_1	Ostwald Coefficient L
298.15	0.00164	1.05
310.15	0.00145	0.96
323.15	0.00140	0.95

The Ostwald coefficient was measured at a very low partial pressure of radon. The carrier gas was nitrogen at a partial pressure of about 100 kPa. The liquid was therefore saturated with nitrogen at the experimental pressure.

The mole fraction solubility of radon at a pressure of 101.325 kPa was calculated by the compiler. It was assumed that the Ostwald coefficient was independent of pressure and that the gram-mole volume of radon is 22,290 cm^3 at 273.15 K and 101.325 kPa.

Smoothed Data: $\Delta G^0/J\ mol^{-1} = -RT\ \ln X_1 = -4,994.5 + 70.204\ T$
Std. Dev. $\Delta G^0 = 66$, Coef. Corr. $= 0.9972$
$\Delta H^0/J\ mol^{-1} = -4,994.5$, $\Delta S^0/J\ K^{-1}\ mol^{-1} = -70.204$

T/K	Mol Fraction	$\Delta G^0/J\ mol^{-1}$
298.15	0.00161	15,937
303.15	0.00156	16,287
308.15	0.00151	16,638
313.15	0.00147	16,990
318.15	0.00142	17,341
323.15	0.00138	17,692

The smoothed data fit was added by the Volume Editor.

AUXILIARY INFORMATION

METHOD:	SOURCE AND PURITY OF MATERIALS:
The concentration of radon was determined by measurement of radio-activity in sample withdrawn from the liquid and from the gas phases. The procedure was stated to be similar in principle to that of Boyle (1).	1. Radon. 2. Formic acid. Eastman Chemical Co. Highest grade.

APPARATUS/PROCEDURE:	ESTIMATED ERROR:
A cylindrical glass vessel with a stopcock at each end. The gamma-rays emitted were estimated by a sodium iodide scintillation counter.	
	REFERENCES: 1. Boyle, R. W. Phil. Mag. 1911, 22, 840.

COMPONENTS:	EVALUATOR:
1. Radon-222; $^{222}_{86}$Rn; 14859-67-7 2. Acetic Acid; $C_2H_4O_2$; 64-19-7	H. L. Clever Chemistry Department Emory University Atlanta, GA 30322 U.S.A. August 1978

CRITICAL EVALUATION:

Hofbauer (1) reported four values of the solubility of radon-222 in acetic acid between 290.65 and 300.25 K, and Nussbaum and Hursh (2) reported three values between 298.15 and 323.15 K. The two smoothed data sets differ by 9-10 percent over the five degree range of common measurement, with the Hofbauer data having the higher values.

The agreement between the two data sets is marginal but within reason, when one considers the 0.1 kPa pressure of the gas in the measurement and the adjustment to the 101.325 kPa pressure for the mole fraction calculation. The two data sets were combined with a weight of 1 to Hofbauer data and a weight of 2 to the Nussbaum and Hursh data in a linear regression of a Gibbs energy equation linear in temperature. The result gives the tentative values for the transfer of one mole of radon from the gas at a partial pressure of 101.325 kPa to the hypothetical unit mole fraction solution of

$$\Delta G^o / J \ mol^{-1} = -RT \ ln \ X_1 = -12,805 + 80.750 \ T$$

Std. Dev. ΔG^o = 121, Coef. corr. = 0.9916

$$\Delta H^o / J \ mol^{-1} = -12,805, \quad \Delta S^o / J \ K^{-1} \ mol^{-1} = -80.750$$

The tentative solubility and Gibbs energy values as a function of temperature are in Table 1.

TABLE 1. The solubility of radon-222 in acetic acid. Tentative values of the mole fraction solubility at 101.325 kPa and the Gibbs energy changes as a function of temperature.

T/K	Mol Fraction $X_1 \times 10^2$	$\Delta G^o / J \ mol^{-1}$
288.15	1.270*	10,462
293.15	1.160*	10,866
298.15	1.060*	11,270
303.15	0.974	11,674
308.15	0.897	12,077
313.15	0.828	12,481
318.15	0.767	12,885
323.15	0.711	13,289

*Rounded to the nearest 0.005×10^{-2}.

REFERENCES.

1. Hofbauer, G. Sitzugsber. Akad. Wiss. Wien, Math. Naturwiss. Kl. Abt. 2A 1914, 123, 2001.

2. Nussbaum, E.; Hursh, J. B. J. Phys. Chem. 1958, 62, 81.

COMPONENTS:	ORIGINAL MEASUREMENTS:
1. Radon; $^{222}_{86}$Rn; 14859-67-7 2. Acetic Acid; $C_2H_4O_2$; 64-19-7	Hofbauer, G. Sitzungsber. Akad. Wiss. Wien, Math. Naturwiss. Kl. Abt. 2A 1914, 123, 2001

VARIABLES:
 	
    T/K: 285.75 - 300.25
	PREPARED BY:
	
	        W. Gerrard

EXPERIMENTAL VALUES:

T/K	Mol Fraction $X_1 \times 10^2$	Ostwald Coefficient L
285.75	–	(4.83)[*]
290.65	1.20	5.01
294.15	1.19	4.98
298.95	1.10	4.72
300.25	1.06	4.53

[*]Operational defect.

The author reported a coefficient of solubility, α', based on the radon concentration ratio in the liquid and gaseous phases at equilibrium by radioactivity. The coefficient of solubility is labelled an Ostwald coefficient above. It was measured at a radon partial pressure of probably less than 0.1 kPa at equilibrium. The liquid was also saturated with air as a carrier gas at an initial pressure of about 100 kPa.

The mole fraction solubility was calculated by the compiler. It was assumed that the Ostwald coefficient was independent of pressure and that the gram-mole volume of radon is 22,290 cm^3 at 273.15 K and 101.325 kPa.

Smoothed Data: $\Delta G^0/J\ mol^{-1} = -RT\ \ln X_1 = -9.508.4 + 69.357\ T$
Std. Dev. $\Delta G^0 = 44.8$, Coef. Corr. = 0.9895
$\Delta H^0/J\ mol^{-1} = -9,508.4$, $\Delta S^0/J\ K^{-1}\ mol^{-1} = -69.357$

T/K	Mol Fraction $X_1 \times 10^2$	$\Delta G^0/J\ mol^{-1}$	
288.15	1.26	10,477	The smoothed data
293.15	1.18	10,823	fit was added by
298.15	1.10	11,170	the Volume Editor.
303.15	1.04	11,517	

AUXILIARY INFORMATION

METHOD:	SOURCE AND PURITY OF MATERIALS:
The method was based on the technique of Kofler (1,2). The radon radio-activity was measured in both the liquid and gaseous phases. The apparatus consists of two glass bulbs connected by a tap. Each bulb carries two separate taps. Both Hofbauer and Kofler give a diagram.	1. Radon. 2. Acetic acid. Purified by distillation. The density was given.
	ESTIMATED ERROR:

See the evaluation on page 291 for the tentative recommended thermodynamic and solubility values.	REFERENCES:
	1. Kofler, M.
	   Sitzungsber. Akad. Wiss. Wien
	   1912, 121, 2169.
	
	2. Kofler, M.
	   Monatsh. 1913, 34, 389.

COMPONENTS:	ORIGINAL MEASUREMENTS:
1. Radon; $^{222}_{86}$Rn; 14859-67-7 2. Acetic Acid; $C_2H_4O_2$; 64-19-7	Nussbaum, E.; Hursh, J. B. J. Phys. Chem. 1958, 62, 81 - 84.

VARIABLES:	PREPARED BY:
T/K: 298.15 - 323.15	W. Gerrard May 1977

EXPERIMENTAL VALUES:

T/K	Mol Fraction $X_1 \times 10^2$	Ostwald Coefficient L
298.15	1.04	4.43
310.15	0.806	3.53
323.15	0.736	3.30

The Ostwald coefficient was measured at a very low partial pressure of radon. The carrier gas was nitrogen at a partial pressure of about 100 kPa The liquid was therefore saturated with nitrogen at the experimental pressure.

The mole fraction solubility of radon at a pressure of 101.325 kPa was calculated by the compiler. It was assumed that the Ostwald coefficient was independent of pressure and that the gram-mole volume of radon is 22,290 cm^3 at 273.15 K and 101.325 kPa.

Smoothed Data: $\Delta G^\circ/J\ mol^{-1} = - RT\ \ln X_1 = -10,938 + 74.895\ T$
Std. Dev. $\Delta G^\circ = 122$, Coef. Corr. = 0.9916
$\Delta H^\circ/J\ mol^{-1} = -10,938$, $\Delta S^\circ/J\ K^{-1}\ mol^{-1} = -74.895$

T/K	Mol Fraction $X_1 \times 10^2$	$\Delta G^\circ/J\ mol^{-1}$	
298.15	1.01	11,392	The smoothed data
303.15	0.939	11,766	fit was added by
308.15	0.875	12,141	the Volume Editor
313.15	0.817	12,515	
318.15	0.765	12,890	
323.15	0.718	13,264	

AUXILIARY INFORMATION

METHOD:	SOURCE AND PURITY OF MATERIALS:
The concentration of radon was determined by measurement of radio-activity in samples withdrawn from the liquid and from the gas phases. The procedure was stated to be similar in principle to that of Boyle (1).	1. Radon. 2. Acetic Acid. Eastman Chemical Co. Highest grade.

APPARATUS/PROCEDURE:	ESTIMATED ERROR:
A cylindrical glass vessel with a stopcock at each end. The gamma-rays emitted were estimated by a sodium iodide scintillation counter. See the evaluation on page 291 for the tentative recommendation of thermodynamic and solubility values.	REFERENCES: 1. Boyle, R. W. Phil. Mag. 1911, 22, 840.

COMPONENTS:	EVALUATOR:
1. Radon-222; $^{222}_{86}$Rn; 14859-67-7 2. Propanoic Acid; $C_3H_6O_2$; 79-09-4	H. L. Clever Department of Chemistry Emory University Atlanta, GA 30322 U.S.A. August 1978

CRITICAL EVALUATION:

Hofbauer (1) reported values of the radon-222 solubility in propanoic acid at temperatures of 293.85 and 302.85 K, and Nussbaum and Hursh (2) reported three values between temperatures of 298.15 and 323.15 K.

The two data sets show more scatter than the same two laboratory's results for acetic acid. It is likely that the increase in solubility shown by the Nussbaum and Hursh data between temperatures of 310.15 and 323.15 K is in error. The 323.15 K solubility value was not used in the linear regression described below.

The two data sets were combined on a one to one weight bases in a linear regression of a Gibbs energy equation linear in temperature. The result gives the tentative values for the thermodynamic changes for the transfer of one mole of radon from the gas at a partial pressure of 101.325 kPa to the hypothetical unit mole fraction solution of

$$\Delta G^o / J \ mol^{-1} = -RT \ ln \ X_1 = -15,919 + 85.697 \ T$$

$$Std. \ Dev. \ \Delta G^o = 159.7, \ Coef. \ corr. = 0.9661$$

$$\Delta H^o / J \ mol^{-1} = -15,919, \ \Delta S^o / J \ K^{-1} \ mol^{-1} = -85.697$$

The tentative solubility and Gibbs energy values as a function of temperature are in Table 1.

TABLE 1. The solubility of radon-222 in propanoic acid. Tentative values of the mole fraction solubility at 101.325 kPa partial pressure radon and the Gibbs energy change as a function of temperature.

T/K	Mol Fraction X_1 x 10^2	$\Delta G^o / J \ mol^{-1}$
293.15	2.29	9,203.1
298.15	2.05	9,631.6
303.15	1.85	10,060
308.15	1.67	10,489
313.15	1.51	10,917
318.15	1.37	11,346
323.15	1.25	11,774

REFERENCES.

1. Hofbauer, G. Sitzugsber. Akad. Wiss. Wien, Math. Naturwiss. Kl. Abt. 2A 1914, 123, 2001.

2. Nussbaum, E.; Hursh, J. B. J. Phys. Chem. 1958, 62, 81.

COMPONENTS:	ORIGINAL MEASUREMENTS:
1. Radon; $^{222}_{86}$Rn; 14859-67-7 2. Propanoic Acid; $C_3H_6O_2$; 79-09-4	Hofbauer, G. Sitzungsber. Akad. Wiss. Wien, Math. Naturwiss. Kl. Abt. 2A 1914, 123, 2001.
VARIABLES: T/K: 283.05 - 302.85	PREPARED BY: W. Gerrard

EXPERIMENTAL VALUES:

T/K	Mol Fraction $X_1 \times 10^2$	Ostwald Coefficient L
283.05	--	(8.74)*
293.85	2.23	7.31
302.85	2.04	6.82

*Operational defect.

The author reported a coefficient of solubility, α', based on the radon concentration ratio in the liquid and gaseous phases at equilibrium by radioactivity. The coefficient of solubility is labelled an Ostwald coefficient above. It was measured at a radon partial pressure of probably less than 0.1 kPa at equilibrium. The liquid was also saturated with air as a carrier gas at an initial pressure of about 100 kPa.

The mole fraction solubility was calculated by the compiler. It was assumed that the Ostwald coefficient was independent of pressure and that the gram-mole volume of radon is 22,290 cm^3 at 273.15 and 101.325 kPa.

Smoothed Data: $\Delta G^O/J\ mol^{-1} = -RT\ ln\ X_1 = -7,321.1 + 56.535\ T$

$\Delta H^O/J\ mol^{-1} = -7,321.1, \quad \Delta S^O/J\ K^{-1}\ mol^{-1} = -56.535$

T/K	Mol Fraction $X_1 \times 10^2$	$\Delta G^O/J\ mol^{-1}$
293.15	2.25	9,252.2
298.15	2.14	9,534.8
303.15	2.03	9,817.5

The smoothed data fit was added by the Volume Editor.

AUXILIARY INFORMATION

METHOD /APPARATUS/PROCEDURE:	SOURCE AND PURITY OF MATERIALS:
The method was based on the technique of Kofler (1,2). The radon radio-activity was measured in both the liquid and gaseous phases. The apparatus consists of two glass bulbs connected by a tap. Each bulb carries two separate taps. Both Hofbauer and Kofler give a diagram.	1. Radon. 2. Propanoic Acid. Purified by distillation. The density was given.
	ESTIMATED ERROR:
See the evaluation on page 294 for the tentative recommendation of solubility and thermodynamic values.	REFERENCES: 1. Kofler, M. Sitzungsber. Akad. Wiss. Wien 1912, 121, 2169. 2. Kofler, M. Monatsh. 1913, 34, 389.

COMPONENTS:	ORIGINAL MEASUREMENTS:

COMPONENTS:

1. Radon; $^{222}_{86}$Rn; 14859-67-7

2. Propanoic Acid; $C_3H_6O_2$; 79-09-4

ORIGINAL MEASUREMENTS:

Nussbaum, E.; Hursh, J. B.

J. Phys. Chem. 1958, 62, 81 - 84.

VARIABLES:

T/K: 298.15 - 323.15

PREPARED BY:

W. Gerrard

EXPERIMENTAL VALUES:

T/K	Mol Fraction $X_1 \times 10^2$	Ostwald Coefficient L
298.15	1.97	6.52
310.15	1.54	5.23
323.15	1.57	5.47

The Ostwald coefficient was measured at a very low partial pressure of radon. The carrier gas was nitrogen at a partial pressure of about 100 kPa. The liquid was therefore saturated with nitrogen at the experimental pressure.

The mole fraction solubility of radon at a pressure of 101.325 kPa was calculated by the compiler. It was assumed that the Ostwald coefficient was independent of pressure and that the gram-mole volume of radon is 22,290 cm^3 at 273.15 K and 101.325 kPa.

Smoothed Data: $\Delta G°/J\ mol^{-1} = - RT\ \ln X_1 = -7,050.2 + 56.694$
   Std. Dev. $\Delta G° = 198$,   Coef. Corr. = 0.9634
   $\Delta H°/J\ mol^{-1} = -7,050.2$, $\Delta S°/J\ K^{-1}\ mol^{-1} = -56.694$

T/K	Mol Fraction $X_1 \times 10^2$	$\Delta G°/J\ mol^{-1}$
298.15	1.88	9,853.6
303.15	1.79	10,137
308.15	1.71	10,420
313.15	1.64	10,704
318.15	1.57	10,987
323.15	1.51	11,271

The smoothed data fit was added by the Volume Editor.

AUXILIARY INFORMATION

METHOD:

   The concentration of radon was determined by measurement of radioactivity in samples withdrawn from the liquid and from the gas phases. The procedure was stated to be similar in principle to that of Boyle (1).

SOURCE AND PURITY OF MATERIALS:

1. Radon.

2. Propanoic Acid. Eastman Chemical Co. Highest Grade.

APPARATUS/PROCEDURE:

   A cylindrical glass vessel with a stopcock at each end. The gamma-rays emitted were estimated by a sodium iodide scintillation counter.

   See the evaluation on page 294 for the tentative recommendation of thermodynamic and solubility values.

ESTIMATED ERROR:

REFERENCES:

1. Boyle, R. W.
   Phil. Mag. 1911, 22, 840.

COMPONENTS:	ORIGINAL MEASUREMENTS:
1. Radon; $^{222}_{86}$Rn; 14859-67-7 2. 2-Propenoic Acid (Acrylic Acid); $C_3H_4O_2$ (CH_2=$CHCO_2H$); 79-10-7	Nussbaum, E.; Hursh, J. B. J. Phys. Chem. 1958, 62, 81 - 84.

VARIABLES:	PREPARED BY:
T/K: 310.15	W. Gerrard

EXPERIMENTAL VALUES:

T/K	Mol Fraction $X_1 \times 10^2$	Ostwald Coefficient L
310.15	1.35	5.01

The Ostwald coefficient was measured at a very low partial pressure of radon. The carrier gas was nitrogen at a partial pressure of about 100 kPa. The liquid was therefore saturated with nitrogen at the experimental pressure.

The mole fraction solubility of radon at a pressure of 101.325 kPa was calculated by the compiler. It was assumed that the Ostwald coefficient was independent of pressure and that the gram-mole volume of radon is 22,290 cm^3 at 273.15 K and 101.325 kPa.

AUXILIARY INFORMATION

METHOD:	SOURCE AND PURITY OF MATERIALS:
The concentration of radon was determined by measurement of radio-activity in samples withdrawn from the liquid and from the gas phases. The procedure was stated to be similar in principle to that of Boyle (1).	1. Radon-222. 2. Acrylic Acid. Eastman Chemical Company. Highest grade.

APPARATUS/PROCEDURE:	ESTIMATED ERROR:
A cylindrical glass vessel with a stopcock at each end. The gamma-rays emitted were estimated by a sodium iodide scintillation counter.	

REFERENCES:

1. Boyle, R. W. Phil. Mag. 1911, 22, 840.

COMPONENTS:	ORIGINAL MEASUREMENTS:

COMPONENTS:

1. Radon; $^{222}_{86}$Rn; 14859-67-7

2. 2-Methyl Propanoic Acid; $C_4H_8O_2$; 79-31-2

ORIGINAL MEASUREMENTS:

Hofbauer, G.

Sitzungsber. Akad. Wiss. Wien, Math. Naturwiss. Kl. Abt. 2A 1914, 123, 2001.

VARIABLES:

T/K: 293.05

PREPARED BY:

W. Gerrard

EXPERIMENTAL VALUES:

T/K	Mol Fraction $X_1 \times 10^2$	Ostwald Coefficient L
293.05	0.0346	9.18

The author reported a coefficient of solubility, α', based on the radon concentration ratio in the liquid and gaseous phases at equilibrium by radioactivity. The coefficient of solubility is labelled an Ostwald coefficient above. It was measured at a radon partial pressure of probably less than 0.1 kPa at equilibrium. The liquid was also saturated with air as a carrier gas at an initial pressure of about 100 kPa.

The mole fraction solubility at a radon partial pressure of 101.325 kPa was calculated by the compiler. It was assumed that the Ostwald coefficient was independent of pressure and that the gram-mole volume of radon is 22,290 cm^3 at 273.15 K and 101.325 kPa.

AUXILIARY INFORMATION

METHOL/APPARATUS/PROCEDURE:

The method was based on the technique of Kofler (1,2). The radon radioactivity was measured in both the liquid and gaseous phases.

The apparatus consists of two glass bulbs connected by a tap. Each bulb carries two separate taps. Both Hofbauer and Kofler give a diagram.

SOURCE AND PURITY OF MATERIALS:

1. Radon.

2. Methyl Propanoic Acid. Purified by distillation. The density was given.

DATA CLASS:

ESTIMATED ERROR:

REFERENCES:
1. Kofler, M.
   Sitzungsber. Akad. Wiss. Wien 1912, 121, 2169.

2. Kofler, M.
   Monatsh. 1913, 34, 389.

COMPONENTS:	ORIGINAL MEASUREMENTS:

COMPONENTS:

1. Radon; $^{222}_{86}$Rn; 14859-67-7

2. Butanoic Acid; $C_4H_8O_2$; 107-92-6

ORIGINAL MEASUREMENTS:

Hofbauer, G.

Sitzungsber. Akad. Wiss. Wien,
Math. Naturwiss. Kl. Abt. 2A 1914,
123, 2001.

VARIABLES:

T/K: 293.05

PREPARED BY:

W. Gerrard

EXPERIMENTAL VALUES:

T/K	Mol Fraction $X_1 \times 10^2$	Ostwald Coefficient L
293.05	3.28	8.78

The author reported a coefficient of solubility, α', based on the radon concentration ratio in the liquid and gaseous phases at equilibrium by radioactivity. The coefficient of solubility is labelled an Ostwald coefficient above. It was measured at a radon partial pressure of probably less than 0.1 kPa at equilibrium. The liquid was also saturated with air as a carrier gas at an initial pressure of about 100 kPa.

The mole fraction solubility at a radon partial pressure of 101.325 kPa was calculated by the compiler. It was assumed that the Ostwald coefficient was independent of pressure and that the gram-mole volume of radon is 22,290 cm^3 at 273.15 K and 101.325 kPa.

AUXILIARY INFORMATION

METHOD /APPARATUS/PROCEDURE:

The method was based on the technique of Kofler (1,2). The radon radioactivity was measured in both the liquid and gaseous phases.

The apparatus consists of two glass bulbs connected by a tap. Each bulb carries two separate taps. Both Hofbauer and Kofler give a diagram.

SOURCE AND PURITY OF MATERIALS:

1. Radon.

2. Butanoic Acid. Purified by distillation. The density was given.

DATA CLASS:

ESTIMATED ERROR:

REFERENCES:

1. Kofler, M.
   Sitzungsber. Akad. Wiss. Wien 1912, 121, 2169.

2. Kofler, M.
   Monatsh. 1913, 34, 389.

COMPONENTS:	ORIGINAL MEASUREMENTS:
1. Radon; $^{222}_{86}$Rn; 14859-67-7 2. Butanoic Acid; $C_4H_8O_2$; 107-92-6	Nussbaum, E.; Hursh, J. B. J. Phys. Chem. 1958, 62, 81 - 84.

VARIABLES:	PREPARED BY:
T/K: 298.15 - 323.15	W. Gerrard

EXPERIMENTAL VALUES:

T/K	Mol Fraction $X_1 \times 10^2$	Ostwald Coefficient L
298.15	2.78	7.52
310.15	2.46	6.82
323.15	2.11	5.99

The Ostwald coefficient was measured at a very low partial pressure of radon. The carrier gas was nitrogen at a partial pressure of about 100 kPa. The liquid was therefore saturated with nitrogen at the experimental pressure.

The mole fraction solubility of radon at a pressure of 101.325 kPa was calculated by the compiler. It was assumed that the Ostwald coefficient was independent of pressure and that the gram-mole volume of radon is 22,290 cm^3 at 273.15 K and 101.325 kPa.

Smoothed Data: $\Delta G°/J \ mol^{-1} = - RT \ ln \ X_1 = - 8,862.9 + 59.467 \ T$

Std. Dev. $\Delta G° = 23$,   Coef. Corr. = 0.9995

$\Delta H°/J \ mol^{-1} = -8,862.9$, $\Delta S°/J \ K^{-1} \ mol^{-1} = -59.467$

T/K	Mol Fraction $X_1 \times 10^2$	$\Delta G°/J \ mol^{-1}$
298.15	2.80	8,867.2
303.15	2.64	9,164.6
308.15	2.49	9,461.9
313.15	2.36	9,759.3
318.15	2.23	10,057
323.15	2.12	10,354

The smoothed data fit was added by the Volume Editor.

AUXILIARY INFORMATION

METHOD:	SOURCE AND PURITY OF MATERIALS:
The concentration of radon was determined by measurement of radio-activity in samples withdrawn from the liquid and from the gas phases. The procedure was stated to be similar in principle to that of Boyle (1).	1. Radon-222. 2. Butanoic Acid. Eastman Chemical Company. Highest grade.

APPARATUS/PROCEDURE:	ESTIMATED ERROR:
A cylindrical glass vessel with a stopcock at each end. The gamma-rays emitted were estimated by a sodium iodide scintillation counter.	
	REFERENCES: 1. Boyle, R. W. Phil. Mag. 1911, 22, 840.

COMPONENTS:	ORIGINAL MEASUREMENTS:
1. Radon; $^{222}_{86}$Rn; 14859-67-7 2. Pentanoic Acid (Valeric Acid); $C_5H_{10}O_2$; 109-52-4	Nussbaum, E.; Hursh, J. B. <u>J</u>. <u>Phys</u>. <u>Chem</u>. 1958, <u>62</u>, 81 - 84.
VARIABLES: T/K: 298.15 - 323.15	PREPARED BY: W. Gerrard

EXPERIMENTAL VALUES:

T/K	Mol Fraction X_1 x 10^2	Ostwald Coefficient L
298.15	3.74	8.64
310.15	2.89	6.82
323.15	2.51	6.06

The Ostwald coefficient was measured at a very low partial pressure of radon. The carrier gas was nitrogen at a partial pressure of about 100 kPa. The liquid was therefore saturated with nitrogen at the experimental pressure.

The mole fraction solubility of radon at a pressure of 101.325 kPa was calculated by the compiler. It was assumed that the Ostwald coefficient was independent of pressure and that the gram-mole volume of radon is 22,290 cm^3 at 273.15 K and 101.325 kPa.

Smoothed Data: $\Delta G°/J\ mol^{-1} = - RT\ \ln X_1 = - 12,679 + 70.021\ T$

Std. Dev. $\Delta G° = 86.9$, Coef. Corr. = 0.9951

$\Delta H°/J\ mol^{-1} = - 12,679$, $\Delta S°/J\ K^{-1}\ mol^{-1} = -70.021$

T/K	Mol Fraction X_1 x 10^2	$\Delta G°/J\ mol^{-1}$
298.15	3.66	8,198.1
303.15	3.37	8,548.2
308.15	3.10	8,898.3
313.15	2.87	9,248.4
318.15	2.65	9,598.5
323.15	2.47	9,948.6

The smoothed data fit was added by the Volume Editor.

AUXILIARY INFORMATION

METHOD:

The concentration of radon was determined by measurement of radio-activity in samples withdrawn from the liquid and from the gas phases. The procedure was stated to be similar in principle to that of Boyle (1).

SOURCE AND PURITY OF MATERIALS:

1. Radon-222.

2. Valeric Acid. Eastman Chemical Company. Highest grade.

APPARATUS/PROCEDURE:

A cylindrical glass vessel with a stopcock at each end. The gamma-rays emitted were estimated by a sodium iodide scintillation counter.

ESTIMATED ERROR:

REFERENCES:

1. Boyle, R. W. <u>Phil</u>. <u>Mag</u>. 1911, <u>22</u>, 840.

COMPONENTS:	ORIGINAL MEASUREMENTS:
1. Radon; $^{222}_{86}Rn$; 14859-67-7 2. Hexanoic Acid; $C_6H_{12}O_2$; 142-62-1	Nussbaum, E.; Hursh, J. B. J. Phys. Chem. 1958, 62, 81 - 84.
VARIABLES: T/K: 298.15 - 323.15	PREPARED BY: W. Gerrard May 1977

EXPERIMENTAL VALUES:

T/K	Mol Fraction X_1 x 10^2	Ostwald Coefficient L
298.15	4.47	9.03
310.15	3.50	7.23
323.15	2.92	6.16

The Ostwald coefficient was measured at a very low partial pressure of radon. The carrier gas was nitrogen at a partial pressure of about 100 kPa. The liquid was therefore saturated with nitrogen at the experimental pressure.

The mole fraction solubility of radon at a pressure of 101.325 kPa was cal-culated by the compiler. It was assumed that the Ostwald coefficient was independent of pressure and that the gram-mole volume of radon is 22,290 cm^3 at 273.15 K and 101.325 kPa.

Smoothed Data: $\Delta G°/J\ mol^{-1} = - RT\ ln\ X_1 = -13,590 + 71.513\ T$

Std. Dev. $\Delta G° = 47.1$, Coef. Corr. = 0.9986

$\Delta H°/J\ mol^{-1} = -13,590$, $\Delta S°/J\ K^{-1}\ mol^{-1} = -71.513$

T/K	Mol Fraction X_1 x 10^2	$\Delta G°/J\ mol^{-1}$
298.15	4.42	7,732.2
303.15	4.04	8,089.3
308.15	3.70	8,447.3
313.15	3.40	8,804.9
318.15	3.13	9,162.5
323.15	2.89	9,520.0

The smoothed data fit was added by the Volume Editor.

AUXILIARY INFORMATION

METHOD:	SOURCE AND PURITY OF MATERIALS:
The concentration of radon was determined by measurement of radio-activity in samples withdrawn from the liquid and from the gas phases. The procedure was stated to be similar in principle to that of Boyle (1).	1. Radon-222. 2. Hexanoic Acid. Eastman Chemical Company. Highest grade.
APPARATUS/PROCEDURE: A cylindrical glass vessel with a stopcock at each end. The gamma-rays emitted were estimated by a sodium iodide scintillation counter.	ESTIMATED ERROR: REFERENCES: 1. Boyle, R. W. Phil. Mag. 1911, 22, 840.

COMPONENTS:	ORIGINAL MEASUREMENTS:
1. Radon; $^{222}_{86}$Rn; 14859-67-7 2. Heptanoic Acid; $C_7H_{14}O_2$; 111-14-8	Nussbaum, E.: Hursh. J. B. J. Phys. Chem. 1958, 62, 81 - 84.
VARIABLES: T/K: 298.15 - 323.15	PREPARED BY: W. Gerrard

EXPERIMENTAL VALUES:

T/K	Mol Fraction X_1 x 10^2	Ostwald Coefficient L
298.15	4.87	8.75
310.15	3.90	7.15
323.15	3.38	6.33

The Ostwald coefficient was measured at a very low partial pressure of radon. The carrier gas was nitrogen at a partial pressure of about 100 kPa. The liquid was therefore saturated with nitrogen at the experimental pressure.

The mole fraction solubility of radon at a pressure of 101.325 kPa was calculated by the compiler. It was assumed that the Ostwald coefficient was independent of pressure and that the gram-mole volume of radon is 22,290 cm^3 at 273.15 K and 101.325 kPa.

Smoothed Data: $\Delta G°/J\ mol^{-1} = -RT\ ln\ X_1 = -11,640 + 64.285\ T$
Std. Dev. $\Delta G° = 58.7$, Coef. Corr. = 0.9975
$\Delta H°/J\ mol^{-1} = -11,640$, $\Delta S°/J\ K^{-1}\ mol^{-1} = -64.285$

T/K	Mol Fraction X_1 x 10^2	$\Delta G°/J\ mol^{-1}$
298.15	4.80	7526.7
303.15	4.44	7848.1
308.15	4.12	8169.4
313.15	3.83	8490.7
318.15	3.57	8812.1
323.15	3.34	9133.4

The smoothed data fit was added by the Volume Editor.

AUXILIARY INFORMATION

METHOD:	SOURCE AND PURITY OF MATERIALS:
The concentration of radon was determined by measurement of radio-activity in samples withdrawn from the liquid and from the gas phases. The procedure was stated to be similar in principle to that of Boyle (1).	1. Radon-222. 2. Heptanoic Acid. Eastman Chemical Company. Highest grade.
APPARATUS/PROCEDURE: A cylindrical glass vessel with a stopcock at each end. The gamma-rays emitted were estimated by a sodium iodide scintillation counter.	ESTIMATED ERROR: REFERENCES: 1. Boyle, R. W. Phil. Mag. 1911, 22, 840.

COMPONENTS:	ORIGINAL MEASUREMENTS:
1. Radon; $^{222}_{86}$Rn; 14859-67-7 2. Octanoic Acid, $C_8H_{16}O_2$; 124-07-2	Nussbaum, E.; Hursh, J. B. *J. Phys. Chem.* 1958, 62, 81 - 84.

VARIABLES:
 	
    T/K: 298.15 - 323.15
	PREPARED BY:
	
	    W. Gerrard

EXPERIMENTAL VALUES:

T/K	Mol Fraction $X_1 \times 10^2$	Ostwald Coefficient L
298.15	5.59	9.03
310.15	4.20	6.89
323.15	3.67	6.16

The Ostwald coefficient was measured at a very low partial pressure of radon. The carrier gas was nitrogen at a partial pressure of about 100 kPa. The liquid was therefore saturated with nitrogen at the experimental pressure.

The mole fraction solubility of radon at a pressure of 101.325 kPa was calculated by the compiler. It was assumed that the Ostwald coefficient was independent of pressure and that the gram-mole volume of radon is 22,290 cm^3 at 273.15 K and 101.325 kPa.

Smoothed Data: $\Delta G°/J\ mol^{-1} = -RT \ln X_1 = -13,354 + 68.995\ T$

           Std. Dev. $\Delta G° = 112$,   Coef. Corr. = 0.9916

           $\Delta H°/J\ mol^{-1} = -13,354$, $\Delta S°/J\ K^{-1}\ mol^{-1} = -68,995$

T/K	Mol Fraction $X_1 \times 10^2$	$\Delta G°/J\ mol^{-1}$
298.15	5.44	7,217.1
303.15	4.98	7,562.0
308.15	4.57	7,907.0
313.15	4.20	8,252.0
318.15	3.88	8,597.0
323.15	3.59	8,941.9

The smoothed data fit was added by the Volume Editor.

AUXILIARY INFORMATION

METHOD:

   The concentration of radon was determined by measurement of radio-activity in samples withdrawn from the liquid and from the gas phases. The procedure was stated to be similar in principle to that of Boyle (1).

SOURCE AND PURITY OF MATERIALS:

1. Radon-222.

2. Octanoic Acid. Eastman Chemical Company. Highest grade.

APPARATUS/PROCEDURE:

   A cylindrical glass vessel with a stopcock at each end. The gamma-rays emitted were estimated by a sodium iodide scintillation counter.

ESTIMATED ERROR:

REFERENCES:

1. Boyle, R. W.
   Phil. Mag. 1911, 22, 840.

COMPONENTS:	ORIGINAL MEASUREMENTS:
1. Radon; $^{222}_{86}$Rn; 14859-67-7 2. Nonanoic Acid; $C_9H_{18}O_2$; 112-05-0	Nussbaum, E.; Hursh, J. B. J. Phys. Chem. 1958, 62, 81 - 84.
VARIABLES: T/K: 298.15 - 323.15	PREPARED BY: W. Gerrard

EXPERIMENTAL VALUES:

T/K	Mol Fraction X_1 x 10^2	Ostwald Coefficient L
298.15	5.67	8.32
310.15	4.61	6.89
323.15	3.945	6.00

The Ostwald coefficient was measured at a very low partial pressure of radon. The carrier gas was nitrogen at a partial pressure of about 100 kPa. The liquid was therefore saturated with nitrogen at the experimental pressure.

The mole fraction solubility of radon at a pressure of 101.325 kPa was calculated by the compiler. It was assumed that the Ostwald coefficient was independent of pressure and that the gram-mole volume of radon is 22,290 cm^3 at 273.15 K and 101.325 kPa.

Smoothed Data: $\Delta G°/J\ mol^{-1} = - RT\ \ln X_1 = -11,584 + 62.791\ T$
Std. Dev. $\Delta G° = 38.0$, Coef. Corr. = 0.9990
$\Delta H°/J\ mol^{-1} = -11,584$, $\Delta S°/J\ K^{-1}\ mol^{-1} = -62.791$

T/K	Mol Fraction X_1 x 10^2	$\Delta G°/J\ mol^{-1}$
298.15	5.62	7,137.2
303.15	5.20	7,451.1
308.15	4.83	7,765.0
313.15	4.49	8,078.9
318.15	4.19	8,392.8
323.15	3.91	8,706.6

The smoothed data fit was added by the Volume Editor.

AUXILIARY INFORMATION

METHOD:	SOURCE AND PURITY OF MATERIALS:
The concentration of radon was determined by measurement of radio-activity in samples withdrawn from the liquid and from the gas phases. The procedure was stated to be similar in principle to that of Boyle (1).	1. Radon-222. 2. Nonanoic Acid. Eastman Chemical Company. Highest grade.

APPARATUS/PROCEDURE:	ESTIMATED ERROR:
A cylindrical glass vessel with a stopcock at each end. The gamma-rays emitted were estimated by a sodium iodide scintillation counter.	
	REFERENCES: 1. Boyle, R. W. Phil. Mag. 1911, 22, 840.

COMPONENTS:	ORIGINAL MEASUREMENTS:

1. Radon; $^{222}_{86}$Rn; 14859-67-7

2. Decanoic Acid; $C_{10}H_{20}O_2$; 334-48-5

Nussbaum, E.; Hursh, J. B.

J. Phys. Chem. 1958, 62, 81 - 84.

VARIABLES:	PREPARED BY:
T/K: 310.15	W. Gerrard

EXPERIMENTAL VALUES:

T/K	Mol Fraction $X_1 \times 10^2$	Ostwald Coefficient L
310.15	5.16	7.13

The Ostwald coefficient was measured at a very low partial pressure of radon. The carrier gas was nitrogen at a partial pressure of about 100 kPa. The liquid was therefore saturated with nitrogen at the experimental pressure.

The mole fraction solubility of radon at a pressure of 101.325 kPa was calculated by the compiler. It was assumed that the Ostwald coefficient was independent of pressure and that the gram-mole volume of radon is 22,290 cm^3 at 273.15 K and 101.325 kPa.

AUXILIARY INFORMATION

METHOD:

The concentration of radon was determined by measurement of radio-activity in samples withdrawn from the liquid and from the gas phases. The procedure was stated to be similar in principle to that of Boyle (1).

SOURCE AND PURITY OF MATERIALS:

1. Radon-222.

2. Decanoic Acid. Eastman Chemical Company. Highest grade.

APPARATUS/PROCEDURE:

A cylindrical glass vessel with a stopcock at each end. The gamma-rays emitted were estimated by a sodium iodide scintillation counter.

ESTIMATED ERROR:

REFERENCES:

1. Boyle, R. W.
    Phil. Mag. 1911, 22, 840.

COMPONENTS:	ORIGINAL MEASUREMENTS:
1. Radon; $^{222}_{86}$Rn; 14859-67-7 2. Undecanoic Acid; $C_{11}H_{22}O_2$; 112-37-8	Nussbaum, E.; Hursh, J. B. J. Phys. Chem. 1958, 62, 81-84.

VARIABLES:	PREPARED BY:
T/K: 310.15	W. Gerrard

EXPERIMENTAL VALUES:

T/K	Mol Fraction $X_1 \times 10^2$	Ostwald Coefficient L
310.15	5.34	6.86

The Ostwald coefficient was measured at a very low partial pressure of radon. The carrier gas was nitrogen at a partial pressure of about 100 kPa. The liquid was therefore saturated with nitrogen at the experimental pressure.

The mole fraction solubility of radon at a pressure of 101.325 kPa was calculated by the compiler. It was assumed that the Ostwald coefficient was independent of pressure and that the gram-mole volume of radon is 22,290 cm^3 at 273.15 K and 101.325 kPa.

AUXILIARY INFORMATION

METHOD:	SOURCE AND PURITY OF MATERIALS:
The concentration of radon was determined by measurement of radio-activity in samples withdrawn from the liquid and from the gas phases. The procedure was stated to be similar in principle to that of Boyle (1).	1. Radon-222. 2. Undecanoic Acid. Eastman Chemical Company. Highest grade.

APPARATUS/PROCEDURE:	ESTIMATED ERROR:
A cylindrical glass vessel with a stopcock at each end. The gamma-rays emitted were estimated by a sodium iodide scintillation counter.	
	REFERENCES: 1. Boyle, R. W. Phil. Mag. 1911, 22, 840.

COMPONENTS:	ORIGINAL MEASUREMENTS:
1. Radon; $^{222}_{86}$Rn; 14859-67-7 2. Dodecanoic Acid (Lauric Acid); $C_{12}H_{24}O_2$; 143-07-7	Nussbaum, E.; Hursh, J. B. J. Phys. Chem. 1958, 62, 81 - 84.

VARIABLES:
   T/K: 323.15
	PREPARED BY:
	   W. Gerrard

EXPERIMENTAL VALUES:

T/K	Mol Fraction X_1 x 10^2	Ostwald Coefficient L
323.15	4.93	5.93

The Ostwald coefficient was measured at a very low partial pressure of radon. The carrier gas was nitrogen at a partial pressure of about 100 kPa. The liquid was therefore saturated with nitrogen at the experimental pressure.

The mole fraction solubility of radon at a pressure of 101.325 kPa was calculated by the compiler. It was assumed that the Ostwald coefficient was independent of pressure and that the gram-mole volume of radon is 22,290 cm^3 at 273.15 K and 101.325 kPa.

AUXILIARY INFORMATION

METHOD:	SOURCE AND PURITY OF MATERIALS:
The concentration of radon was determined by measurement of radio-activity in samples withdrawn from the liquid and from the gas phases. The procedure was stated to be similar in principle to that of Boyle (1).	1. Radon-222. 2. Lauric Acid. Eastman Chemical Co. Highest grade.

APPARATUS/PROCEDURE:	ESTIMATED ERROR:
A cylindrical glass vessel with a stopcock at each end. The gamma-rays emitted were estimated by a sodium iodide scintillation counter.	
	REFERENCES: 1. Boyle, R. W. Phil. Mag. 1911, 22, 840.

COMPONENTS:	ORIGINAL MEASUREMENTS:
1. Radon; $^{222}_{86}$Rn; 14859-67-7 2. Tridecanoic Acid; $C_{13}H_{26}O_2$; 638-53-9	Nussbaum, E; Hursh, J. B. J. Phys. Chem. 1958, 62, 81 - 84.
VARIABLES: T/K: 323.15	PREPARED BY: W. Gerrard

EXPERIMENTAL VALUES:

T/K	Mol Fraction X_1 x 10^2	Ostwald Coefficient L
323.15	5.28	5.95

The Ostwald coefficient was measured at a very low partial pressure of
radon. The carrier gas was nitrogen at a partial pressure of about 100 kPa.
The liquid was therefore saturated with nitrogen at the experimental
pressure.

The mole fraction solubility of radon at a pressure of 101.325 kPa was
calculated by the compiler. It was assumed that the Ostwald coefficient was
independent of pressure and that the gram-mole volume of radon is 22,290 cm³
at 273.15 K and 101.325 kPa.

AUXILIARY INFORMATION

METHOD:	SOURCE AND PURITY OF MATERIALS:
The concentration of radon was determined by measurement of radio-activity in samples withdrawn from the liquid and from the gas phases. The procedure was stated to be similar in principle to that of Boyle (1).	1. Radon-222. 2. Tridecanoic Acid. Eastman Chemi-Cal Co. Highest grade.

APPARATUS/PROCEDURE:	ESTIMATED ERROR:
A cylindrical glass vessel with a stopcock at each end. The gamma-rays emitted were estimated by a sodium iodide scintillation counter.	
	REFERENCES: 1. Boyle, R. W. Phil. Mag. 1911, 22, 840.

COMPONENTS:	ORIGINAL MEASUREMENTS:

1. Radon; $^{222}_{86}$Rn; 14859-67-7
 	
2. (ZZ)-9,12- Octadecadienoic Acid
   (Linoleic Acid); $C_{18}H_{32}O_2$;
   60-33-3
	Nussbaum, E.; Hursh, J. B.
	
	J. Phys. Chem. 1958, 62, 81 - 84.

VARIABLES:
        T/K: 298.15 - 323.15
	PREPARED BY:
	        W. Gerrard

EXPERIMENTAL VALUES:

T/K	Mol Fraction X_1 x 10^2	Ostwald Coefficient L
298.15	9.27	7.96
323.15	7.30	6.32

The Ostwald coefficient was measured at a very low partial pressure of radon. The carrier gas was nitrogen at a partial pressure of about 100 kPa. The liquid was therefore saturated with nitrogen at the experimental pressure.

The mole fraction solubility of radon at a pressure of 101.325 kPa was calculated by the compiler. It was assumed that the Ostwald coefficient was independent of pressure and that the gram-mole volume of radon is 22,290 cm^3 at 273.15 K and 101.325 kPa.

Smoothed Data: $\Delta G°/J\ mol^{-1} = -\ RT\ ln\ X_1 = -15,307 + 71.114\ T$

$\Delta H°/J\ mol^{-1} = -15,307, \quad \Delta S°/J\ K^{-1}\ mol^{-1} = -71.114$

T/K	Mol Fraction X_1 x 10^2	$\Delta G°/J\ mol^{-1}$
298.15	9.27	5,895.8
303.15	8.37	6,251.4
308.15	7.59	6,606.9
313.15	6.90	6,962.5

The smoothed data fit was added by the Volume Editor.

AUXILIARY INFORMATION

METHOD:	SOURCE AND PURITY OF MATERIALS:
The concentration of radon was determined by measurement of radio-activity in samples withdrawn from the liquid and from the gas phases. The procedure was stated to be similar in principle to that of Boyle (1).	1. Radon-222. 2. Linoleic Acid. Eastman Chemical Co. Highest grade.
APPARATUS/PROCEDURE:	ESTIMATED ERROR:
A cylindrical glass vessel with a stopcock at each end. The gamma-rays emitted were estimated by a sodium iodide scintillation counter.	REFERENCES: 1. Boyle, R. W. Phil. Mag. 1911, 22, 840.

COMPONENTS:	ORIGINAL MEASUREMENTS:
1. Radon; $^{222}_{86}$Rn; 14859-67-7 2. (Z)-9-Octadecenoic Acid (Oleic Acid); $C_{18}H_{34}O_2$; 112-80-1	Nussbaum, E.; Hursh, J. B. J. Phys. Chem. 1958, 62, 81 - 84.
VARIABLES: T/K: 298.15 - 323.15	PREPARED BY: W. Gerrard

EXPERIMENTAL VALUES:

T/K	Mol Fraction X_1 x 10^2	Ostwald Coefficient L
298.15	9.59	8.10
310.15	7.85	6.72
323.15	6.73	5.86

The Ostwald coefficient was measured at a very low partial pressure of radon. The carrier gas was nitrogen at a partial pressure of about 100 kPa. The liquid was therefore saturated with nitrogen at the experimental pressure.

The mole fraction solubility of radon at a pressure of 101.325 kPa was calculated by the compiler. It was assumed that the Ostwald coefficient was independent of pressure and that the gram-mole volume of radon is 22,290 cm^3 at 273.15 K and 101.325 kPa.

Smoothed Data: ΔG°/J mol^{-1} = - RT ln X_1 =-11,308 + 57.489 T

Std. Dev. ΔG° = 34.4, Coef. Corr. = 0.9989

ΔH°/J mol^{-1} = -11,308, ΔS°/J K^{-1} mol^{-1} = -57.489

T/K	Mol Fraction X_1 x 10^2	ΔG°/J mol^{-1}
298.15	9.51	5,832.3
303.15	8.82	6,119.8
308.15	8.20	6,407.2
313.15	7.64	6,694.6
318.15	7.14	6,982.1
323.15	6.68	7,269.5

The smoothed data fit was added by the Volume Editor.

AUXILIARY INFORMATION

METHOD:	SOURCE AND PURITY OF MATERIALS:
The concentration of radon was determined by measurement of radio-activity in samples withdrawn from the liquid and from the gas phases. The procedure was stated to be similar in principle to that of Boyle (1).	1. Radon-222. 2. Oleic Acid. Eastman Chemical Co. Highest grade.

APPARATUS/PROCEDURE:	ESTIMATED ERROR:
A cylindrical glass vessel with a stopcock at each end. The gamma-rays emitted were estimated by a sodium iodide scintillation counter.	REFERENCES: 1. Boyle, R. W. Phil. Mag. 1911, 22, 840.

COMPONENTS:	ORIGINAL MEASUREMENTS:
1. Radon; $^{222}_{86}$Rn; 14859-67-7 2. Acetic Acid Ethyl Ester (Ethyl Acetate); $C_4H_8O_2$; 141-78-6	Ramstedt, E. Le Radium 1911, 8, 253-256.
VARIABLES: T/K: 255.15 - 291.15	PREPARED BY: W. Gerrard

EXPERIMENTAL VALUES:

T/K	Mol Fraction X_i x 10^2	Ostwald Coefficient L
255.15	5.74	13.6
273.15	3.87	9.41
291.15	2.93	7.35

The Ostwald coefficient was measured at a radon partial pressure of less than 0.1 kPa at equilibrium. The radon was present in a carrier gas (air or nitrogen) at an initial pressure of about 101 kPa.

The mole fraction solubility was calculated by the compiler. It was assumed that the Ostwald coefficient was independent of pressure and that the gram-mole volume of radon is 22,290 cm^3 at 273.15 K and 101.325 kPa.

Smoothed Data: ΔG^o/J mol^{-1} = -RT ln X_1 = -11,510 + 68.977 T

Std. Dev. ΔG = 47.0, Coef. Corr. = 0.9993

ΔH^o/J mol^{-1} = -11,510, ΔS^o/J K^{-1} mol^{-1} = -68.977

T/K	Mol Fraction X_1 x 10^2	ΔG^o/J mol^{-1}
258.15	5.32	6,296.4
263.15	4.81	6,641.3
268.15	4.36	6,986.1
273.15	3.96	7,331.0
278.15	3.62	7,675.9
283.15	3.31	8,020.8
288.15	3.04	8,365.7
293.15	2.80	8,710.6
298.15	2.59	9,055.5

The smoothed data fit was added by the Volume Editor.

AUXILIARY INFORMATION

METHOD: Measurement of radioactivity in the liquid and in the gaseous phase.	SOURCE AND PURITY OF MATERIALS: 1. Radon-222. 2. Acetic Acid Ethyl Ester. Dried and distilled.
APPARATUS/PROCEDURE: Two tubes connected by a wide tap. To determine concentration of radon, each tube is placed in a condenser.	ESTIMATED ERROR: REFERENCES:

COMPONENTS:	ORIGINAL MEASUREMENTS:
1. Radon-222; $^{222}_{86}$Rn; 14859-67-7 2. Acetic Acid Pentyl Ester (Amyl Acetate); $C_7H_{14}O_2$; 628-63-7	Lurie, A. Thesis University of Grenoble, 1910 Microfilm available. See also Tables annuelles de constantes et donnees numeriques de chemie, de physique et de technologie 1913 (for 1911), 2, 401.

VARIABLES:	PREPARED BY:
T/K: 253.15 - 343.15	W. Gerrard

EXPERIMENTAL VALUES:

T/K	Mol Fraction $X_1 \times 10^2$	Ostwald Coefficient L
253.15	-	39.5
273.15	11.5	19.4
293.15	8.58	15.1
323.15	-	7.2
343.15	-	4.8

The author reported a coefficient of absorption, α, which appears to have the same meaning as the ratio of concentrations: concentration of radon in the liquid/concentration of radon in the gas. The coefficient of absorption has been labelled as the Ostwald coefficient.

The coefficient of absorption was measured at a radon partial pressure of less than 0.1 kPa at equilibrium. The radon was present in a carrier gas (air or nitrogen) at an initial pressure of about 101 kPa.

The mole fraction solubility at 101.325 kPa was calculated by the compiler. It was assumed that the Ostwald coefficient was independent of pressure and that the gram-mole volume of radon is 22,290 cm^3 at 273.15 K and 101.325 kPa.

AUXILIARY INFORMATION

METHOD /APPARATUS/PROCEDURE:	SOURCE AND PURITY OF MATERIALS:
The concentration of radon in the gas and liquid phases determined by measurements of radioactivity. An aluminum foil electroscope was used to measure the radioactivity. Diagrams given by Lurie.	1. Radon-222. 2. Acetic Acid Pentyl Ester (Amyl Acetate). Purity of liquid not specified.
	ESTIMATED ERROR:
	REFERENCES:

COMPONENTS:	ORIGINAL MEASUREMENTS:
1. Radon; $^{222}_{86}$Rn; 14859-67-7 2. 1,2,3-Propanetriol, Triacetate (Triacetin); $C_9H_{14}O_6$; 102-76-1	Nussbaum, E.; Hursh, J. B. <u>J</u>. <u>Phys</u>. <u>Chem</u>. 1958, <u>62</u>, 81 - 84.
VARIABLES: T/K: 298.15 - 310.15	PREPARED BY: W. Gerrard

EXPERIMENTAL VALUES:

T/K	Mol Fraction X_1 x 10^2	Ostwald Coefficient L
298.15	2.22	3.42
310.15	2.13	2.88

The Ostwald coefficient was measured at a very low partial pressure of radon. The carrier gas was nitrogen at a partial pressure of about 100 kPa. The liquid was therefore saturated with nitrogen at the experimental pressure.

The mole fraction solubility of radon at a pressure of 101.325 kPa was calculated by the compiler. It was assumed that the Ostwald coefficient was independent of pressure and that the gram-mole volume of radon is 22,290 cm^3 at 273.15 K and 101.325 kPa.

Smoothed Data: $\Delta G°/J \ mol^{-1} = - RT \ ln \ X_1 = -2,651.5 + 40.551 \ T$

$\Delta H°/J \ mol^{-1} = -2,651.5$, $\Delta S°/J \ K^{-1} \ mol^{-1} = -40.551$

T/K	Mol Fraction X_1 x 10^2	$\Delta G°/J \ mol^{-1}$
298.15	2.22	9,438.8
303.15	2.18	9,641.6
308.15	2.14	9,844.4
313.15	2.11	10,047

The smoothed data fit was added by the Volume Editor.

AUXILIARY INFORMATION

METHOD:	SOURCE AND PURITY OF MATERIALS:
The concentration of radon was determined by measurement of radio-activity in samples withdrawn from the liquid and from the gas phases. The procedure was stated to be similar in principle to that of Boyle (1).	1. Radon-222. 2. Triacetin. Eastman Chemical Company. Highest grade.

APPARATUS/PROCEDURE:	ESTIMATED ERROR:
A cylindrical glass vessel with a stopcock at each end. The gamma-rays emitted were estimated by a sodium iodide scintillation counter.	
	REFERENCES: 1. Boyle, R. W. <u>Phil</u>. <u>Mag</u>. 1911, <u>22</u>, 840.

COMPONENTS:	ORIGINAL MEASUREMENTS:
1. Radon; $^{222}_{86}$Rn; 14859-67-7 2. Butanoic Acid, 1,2,3-propanetriyl ester (Tributyrin); $C_{15}H_{26}O_6$; 60-01-5	Nussbaum, E.; Hursh, J. B. J. Phys. Chem 1958, 62, 81 - 84.
VARIABLES: T/K: 298.15 - 323.15	PREPARED BY: W. Gerrard

EXPERIMENTAL VALUES:

T/K	Mol Fraction X_1 x 10^2	Ostwald Coefficient L
298.15	7.19	6.42
310.15	5.54	5.01

The Ostwald coefficient was measured at a very low partial pressure of radon. The carrier gas was nitrogen at a partial pressure of about 100 kPa. The liquid was therefore saturated with nitrogen at the experimental pressure.

The mole fraction solubility of radon at a pressure of 101.325 kPa was calculated by the compiler. It was assumed that the Ostwald coefficient was independent of pressure and that the gram-mole volume of radon is 22,290 cm^3 at 273.15 K and 101.325 kPa.

Smoothed Data: $\Delta G°/J\ mol^{-1} = - RT\ ln\ X_1 = -16,703 + 77.908\ T$

$\Delta H°/J\ mol^{-1} = -16,718,\quad \Delta S°/J\ K^{-1}\ mol^{-1} = -77.908$

T/K	Mol Fraction X_1 x 10^2	$\Delta G°/J\ mol^{-1}$
298.15	7.19	6,525.7
303.15	6.43	6,915.2
308.15	5.78	7,304.8
313.15	5.21	7,694.3

The smoothed data fit was added by the Volume Editor.

AUXILIARY INFORMATION

METHOD:	SOURCE AND PURITY OF MATERIALS:
The concentration of radon was determined by measurement of radio-activity in samples withdrawn from the liquid and from the gas phases. The procedure was stated to be similar in principle to that of Boyle (1).	1. Radon-222. 2. Tributyrin. Eastman Chemical Co. Highest grade.
APPARATUS/PROCEDURE: A cylindrical glass vessel with a stopcock at each end. The gamma-rays emitted were estimated by a sodium iodide scintillation counter.	ESTIMATED ERROR: REFERENCES: 1. Boyle, R. W. Phil. Mag. 1911, 22, 840.

COMPONENTS:	ORIGINAL MEASUREMENTS:
1. Radon; $^{222}_{86}$Rn; 14859-67-7 2. Hexanoic acid; 1,2,3-propanetriyl ester (Trihexanoin); $C_{21}H_{38}O_6$; 621-70-5	Nussbaum, E.; Hursh, J. B. J. Phys. Chem. 1958, 62, 81 - 84.

VARIABLES:	PREPARED BY:
T/K: 298.15 - 323.15	W. Gerrard

EXPERIMENTAL VALUES:

T/K	Mol Fraction X_1 x 10^2	Ostwald Coefficient L
298.15	10.5	7.25
310.15	8.75	6.10
323.15	7.34	5.17

The Ostwald coefficient was measured at a very low partial pressure of radon. The carrier gas was nitrogen at a partial pressure of about 100 kPa. The liquid was therefore saturated with nitrogen at the experimental pressure.

The mole fraction solubility of radon at a pressure of 101.325 kPa was calculated by the compiler. It was assumed that the Ostwald coefficient was independent of pressure and that the gram-mole volume of radon is 22,290 cm^3 at 273.15 K and 101.325 kPa.

Smoothed Data: $\Delta G°/J\ mol^{-1} = -\ RT\ \ln X_1 = -11,467 + 57.208$

Std. Dev. $\Delta G° = 4.9$,   Coef. Corr. = 0.9999

$\Delta H°/J\ mol^{-1} = -11,467$, $\Delta S°/J\ K^{-1}\ mol^{-1} = -57.208$

T/K	Mol Fraction X_1 x 10^2	$\Delta G°/J\ mol^{-1}$	
298.15	10.5	5,589.9	The smoothed data
303.15	9.72	5,875.9	fit was added by
308.15	9.03	6,161.9	the Volume Editor.
313.15	8.40	6,448.0	
318.15	7.84	6,734.0	
323.15	7.33	7,020.1	

AUXILIARY INFORMATION

METHOD:	SOURCE AND PURITY OF MATERIALS:
The concentration of radon was determined by measurement of radio-activity in samples withdrawn from the liquid and from the gas phases. The procedure was stated to be similar in principle to that of Boyle (1).	1. Radon-222. 2. Trihexanoin. Eastman Chemical Co. Highest Grade.

APPARATUS/PROCEDURE:	ESTIMATED ERROR:
A cylindrical glass vessel with a stopcock at each end. The gamma-rays emitted were estimated by a sodium iodide scintillation counter.	
	REFERENCES: 1. Boyle, R. W. Phil. Mag. 1911, 22, 840.

COMPONENTS:	ORIGINAL MEASUREMENTS:
1. Radon; $^{222}_{86}$Rn; 14859-67-7 2. Octanoic acid; 1,2,3-propanetriyl ester (Trioctanoin); $C_{27}H_{50}O_6$; 538-23-8	Nussbaum, E.; Hursh, J. B. J. Phys. Chem. 1958, 62, 81 - 84.

VARIABLES:	PREPARED BY:
T/K: 298.15 - 323.15	W. Gerrard

EXPERIMENTAL VALUES:

T/K	Mol Fraction $X_1 \times 10^2$	Ostwald Coefficient L
298.15	13.3	7.55
310.15	10.8	6.12
323.15	9.8	5.63

The Ostwald coefficient was measured at a very low partial pressure of radon. The carrier gas was nitrogen at a partial pressure of about 100 kPa. The liquid was therefore saturated with nitrogen at the experimental pressure.

The mole fraction solubility of radon at a pressure of 101.325 kPa was calculated by the compiler. It was assumed that the Ostwald coefficient was independent of pressure and that the gram-mole volume of radon is 22,290 cm^3 at 273.15 K and 101.325 kPa.

Smoothed Data: $\Delta G°/J \, mol^{-1} = - RT \ln X_1 = -9,690.1 + 49.440 \, T$

Std. Dev. $\Delta G° = 82.6$, Coef. Corr. = 0.9912

$\Delta H°/J \, mol^{-1} = -9,690.1$, $\Delta S°/J \, K^{-1} \, mol^{-1} = -49.440$

T/K	Mol Fraction $X_1 \times 10^2$	$\Delta G°/J \, mol^{-1}$
298.15	13.0	5,050.5
303.15	12.2	5,297.7
308.15	11.5	5,544.9
313.15	10.8	5,792.1
318.15	10.2	6,039.4
323.15	9.63	6,286.6

The smoothed data fit was added by the Volume Editor.

AUXILIARY INFORMATION

METHOD:

The concentration of radon was determined by measurement of radioactivity in samples withdrawn from the liquid and from the gas phases. The procedure was stated to be similar in principle to that of Boyle (1).

SOURCE AND PURITY OF MATERIALS:

1. Radon-222.

2. Trioctanoin. Eastman Chemical Co. Highest grade.

APPARATUS/PROCEDURE:

A cylindrical glass vessel with a stopcock at each end. The gamma-rays emitted were estimated by a sodium iodide scintillation counter.

ESTIMATED ERROR:

REFERENCES:

1. Boyle, R. W. Phil. Mag. 1911, 22, 840.

COMPONENTS:	ORIGINAL MEASUREMENTS:
1. Radon; $^{222}_{86}$Rn; 14859-67-7 2. Trichloromethane (Chloroform); CHCl$_3$; 67-66-3	Ramstedt, E. Le Radium 1911, **8**, 253-256.

VARIABLES: T/K: 255.15 - 291.15	PREPARED BY: W. Gerrard

EXPERIMENTAL VALUES:

T/K	Mol Fraction X_1 x 10^2	Ostwald Coefficient L
255.15	9.18	28.5
273.15	6.71	20.5
291.15	4.83	15.08

The Ostwald coefficient was measured at a radon partial pressure of less than 0.1 kPa at equilibrium. The radon was present in a carrier gas (air or nitrogen) at an initial pressure of about 101 kPa.

The mole fraction solubility was calculated by the compiler. It was assumed that the Ostwald coefficient was independent of pressure and that the gram-mole volume of radon is 22,290 cm^3 at 273.15 K and 101.325 kPa.

Smoothed Data: ΔG^O/J mol^{-1} = -RT ln X_1 = -11,040 + 63.037 T

Std. Dev. ΔG = 37.8, Coef. Corr. = 0.9994

ΔH^O/J mol^{-1} = -11,040, ΔS^O/J K^{-1} mol^{-1} = -63.037

T/K	Mol Fraction X_1 x 10^2	ΔG^O/J mol^{-1}
258.15	8.73	5,233.5
263.15	7.92	5,548.7
268.15	7.21	5,863.9
273.15	6.58	6,179.0
278.15	6.03	6,494.2
283.15	5.54	6,809.4
288.15	5.11	7,124.6
293.15	4.72	7,439.8
298.15	3.79	7,755.0

The smoothed data fit was added by the Volume Editor.

AUXILIARY INFORMATION

METHOD: Measurement of radioactivity in the liquid and in the gaseous phase.	SOURCE AND PURITY OF MATERIALS: 1. Radon-222. 2. Trichloromethane (Chloroform). Dried and distilled.
APPARATUS/PROCEDURE: Two tubes connected by a wide tap. To determine concentration of radon, each tube is placed in a condenser.	ESTIMATED ERROR: REFERENCES:

COMPONENTS:	ORIGINAL MEASUREMENTS:
1. Radon; $^{222}_{86}$Rn; 14859-67-7 2. Carbon Disulfide; CS_2; 75-15-0	Ramstedt, E. Le Radium 1911, 8, 253-256.

VARIABLES:
     T/K: 255.15 - 291.15
	PREPARED BY:
	
	     W. Gerrard

EXPERIMENTAL VALUES:

T/K	Mol Fraction X_1 x 10^2	Ostwald Coefficient L
255.15	12.2	50.3
273.15	8.10	33.4
291.15	5.53	23.14

The Ostwald coefficient was measured at a radon partial pressure of less than 0.1 kPa at equilibrium. The radon was present in a carrier gas (air or nitrogen) at an initial pressure of about 101 kPa.

The mole fraction solubility was calculated by the compiler. It was assumed that the Ostwald coefficient was independent of pressure and that the gram-mole volume of radon is 22,290 cm^3 at 273.15 K and 101.325 kPa.

Smoothed Data: ΔG^o/J mol^{-1} = -RT ln X_1 = -13,584 + 70.696 T

         Std. Dev. ΔG = 15.9, Coef. Corr. = 0.9999

      ΔH^o/J mol^{-1} = -13,584, ΔS^o/J K^{-1} mol^{-1} = -70.696

T/K	Mol Fraction X_1 x 10^2	ΔG^o/J mol^{-1}	
258.15	11.4	4,665.8	
263.15	10.1	5,019.2	
268.15	8.98	5,372.7	The smoothed
273.15	8.03	5,726.2	data fit was
278.15	7.22	6,079.7	added by the
283.15	6.50	6,433.2	Volume Editor.
288.15	5.88	6,786.6	
293.15	5.34	7,140.1	
298.15	4.87	7,493.6	

AUXILIARY INFORMATION

METHOD: Measurement of radioactivity in the liquid and in the gaseous phase.	SOURCE AND PURITY OF MATERIALS: 1. Radon-222. 2. Carbon Disulfide. Dried and distilled.

APPARATUS/PROCEDURE:
   Two tubes connected by a wide tap. To determine concentration of radon, each tube is placed in a condenser.
 	
   The mole fraction solubilities calculated above are for a radon partial pressure of 101.325 kPa.
	ESTIMATED ERROR:
	
	 REFERENCES:

COMPONENTS:	ORIGINAL MEASUREMENTS:
1. Radon; $^{222}_{86}$Rn; 14859-67-7 2. Benzenamine (Aniline); C_6H_7N; 62-53-3	Ramstedt, E. Le Radium 1911, 8, 253-356.

VARIABLES:	PREPARED BY:
T/K: 273.15 - 291.15	W. Gerrard

EXPERIMENTAL VALUES:

T/K	Mol Fraction $X_1 \times 10^2$	Ostwald Coefficient L
273.15	1.75	4.43
291.15	1.44	3.80

The Ostwald coefficient was measured at a radon partial pressure of less than 0.1 kPa at equilibrium. The radon was present in a carrier gas (air or nitrogen) at an initial pressure of about 101 kPa.

The mole fraction solubility was calculated by the compiler. It was assumed that the Ostwald coefficient was independent of pressure and that the gram-mole volume of radon is 22,290 cm^3 at 273.15 K and 101.325 kPa.

Smoothed Data: $\Delta G^O / J\ mol^{-1} = -RT\ \ln X_1 = -7162.2 + 59.857\ T$

$$\Delta H^O / J\ mol^{-1} = -7,162.2, \quad \Delta S^O / J\ K^{-1}\ mol^{-1} = -59.857$$

T/K	Mol Fraction $X_1 \times 10^2$	$\Delta G^O / J\ mol^{-1}$
273.15	1.75	9,187.7
278.15	1.65	9,486.9
283.15	1.57	9,786.2
288.15	1.48	10,086
293.15	1.41	10,385
298.15	1.34	10,684

The smoothed data fit was added by the Volume Editor.

AUXILIARY INFORMATION

METHOD:	SOURCE AND PURITY OF MATERIALS:
Measurement of radioactivity in the liquid and in the gaseous phase.	1. Radon-222. 2. Benzenamine. Dried and distilled.

APPARATUS/PROCEDURE:	ESTIMATED ERROR:
Two tubes connected by a wide tap. To determine concentration of radon, each tube is placed in a condenser. The mole fraction solubilities calculated above are for a radon partial pressure of 101.325 kPa.	
	REFERENCES:

321

COMPONENTS:	ORIGINAL MEASUREMENTS:
1. Radon-222; $^{222}_{86}$Rn; 14859-67-7 2. Nitrobenzene; $C_6H_5NO_2$; 98-95-3	Traubenberg, H.F.R. Phys. Z. 1904, 5, 130 - 134.

VARIABLES:	PREPARED BY:
	W. Gerrard

EXPERIMENTAL VALUES:

T/K	Ratio compared with water*	Ostwald Coefficient L
"Room temperature"	18.06	5.42

*water was taken as 0.3

The author reported his solubility value as an absorption coefficient based
on the concentration ratio of radon in the gas and in the liquid phases.
It has been labelled as an Ostwald coefficient above. The solubility was
measured at a radon partial pressure of less than 0.1 kPa. The liquid was
also saturated with air as a carrier gas at a pressure of about 100 kPa.

AUXILIARY INFORMATION

METHOD: Measurement of radioactivity.	SOURCE AND PURITY OF MATERIALS: Not stated.
APPARATUS/PROCEDURE: Electroscope for the measurement of radioactivity.	ESTIMATED ERROR: Traubenberg's values appear to be low when compared with values reported by other workers.
	REFERENCES:

COMPONENTS:	EVALUATOR:
1. Radon-222; $^{222}_{86}$Rn; 14859-67-7 2. Petroleum Products	William Gerrard Department of Chemistry The Polytechnic of North London Holloway, London N7 8DB U.K. May 1977

CRITICAL EVALUATION:

Petroleum.

There are three reports of the solubility of radon-222 in petroleum.
Hofman's (1) values were based on a primitive electroscopic technique and
are of doubtful accuracy. Lurie (2) cited Hofman's values and gave his own
as being in approximate agreement. Traubenberg's (3) value appears to be
much too small.

Paraffin oil.

There are three reports of the solubility of radon-222 in paraffin oil.
Two of the reports give the temperature as only "room temperature."
Schrodt's (4) value is too large and Traubenberg's (3) value is too small.
Ramstedt's (5) values of the radon Ostwald coefficient at 273.15 and
291.15 K are classed as tentative.

Petroleum ether.

Traubenberg's (3) value is too small.

REFERENCES.

1. Hofman, R. Phys. Z. 1905, 6, 339.

2. Lurie, A. Thesis 1910, University of Grenoble.

3. Traubenberg, H. F. R. Phys. Z. 1904, 5, 130.

4. Schrodt, O. Roentgenpraxis (Leipzig) 1938, 10, 743.

5. Ramstedt, E. Le Radium 1911, 8, 253.

COMPONENTS:	ORIGINAL MEASUREMENTS:
1. Radon-222; $^{222}_{86}$Rn; 14859-67-7 2. Petroleum.	Lurie, A. Thesis University of Grenoble, 1910 Microfilm available. See also Tables annuelles de con- stantes et donnees numeriques de chemie, de physique et de technologie 1913 (for 1911), 2, 401.
VARIABLES: T/K: 255.15 - 333.15	PREPARED BY: W. Gerrard

EXPERIMENTAL VALUES:

T/K	Ostwald Coefficient L
255.15	21.4
276.15	12.6
293.15	9.01
313.15	8.7
333.15	6.9

The author reported a coefficient of absorption, α, which appears to have the same meaning as the ratio of concentrations: concentration of radon in the liquid/concentration of radon in the gas. The coefficient of absorption has been labelled as the Ostwald coefficient.

The coefficient of absorption was measured at a radon partial pressure of less than 0.1 kPa at equilibrium. The radon was present in a carrier gas (air or nitrogen) at an initial pressure of about 101 kPa.

AUXILIARY INFORMATION

METHOD/APPARATUS/PROCEDURE:	SOURCE AND PURITY OF MATERIALS:
The concentration of radon in the gas and liquid phases determined by measurements of radioactivity. An aluminum foil electroscope was used to measure the radioactivity. Diagrams given by Lurie.	1. Radon-222. 2. Petroleum. Purity of liquid not specified.
	DATA CLASS:
	ESTIMATED ERROR:
	REFERENCES:

COMPONENTS:	ORIGINAL MEASUREMENTS:
1. Radon-222; $^{222}_{86}$Rn; 14859-67-7 2. Petroleum	Hofmann, R. Phys. Z. 1905, 6, 339-340.

VARIABLES:	PREPARED BY:
T/K: 255.15 - 333.15	W. Gerrard

EXPERIMENTAL VALUES:

T/K	Ostwald Coefficient L
255.15	22.7
276.15	12.87
293.15	9.55
313.15	8.13
333.15	7.01

The author reported an Absorption coefficient which appears equivalent to the Ostwald coefficient. The coefficient was measured at a radon-222 partial pressure of less than 0.1 kPa at equilibrium. The radon was present in a carrier gas (air or nitrogen) at an initial pressure of about 101 kPa.

AUXILIARY INFORMATION

METHOD:	SOURCE AND PURITY OF MATERIALS:
Determination of concentration of radon in the gas and liquid phases by means of radioactivity.	No specific information.

APPARATUS/PROCEDURE:	ESTIMATED ERROR:
Electroscope, and glass containers for measurement of radioactivity.	Of doubtful accuracy because of the primitive electroscope technique.
	REFERENCES: 1. Traubenberg, H.F.R. Phys. Z. 1904, 5, 130.

COMPONENTS:	ORIGINAL MEASUREMENTS:
1. Radon-222; $^{222}_{86}$Rn; 14859-67-7 2. Petroleum Products	Traubenberg, H. F. R. Phys. Z. 1904, 5, 130 - 134.

VARIABLES:	PREPARED BY:
T/K: "Room Temperature"	W. Gerrard

EXPERIMENTAL VALUES:

T/K	Ratio Compared With Water[*]	Ostwald Coefficient L
Petroleum		
Room Temperature	25.2	7.56
Paraffin Oil		
Room Temperature	14.46	4.34
Petroleum Ether		
Room Temperature	16.19	4.86
"Kaiserol"		
Room Temperature	20.58	6.17

[*]Water taken as 0.3.

The author reported his solubility values as an Absorption coefficient based on the concentration of radon in the gas and liquid phases. We have labelled it as an Ostwald coefficient above. The solubility was measured at a radon partial pressure of less than 0.1 kPa. The liquid was also saturated with air as a carrier gas at a pressure of about 100 kPa.

AUXILIARY INFORMATION

METHOD:	SOURCE AND PURITY OF MATERIALS:
Measurement of radioactivity.	Not stated.

APPARATUS/PROCEDURE:	ESTIMATED ERROR:
Electroscope used for the measurement of radioactivity.	Traubenberg's values appear to be low when compared with values reported by other workers.
	REFERENCES:

COMPONENTS:	ORIGINAL MEASUREMENTS:
1. Radon-222; $^{222}_{86}$Rn; 14859-67-7 2. Paraffin oil.	Ramstedt, E. Le Radium 1911, 8, 253-256.

VARIABLES:	PREPARED BY:
T/K: 273.15 - 291.15	W. Gerrard

EXPERIMENTAL VALUES:

T/K	Ostwald Coefficient L
273.15	12.6
291.15	9.2

The author reported a coefficient of solubility, S, which we have labeled an Ostwald coefficient. The coefficient was measured at a radon-222 partial pressure of less than 0.1 kPa at equilibrium. The radon was present in a carrier gas (air or nitrogen) at an initial pressure of about 101 kPa.

AUXILIARY INFORMATION

METHOD:	SOURCE AND PURITY OF MATERIALS:
Measurement of radioacitivty in the liquid and in the gaseous phase.	1. Radon-222. 2. Paraffin oil. Dried and redistilled.

APPARATUS/PROCEDURE:	ESTIMATED ERROR:
Two tubes connected by a wide tap. To determine concentration of radon, each tube is placed in a condenser.	REFERENCES:

COMPONENTS:	ORIGINAL MEASUREMENTS:
1. Radon-222; $^{222}_{86}$Rn; 14859-67-7 2. Turpentine	Lurie, A. Thesis University of Grenoble, 1910 Microfilm available. See also Tables annuelles de constantes et donnees numeriques de chemie, de physique et de technologie 1913 (for 1911), 2, 401.

VARIABLES:	PREPARED BY:
T/K: 252.15 - 338.15	W. Gerrard May 1977

EXPERIMENTAL VALUES:

T/K	Mol Fraction $X_1 \cdot x\ 10^2$	Ostwald Coefficient L
252.15	–	45.5
273.15	14.1	23.1
291.15	–	16.6
293.15	–	15.9
323.15	–	7.5
338.15	–	4.08

The author reported a coefficient of absorption, α, which appears to have the same meaning as the ratio of concentrations: concentration of radon in the liquid/concentration of radon in the gas.   The coefficient of absorption has been labelled as the Ostwald coefficient.

The coefficient of absorption was measured at a radon partial pressure of less than 0.1 kPa at equilibrium.   The radon was present in a carrier gas (air or nitrogen) at an initial pressure of about 101 kPa.

The mole fraction solubility at 101.325 kPa was calculated by the compiler. It was assumed that the Ostwald coefficient was independent of pressure and that the gram-mole volume of radon is 22,290 cm^3 at 273.15 K and 101.325 kPa.   The turpentine was taken as equivalent to α-pinene, $C_{10}H_{16}$, for the mole fraction calculation.

AUXILIARY INFORMATION

METHOD /APPARATUS/PROCEDURE:	SOURCE AND PURITY OF MATERIALS:
The concentration of radon in the gas and liquid phases determined by measurements of radioactivity. An aluminum foil electroscope was used to measure the radioactivity. Diagrams given by Lurie.	1. Radon-222. 2. Turpentine. Purity of liquid not specified.
	DATA CLASS:
	ESTIMATED ERROR:
	REFERENCES:

COMPONENTS:	EVALUATOR:
1. Radon-222; $^{222}_{86}$Rn; 14859-67-7 2. Olive Oil	William Gerrard Department of Chemistry The Polytechnic of North London Holloway, London N7 8DB U.K. May 1977

CRITICAL EVALUATION:

Lurie (1) reports Ostwald coefficients for radon-222 in olive oil at temperatures of 273.15, 288.15, 313.15 and 333.15 K. Both Strasburger (2) and Schrodt (3) report a "room temperature" Ostwald coefficient which agrees with the Lurie 288.15 K value. Nussbaum and Hursh (4) report Ostwald coefficients at 298.15 and 310.15 K which are about one-third the Lurie values, and may be taken as reliable.

The mole fraction solubility of radon-222 in olive oil was calculated assuming that olive oil is 1,2,3-propanetriyl ester of (Z)-9-octadecenoic acid, or triolein, of molecular weight 885.46. The Nussbaum and Hursh solubility values are preferred.

The tentative values of the thermodynamic properties of solution for the transfer of one mole of radon from the gas at 101.325 kPa (1 atm) to the hypothetical unit mole fraction solution are

$$\Delta G^{o}/J\ mol^{-1} = -RT\ ln\ X_1 = -11.914 + 51.962\ T$$

$$\Delta H^{o}/J\ mol^{-1} = -11.914,\ \Delta S^{o}/J\ K^{-1}\ mol^{-1} = -51.962$$

A table of tentative mole fraction solubility and Gibbs energy values as a function of temperature appears on the radon + olive oil data sheet of Nussbaum and Hursh, and in Table 1 below.

Lawrence, Loomis, Tobias and Turpin (5) cite Bunsen coefficients for radon-222 in olive oil which they attribute to Lurie and state to be given in the International Critical Tables (6). The values are 0.01 times the values reported by Lurie. They are not found in the International Critical Tables, but they are in Seidell and Linke (7) which cite no other of Lurie's data. The radon in olive oil data cited in these references (5,7) are in error and should not be used.

REFERENCES.

1. Lurie, A. Thesis 1910, University of Grenoble.

2. Strasburger, J. Deut. Med. Wochsehr. 1923, 49, 1459.

3. Schrodt, O. Roentgenpraxis (Leipzig) 1938, 10, 743.

4. Nussbaum, E.; Hursh, J. B. J. Phys. Chem. 1958, 62, 81.

5. Lawrence, J. H.; Loomis, W. F.; Tobias, C. A.; Turpin, F. H. J. Physiol. 1946, 105, 197.

6. Washburn, E. W. Editor, International Critical Tables, McGraw Hill Co., New York 1928.

7. Seidell, A.; Linke, W. F. Solubilities of Inorganic and Organic Compounds, American Chemical Society, 1958, 1965.

TABLE 1. The solubility of radon-222 in olive oil. Tentative values of the mole fraction solubility at a radon partial pressure of 101.325 kPa and the Gibbs energy change as a function of temperature.

T/K	Mol Fraction $X_1 \times 10^2$	$\Delta G^{o}/J\ mol^{-1}$
298.15	23.6	3,578.7
303.15	21.8	3,838.5
308.15	20.2	4,098.3

COMPONENTS:	ORIGINAL MEASUREMENTS:

COMPONENTS:

1. Radon-222; $^{222}_{86}Rn$; 14859-67-7

2. Olive oil

ORIGINAL MEASUREMENTS:

Lurie, A.
Thesis
University of Grenoble, 1910
Microfilm available.
See also Tables annuelles de constantes et donnees numeriques de chemie, de physique et de technologie 1913 (for 1911), 2, 401.

VARIABLES:

T/K: 273.15 - 333.15

PREPARED BY:

W. Gerrard

EXPERIMENTAL VALUES:

T/K	Mol Fraction $X_1 \times 10^2$	Ostwald Coefficient L
273.15	66.8	45.9
288.15	-	28.6
313.15	-	18.6
333.15	-	11.1

The author reported a coefficient of absorption, α, which appears to have the same meaning as the ratio of concentrations: concentration of radon in the liquid/concentration of radon in the gas. The coefficient of absorption has been labelled as the Ostwald coefficient.

The coefficient of absorption was measured at a radon partial pressure of less than 0.1 kPa at equilibrium. The radon was present in a carrier gas (air or nitrogen) at an initial pressure of about 101 kPa.

The mole fraction solubility at 101.325 kPa was calculated by the compiler. It was assumed that the Ostwald coefficient was independent of pressure and that the gram-mole volume of radon is 22,290 cm^3 at 273.15 K and 101.325 kPa. The olive oil was taken as equivalent to triolein for the mole fraction calculation.

AUXILIARY INFORMATION

METHOD/APPARATUS/PROCEDURE:

The concentration of radon in the gas and liquid phases determined by measurements of radioactivity. An aluminum foil electroscope was used to measure the radioactivity. Diagrams given by Lurie.

SOURCE AND PURITY OF MATERIALS:

1. Radon-222.

2. Olive oil. Purity of liquid not specified.

DATA CLASS:

ESTIMATED ERROR:

REFERENCES:

COMPONENTS:	ORIGINAL MEASUREMENTS:
1. Radon-222; $^{222}_{86}$Rn; 14859-67-7 2. Olive oil	Nussbaum, E.; Hursh, J.B. J. Phys. Chem. 1958, 62, 81-84.

VARIABLES:
 	
    T/K: 298.15 - 310.15
	PREPARED BY:
	
	    W. Gerrard

EXPERIMENTAL VALUES:

T/K	Mol Fraction X_1 x 10^2	Ostwald Coefficient L
298.15	23.6	7.70
310.15	19.6	6.24

The Ostwald coefficient was measured at a very low partial pressure of radon-222. The carrier gas was nitrogen at a partial pressure of about 100 kPa. The liquid was therefore saturated with nitrogen at the experimental pressure.

The mole fraction solubility of radon-222 at a pressure of 101.325 kPa was calculated by the compiler. It was assumed that the Ostwald coefficient was independent of pressure and that the gram-mole volume of radon is 22,290 cm^3 at 273.15 K and 101.325 kPa. Olive oil was taken as 1,2,3-propanetriyl ester of (Z)-9-octadecenoic acid (triolein); $C_{57}H_{104}O_6$; 122-32-7 of molecular weight 885.46.

Smoothed Data: ΔH^o/J mol^{-1} = -11,914, ΔS^o/J K^{-1} mol^{-1} = -51.962

T/K	Mol Fraction X_1 x 10^2	ΔG^o/J mol^{-1}
298.15	23.6	3,578.7
303.15	21.8	3,838.5
308.15	20.2	4,098.3

$$\Delta G^o/\text{J mol}^{-1} = -11,914 + 51.962\ T$$

The smoothed data table was added by the Editor.

AUXILIARY INFORMATION

METHOD: The concentration or radon was determined by measurement of radioactivity in samples withdrawn from the liquid and from the gas phases. The procedure was stated to be similar in principle to that of Boyle (1).	SOURCE AND PURITY OF MATERIALS: 1. Radon-222.
APPARATUS/PROCEDURE: A cylindrical glass vessel with a stopcock at each end. The gamma-rays emitted were estimated by a sodium iodide scintillation counter.	ESTIMATED ERROR: REFERENCES: 1. Boyle, R.W. Phil. Mag. 1911, 22, 840.

COMPONENTS:	ORIGINAL MEASUREMENTS:
1. Radon-222; $^{222}_{86}$Rn; 14859-67-7 2. Fats and Oils for Ointment Base.	Schrodt, O. Roentgenpraxis (Leipzig) 1938, 10, 743.

VARIABLES:
 	
  T/K: "Room temperature" (?)
	PREPARED BY:
	    W. Gerrard

EXPERIMENTAL VALUES:

T/K	Ostwald Coefficient L
	Vaseline
-	6.68
	Lard
-	7.73
	Beeftallow
-	4.1
	Lanolin
-	7.3
	Eucerin anhydricum
-	6.8
	Eucerin cum aqua
-	7.27
	Unguentolan
-	9.26
	Cocoabutter
-	5.5
	Olive oil
-	28.1
	Paraffin oil
-	11.7

The absorption coefficient (Ostwald coefficient) was based on the ratio of concentrations of radon in the liquid and gaseous phases at very small partial pressures of radon-222.

AUXILIARY INFORMATION

METHOD:

Concentrations of radon were measured by the determination of radioactivity at very small partial pressures of radon.

SOURCE AND PURITY OF MATERIALS:
  No specific statement.

APPARATUS/PROCEDURE:

Glass bulbs and electroscope. Diagrams given by Schrodt.

ESTIMATED ERROR:

REFERENCES:

COMPONENTS:	ORIGINAL MEASUREMENTS:
1. Radon-222; $^{222}_{86}$Rn; 14859-67-7 2. Colza oil	Lurie, A. Thesis University of Grenoble, 1910 Microfilm available. See also Tables annuelles de constantes et donnees numeriques de chemie, de physique et de technologie 1913 (for 1911), 2, 401.

VARIABLES:
 	
    T/K: 270.15 - 473.15
	PREPARED BY:
	    W. Gerrard

EXPERIMENTAL VALUES:

T/K	Ostwald Coefficient L
270.15	51.2
283.15	35.3
293.15	26.1
323.15	16.75
373.15	6.25
473.15	3.3

The author reported a coefficient of absorption, α, which appears to have the same meaning as the ratio of concentrations: concentration of radon in the liquid/concentration of radon in the gas. The coefficient of absorption has been labelled as the Ostwald coefficient.

The coefficient of absorption was measured at a radon partial pressure of less than 0.1 kPa at equilibrium. The radon was present in a carrier gas (air or nitrogen) at an initial pressure of about 101 kPa.

AUXILIARY INFORMATION

METHOD /APPARATUS/PROCEDURE:	SOURCE AND PURITY OF MATERIALS:
The concentration of radon in the gas and liquid phases determined by measurements of radioactivity. An aluminum foil electroscope was used to measure the radioactivity. Diagrams given by Lurie.	1. Radon-222. 2. Colza oil. Purity of liquid not specified.
	ESTIMATED ERROR:
	REFERENCES:

COMPONENTS:	ORIGINAL MEASUREMENTS:
1. Radon-222; $^{222}_{86}$Rn; 14859-67-7 2. Poppy oil	Lurie, A. Thesis University of Grenoble, 1910 Microfilm available. See also Tables annuelles de con- stantes et donnees numeriques de chemie, de physique et de technologie 1913 (for 1911), 2, 401.
VARIABLES: T/K: 268.15 - 363.15	PREPARED BY: W. Gerrard

EXPERIMENTAL VALUES:

T/K	Ostwald Coefficient L
268.15	50.7
289.15	30.2
313.15	19.05
338.15	12.4
363.15	8.4

The author reported a coefficient of absorption, α, which appears to have the same meaning as the ratio of concentrations: concentration of radon in the liquid/concentration of radon in the gas. The coefficient of absorption has been labelled as the Ostwald coefficient.

The coefficient of absorption was measured at a radon partial pressure of less than 0.1 kPa at equilibrium. The radon was present in a carrier gas (air or nitrogen) at an initial pressure of about 101 kPa.

AUXILIARY INFORMATION

METHOD /APPARATUS/PROCEDURE:	SOURCE AND PURITY OF MATERIALS:
The concentration of radon in the gas and liquid phases determined by measurements of radioactivity. An aluminum foil electroscope was used to measure the radioactivity. Diagrams given by Lurie.	1. Radon-222. 2. Poppy oil. Purity of liquid not specified.
	ESTIMATED ERROR:
	REFERENCES:

COMPONENTS:	ORIGINAL MEASUREMENTS:
1. Radon-222; $^{222}_{86}Rn$; 14859-67-7 2. Vaseline oil	Lurie, A. Thesis University of Grenoble, 1910 Microfilm available. See also Tables annuelles de constantes et donnees numeriques de chemie, de physique et de technologie 1913 (for 1911), 2, 401.
VARIABLES: T/K: 263.15 - 323.15	PREPARED BY: W. Gerrard

EXPERIMENTAL VALUES:

T/K	Ostwald Coefficient L
263.15	23.67
273.15	15.16
288.15	11.1
298.15	8.36
323.15	6.6

The author reported a coefficient of absorption, α, which appears to have the same meaning as the ratio of concentrations: concentration of radon in the liquid/concentration of radon in the gas. The coefficient of absorption has been labelled as the Ostwald coefficient.

The coefficient of absorption was measured at a radon partial pressure of less than 0.1 kPa at equilibrium. The radon was present in a carrier gas (air or nitrogen) at an initial pressure of about 101 kPa.

AUXILIARY INFORMATION

METHOD/APPARATUS/PROCEDURE:	SOURCE AND PURITY OF MATERIALS:
The concentration of radon in the gas and liquid phases determined by measurements of radioactivity. An aluminum foil electroscope was used to measure the radioactivity. Diagrams given by Lurie.	1. Radon-222. 2. Vaseline oil. Purity of liquid not specified.
	ESTIMATED ERROR:
	REFERENCES:

COMPONENTS:	ORIGINAL MEASUREMENTS:
1. Radon-222; $^{222}_{86}$Rn; 14859-67-7 2. Animal Fats	Nussbaum, E.; Hursh, J.B. J. Phys. Chem. 1958, 62, 81-84.

| VARIABLES:
 T/K: 310.15 | PREPARED BY:
 W. Gerrard |

EXPERIMENTAL VALUES:

T/K	Ostwald Coefficient L
Butter fat	
310.15	5.91
Rat fatty acids (extracted)	
310.15	5.85
Human fat (extracted)	
310.15	6.33

The Ostwald coefficient was measured at a very low partial pressure of radon-222. The carrier gas was nitrogen at a partial pressure of about 100 kPa. The liquid was therefore saturated with nitrogen at the experimental pressure.

AUXILIARY INFORMATION

METHOD: The concentration of radon was determined by measurement of radioactivity in samples withdrawn from the liquid and from the gas phases. The procedure was stated to be similar in principle to that of Boyle (1).	SOURCE AND PURITY OF MATERIALS: 1. Radon-222.
APPARATUS/PROCEDURE: A cylindrical glass vessel with a stopcock at each end. The gamma-rays emitted were estimated by a sodium iodide scintillation counter.	ESTIMATED ERROR: REFERENCES: 1. Boyle, R.W. Phil. Mag. 1911, 22, 840.

COMPONENTS:	ORIGINAL MEASUREMENTS:
1. Radon-222; $^{222}_{86}$Rn; 14859-67-7 2. Rat Tissues	Nussbaum, E.; Hursh, J.B. Science 1957, 125, 552.
VARIABLES: T/K: 310.15	PREPARED BY: W. Gerrard

EXPERIMENTAL VALUES:

T/K	Ostwald Coefficient L ± Std. Dev.
Omental Fat	
310.15	4.83 ± 0.07
Venous Blood	
310.15	0.405 ± 0.016
Brain	
310.15	0.309 ± 0.008
Kidney	
310.15	0.285 ± 0.012
Liver	
310.15	0.306 ± 0.004
Heart	
310.15	0.221 ± 0.013
Testis	
310.15	0.184 ± 0.007
Muscle	
310.15	0.154 ± 0.005

The Ostwald coefficient was measured at a very low partial pressure of radon. The carrier gas was nitrogen at a partial pressure of about 100 kPa. The liquid was therefore saturated with nitrogen at the experimental pressure.

AUXILIARY INFORMATION

METHOD:

The concentration of radon was determined by measurement of radioactivity in samples withdrawn from the liquid and from the gas phases. The procedure was stated to be similar in principle to that of Boyle (1).

SOURCE AND PURITY OF MATERIALS:

1. Radon-222.

2. Rat Tissues. Adult Rochester Wistar rats breathed air containing 0.5 - 5 µc dm^{-3} radon-222 from 30 minues to 48 hours. Rats were sacrificed by the introduction of CO to the breathing chamber. Specified rat tissues were dissected, placed in tarred test tubes and sealed.

APPARATUS/PROCEDURE:

A cylindrical glass vessel with a stopcock at each end. The gamma-rays emitted were estimated by a sodium iodide scintillation counter.

ESTIMATED ERROR:

REFERENCES:

1. Boyle, R.W.
 Phil. Mag. 1911, 22, 840.

COMPONENTS:	ORIGINAL MEASUREMENTS:
1. Radon-222; $^{222}_{86}$Rn; 14859-67-7 2. Blood	Knaffle-Lenz, E. Z. Balneolgie 1912, 5, nr. 14.
VARIABLES: T/K: 310.15	PREPARED BY: W. Gerrard

EXPERIMENTAL VALUES:

Knaffle-Lenz reported that at 310.15 K the
absorption coefficient for blood is 0.42
times that for water.

AUXILIARY INFORMATION

METHOD:	SOURCE AND PURITY OF MATERIALS:
APPARATUS/PROCEDURE:	ESTIMATED ERROR:
	REFERENCES:

COMPONENTS:	ORIGINAL MEASUREMENTS:
1. Radon-222; $^{222}_{86}$Rn; 14859-67-7 2. Egg Lecithin	Tasca, M. Radiol. Med. (Milan) 1940, 27, 401.

VARIABLES:	PREPARED BY:
	W. Gerrard

EXPERIMENTAL VALUES:

Tasca reported the the "quantity" of radon retained by egg
lecithin is 40 time that retained by an equal volume of
water, at a temperature which was not clearly stated, but
could have been room temperature.

Nussbaum and Hursh (1) cite Tasca and express this solubility
as the Ostwald coefficient, 6.4 at 310.15 K, presumably obtained
from 40 x 0.16, where 0.16 is the radon Ostwald coefficient in
water at 310.15 K.

AUXILIARY INFORMATION

METHOD:	SOURCE AND PURITY OF MATERIALS:

APPARATUS/PROCEDURE:	ESTIMATED ERROR:
	REFERENCES: 1. Nussbaum, E.; Hursh, J. B. J. Phys. Chem. 1958, 62, 81.

COMPONENTS:	EVALUATOR:
1. Radon-219 (Actinium Emanation); $^{219}_{86}$Rn; 14835-02-0 2. Liquids	William Gerrard Department of Chemistry The Polytechnic of North London Holloway, London N7 8DB United Kingdom

CRITICAL EVALUATION:

The half life of radon-219 (actinon) is only 3.92 seconds. The mean values of the partition coefficient show a variation from 5 to 10 % (one is 20 %) from the observed maximum. The partition coefficient for the non-aqueous liquid, or the aqueous solution, was given as that compared with water, taken to be 2. The evaluator contends that it is much more reliable to take the observed radon-222 values as those for radon-219 (actinon).

Hevesy's (1,2) results for the solubility of radon-219 in various liquids are on the following data sheets.

1. Hevesy, G. Phys. Z. 1911, 12, 1214.

2. Hevesy, G. J. Phys. Chem. 1912, 16, 429.

COMPONENTS:	ORIGINAL MEASUREMENTS:
1. Radon-219 (Actinium Emanation); $^{219}_{86}$Rn; 14835-02-0 2. Water; H_2O; 7732-18-5 3. Electrolytes	Hevesy, G. Phys. Z. 1911, 12, 1214 J. Phys. Chem. 1912, 16, 429-453.

VARIABLES:
     T/K: "Ordinary Temperatures"
	PREPARED BY:
	       W. Gerrard

EXPERIMENTAL VALUES:

T/K	Absorption Relative to Water	Ostwald Coefficient L
Water; H_2O; 7732-18-5		
"Ordinary Temperatures"	(1)	2
Potassium chloride; KCl; 7447-40-7		
"Ordinary Temperatures"	0.9	1.8
Sulfuric Acid (conc.); H_2SO_4; 7664-93-9		
"Ordinary Temperatures"	0.95	1.9

The author reported a partition coefficient of radon-219 between liquid and air relative to the partition coefficient of radon-219 between air and water, which was taken as 2. We have called the partition coefficient an Ostwald coefficient.

The Ostwald coefficient was measured at a radon partial pressure of less than 0.1 kPa at equilibrium. The radon was present in air as a carrier gas.

AUXILIARY INFORMATION

METHOD: Dried air was passed over a radon-219 preparation, and then through a column of liquid in which part was absorbed, and the remainder passed to the ionization chamber of an electroscope.	SOURCE AND PURITY OF MATERIALS: No details.
APPARATUS/PROCEDURE: Assembly for the "dynamical flow method" as distinct from the shaking method. Radioactivity was measured. The partial pressure of radon-219 was extremely small.	ESTIMATED ERROR: $\delta L/L = 0.05 - 0.10$ REFERENCES:

COMPONENTS:	ORIGINAL MEASUREMENTS:
1. Radon-219 (Actinium Emanation); $^{219}_{86}$Rn; 14835-02-0 2. Hydrocarbons	Hevesy, G. Phys. Z. 1911, 12, 1214 J. Phys. Chem. 1912, 16, 429-453.
VARIABLES: T/K: "Ordinary Temperatures"	PREPARED BY: W. Gerrard

EXPERIMENTAL VALUES:

T/K	Absorption Relative to Water	Ostwald Coefficient L
Benzene; C_6H_6; 71-43-2		
"Ordinary Temperatures"	1.7	3.4
Toluene; C_7H_8; 108-88-3		
"Ordinary Temperatures"	1.8	3.6
Petroleum		
"Ordinary Temperatures"	1.9	3.8

The author reported a partition coefficient of radon-219 between liquid and air relative to the partition coefficient of radon-219 between air and water, which was taken as 2. We have called the partition coefficient an Ostwald coefficient.

The Ostwald coefficient was measured at a radon partial pressure of less than 0.1 kPa at equilibrium. The radon was present in air as a carrier gas.

AUXILIARY INFORMATION

METHOD: Dried air was passed over a radon-219 preparation, and then through a column of liquid in which part was absorbed, and the remainder passed to the ionization chamber of an electroscope.	SOURCE AND PURITY OF MATERIALS: No details.
APPARATUS/PROCEDURE: Assembly for the "dynamical flow method" as distinct from the shaking method. Radioactivity was measured. The partial pressure of radon-219 was extremely small.	ESTIMATED ERROR: $\delta L/L = 0.05 - 0.10$ REFERENCES:

COMPONENTS:	ORIGINAL MEASUREMENTS:
1. Radon-219 (Actinium Emanation); $^{219}_{86}Rn$; 14835-02-0 2. Oxygen or sulfur containing organic compounds	Hevesy, G. Phys. Z. 1911, 12, 1214 J. Phys. Chem. 1912, 16, 429-453.
VARIABLES: T/K: "Ordinary Temperatures"	PREPARED BY: W. Gerrard

EXPERIMENTAL VALUES:

T/K	Absorption Relative to Water	Ostwald Coefficient L
Ethanol; C_2H_6O; 64-17-5		
"Ordinary Temperatures"	1.1	2.2
Pentanol-1 (Amyl Alcohol); $C_5H_{12}O$; 71-41-0		
"Ordinary Temperatures"	1.6	3.2
Benzaldehyde; C_7H_6O; 100-52-7		
"Ordinary Temperatures"	1.7	3.4
Carbon Disulfide; CS_2; 75-15-0		
"Ordinary Temperatures:	2.1	4.2

The author reported a partition coefficient of radon-219 between liquid and air relative to the partition coefficient of radon-219 between air and water, which was taken as 2. We have called the partition coefficient an Ostwald coefficient.

The Ostwald coefficient was measured at a radon partial pressure of less than 0.1 kPa at equilibrium. The radon was present in air as a carrier gas.

AUXILIARY INFORMATION

METHOD:	SOURCE AND PURITY OF MATERIALS:
Dried air was passed over a radon-219 preparation, and then through a column of liquid in which part was absorbed, and the remainder passed to the ionization chamber of an electroscope.	No details.

APPARATUS/PROCEDURE:	ESTIMATED ERROR: $\delta L/L = 0.05 - 0.10$
Assembly for the "dynamical flow method" as distinct from the shaking method. Radioactivity was measured. The partial pressure of radon-219 was extremely small.	REFERENCES:

COMPONENTS:	EVALUATOR:
1. Radon-220 (Thorium Emanation); $^{220}_{86}$Rn; 22481-88-7 2. Liquids	William Gerrard Department of Chemistry The Polytechnic of North London Holloway, London N7 8DB United Kingdom

CRITICAL EVALUATION:

The half life of radon-220 (thoron), $^{220}_{86}$Rn, is 54.5 seconds. Therefore the reported absorption coefficients must be accepted with caution. Boyle (1) used a streaming method (diagram given in the original paper) and placed the following liquids in the order of increasing solutility: but did not give actual values: $CuSO_4$ (aqueous) < $CaCl_2$ (aqueous) < water < H_2SO_4 < alcohol < petroleum; "the same as for radon (radon-222)."

Klaus (2) reported the solubility of radon-220 in water and petroleum (see the following two data sheets).

The evaluator contends that it is much more reliable to take the observed values for radon-222 as the values for radon-220 (thoron).

References: 1. Boyle, R.W., Le Radium, 1910, 7, 200. See also

Boyle, R.W., Bull. Macdonald Physics Bldg., 1910, 1, 52.

2. Klaus, A. Phys. Z. 1905, 6, 820.

COMPONENTS:	ORIGINAL MEASUREMENTS:
1. Radon-220 (Thorium Emanation); $^{220}_{86}$Rn; 22481-88-7 2. Water; H_2O; 7732-18-5	Klaus, A. Phys. Zeit., 1905, 6, 820.

VARIABLES:	PREPARED BY:
T/K: about 292	W. Gerrard

EXPERIMENTAL VALUES:

T/K	Ostwald Coefficient L
about 292	1.052

The author reported an Absorption coefficient as (concentration radon-220 in liquid phase)/(concentration radon-220 in gas phase) which we have called an Ostwald coefficient.

The Ostwald coefficient was measured at a radon partial pressure of less than 0.1 kPa at equilibrium. The radon was present in a carrier gas (air or nitrogen) at an initial pressure of about 101 kPa.

AUXILIARY INFORMATION

METHOD:	SOURCE AND PURITY OF MATERIALS:
Measurement of radioactivity by an electroscope.	Not specified.

APPARATUS/PROCEDURE:	ESTIMATED ERROR:
Gas cylinder, 500 cm^3. Absorption vessel, 1.25 dm^3, which is shaken. Author gave a diagram.	
	REFERENCES:

COMPONENTS:	ORIGINAL MEASUREMENTS:
1. Radon-220 (Thorium Emanation); $^{220}_{86}$Rn; 22481-88-7 2. Petroleum	Klaus, A. Phys. Zeit., 1905, 6, 820.
VARIABLES: $\quad$ T/K: about 292	PREPARED BY: $\quad$ W. Gerrard

EXPERIMENTAL VALUES:

T/K	Ostwald Coefficient L
about 292	4.97

The author reported an Absorption coefficient as (concentration radon-220 in liquid phase)/(concentration radon-220 in gas phase) which we have called an Ostwald coefficient.

The Ostwald coefficient was measured at a radon partial pressure of less than 0.1 kPa at equilibrium. The radon was present in a carrier gas (air or nitrogen) at an initial pressure of about 101 kPa.

AUXILIARY INFORMATION

METHOD:	SOURCE AND PURITY OF MATERIALS:
Measurement of radioactivity by an electroscope.	Not specified.
APPARATUS/PROCEDURE: Gas cylinder, 500 cm^3. Absorption vessel, 1.25 dm^3, which is shaken. Author gave a diagram.	ESTIMATED ERROR:
	REFERENCES:

SYSTEM INDEX

Underlined page numbers refer to the evaluation text and those not underlined to the compiled tables for that system.   The compounds are listed
in the order as in the Chemical Abstract indexes, for example toluene is
listed as benzene, methyl- and dimethylsulfoxide is listed as methane,
sulfinylbis-.

A

E

Eucerin cum aqua	+ radon-222	331
Eyes, guinea pig, see guinea pig eyes		

F

Fat, butter, see butter fat
    dog, see dog fat
    human, see human fat
    rat, see rat fat
    subcutaneous, guinea pig, see guinea pig subcutaneous
        fat

Ferrate(4-) hexakis (cyano-C)-, tetrapotassium (OC-9-11)-		
(aqueous)	+ radon-222	<u>238 - 241</u>, 253
Ferrocyanide, potassium, see ferrate (4-), hexakis		
(cyano-C)-tetrapotassium (OC-9-11)		
Ferrous sulfate, see sulfuric acid iron (2+) salt		
Fluorobenzene, see benzene, fluoro-		
Formic acid	+ radon-222	290
Freon 113, see ethane,1,1,2-trichloro-1,2,2-trifluoro-		
Freon 12, see methane, dichlorodifluoro-		

G

Gasoline	+ krypton	68
Glycerol, see 1,2,3-propanetriol		
hexanoate, see hexanoic acid, 1,2,3-propanetriyl ester		
octanoate, see octanoic acid, 1,2,3-propanetriyl ester		
triacetate, see 1,2,3-propanetriol, triacetate		
tributurate, see butanoic acid, 1,2,3-propanetriyl		
ester		
Glycine, N,N'-1,2-ethanediylbis(N-carboxymethyl)-,		
(aqueous)	+ krypton-85	125
Guinea pig adrenals	+ krypton-85	129
blood	+ krypton-85	125, 130
bone marrow	+ krypton-85	129
brain (in sodium chloride solution)		
	+ krypton-85	130
brain	+ krypton-85	130
brain (in sodium chloride solution)		
	+ krypton	133
eyes (whole)	+ krypton-85	130
heart	+ krypton-85	130
kidneys	+ krypton-85	130
large intestine	+ krypton-85	129
liver	+ krypton-85	129
lungs	+ krypton-85	129
lymph nodes	+ krypton-85	129
muscle	+ krypton-85	130
omental fat	+ krypton-85	130
ovaries	+ krypton-85	130
seminal vesicles	+ krypton-85	130
small intestine	+ krypton-85	129
spleen	+ krypton-85	130
stomach	+ krypton-85	130
subcutaneous fat	+ krypton-85	129
testes	+ krypton-85	130
thymus	+ krypton-85	129
thyroids	+ krypton-85	129
uterus	+ krypton-85	130

H

Hamster blood, Chinese, see Chinese hamster blood		
Heart, guinea pig, see guinea pig heart		
Hemoglobin, see bovine hemoglobin, dog hemoglobin, human		
hemoglobin		
Heptane	+ krypton	30
	+ xenon	157
3-methyl-	+ krypton	31
Heptanoic acid	+ radon-222	303
Hexadecane	+ krypton	49

H

Hexadecane	+ xenon	161
Hexafluorobenzene, see benzene, hexafluoro-		
Hexane	+ krypton	<u>26</u>, 27 – 29
	+ radon-222	262
	+ xenon	156
2,3-dimethyl-	+ krypton	32
2,4-dimethyl-	+ krypton	33
Hexanoic acid	+ radon-222	302
acid, 1,2,3-propanetriyl ester	+ radon-222	316
Horse heart cyanometmyoglobin, metmyoglobin and		
myoglobin (aqueous)	+ xenon	220
Human albumin (in sodium chloride aqueous solution)		
	+ krypton	126
albumin (in sodium chloride aqueous solution)		
	+ xenon	208, 216, 217
blood	+ krypton	123
blood	+ radon-222	337
blood	+ xenon	211
blood	+ xenon-133	212– 214
brain grey matter	+ xenon-133	225, 226
brain white matter	+ xenon-133	225, 226
erythrocytes	+ xenon-133	215
fat	+ krypton	<u>114–115</u>, 116 – 117
fat	+ radon-222	335
fat	+ xenon	<u>199</u>, 202
hemaglobin (aqueous)	+ krypton	123
hemaglobin (aqueous)	+ krypton-85	124
hemaglobin (aqueous)	+ xenon	199 – 200
		208, <u>209</u>, 211
liver tissue	+ krypton-85	131
liver tissue	+ xenon-133	223
methemoglobin (aqueous)	+ xenon	199 – 200, 208
methemoglobin (saline)	+ xenon-133	<u>199 – 200</u>, 219
red blood cells	+ krypton-85	121
red blood cells	+ xenon-133	213, 214, 218
Hydrochloric acid	+ krypton	<u>12 – 13</u>, 16

I

Iodide, potassium, see potassium iodide
        tetramethyl ammonium, see methanaminium, N,N,N-
           trimethyl iodide
Iodobenzene, see benzene, iodo-
*Iso*butanol, see 1-propanol, 2-methyl-

K

Kaiserol	+ radon-222	325
Kidneys, guinea pig, see guinea pig kidneys		
Koppers emulsion K-900	+ xenon	191

L

Lanolin	+ radon-222	331
Lard	+ radon-222	331
Large intestine, see guinea pig large intestine		
Lauric acid, see dodecanoic acid		
Lead nitrate, see nitric acid, lead salt		
Lecithin	+ krypton-85	120
	+ radon-222	338
	+ xenon-133	206
Linoleic acid, see 9,12-octadecadienoic acid		
Lithium chloride (aqueous)	+ krypton	<u>12 – 13</u>, 21
Liver, see guinea pig liver		
Lungs, see guinea pig lungs		
Lymph nodes, see guinea pig lymph nodes		

M

Mercury chloride (aqueous)	+ radon-222	<u>238 – 241</u>, 247
Mesitylene, see benzene 1,3,5-trimethyl-		
Methanaminium, N,N,N-trimethyl-, iodide	+ krypton	<u>12 – 13</u>, 15

P

Pentadecane	+ krypton	49
	+ xenon	161
Pentane	+ krypton	25
	+ xenon	155
2,2,4-trimethyl-	+ krypton	34
2,2,4-trimethyl-	+ xenon	158
Pentanoic acid	+ radon-222	301
1-Pentanol	+ krypton	78
	+ radon-222	284, 285
	+ radon-219	339, 342

Perfluorohexane, see benzene, hexafluoro-
       methylcyclohexane, see cyclohexane, undecafluoro-
        (trifluoromethyl)-
       tributylamine, see 1-butanamine, 1,1,2,2,3,3,4,4,4-
        nonafluoro-N,N-bis (nonafluorobutyl)-
Permanganate, potassium, see permanganic acid, potassium
       salt

Permanganic acid, potassium salt (aqueous)	+ radon-222	238 - 241, 254
Petroleum	+ krypton	68
	+ radon-219	339, 341
	+ radon-220	343, 345
	+ radon-222	322, 323 - 325
ether	+ radon-222	322, 325

Phosphate, tricresyl-, see phosphoric acid, tris(methyl-
       phenyl) ester

Phosphoric acid, sodium salt (phosphate buffer in water)	+ krypton	12 - 13, 23
sodium salt (phosphate buffer in water)	+ xenon	149, 154
tris (methylphenyl) ester	+ krypton	106
Pine oil	+ xenon	191
oil (ternary)	+ xenon	191
Plasma	+ krypton-85	121, 124, 127
	+ xenon-133	207, 213 - 215 218
Poppy oil	+ radon-222	333
Potassium bromide (aqueous)	+ krypton	12 - 13 21
chloride (aqueous)	+ krypton	12 - 13, 21
chloride (aqueous)	+ radon-219	349, 340
chloride (aqueous)	+ radon-222	238 - 241, 255

     ferrocyanide, see ferrate (4-), hexakis
      (cyano-C)-, tetrapotassium (OC-9-11)-

iodide (aqueous)	+ krypton	12 - 13, 21
iodide (aqueous)	+ xenon	149, 151

     nitrate, see nitric acid, potassium salt
     permanganate, see permanganic acid,
      potassium salt

1,2,3-Propanetriol	+ krypton	81
	+ radon-222	286, 287
triacetate	+ radon-222	314
Propanoic acid	+ radon-222	294, 295 - 296
2-methyl-	+ radon-222	298
1-Propanol	+ krypton	75
	+ radon-222	278
	+ xenon	170
2-methyl-	+ krypton	77
2-methyl-	+ radon-222	282
2-Propanol	+ radon-222	279
2-Propanone	+ krypton	83
	+ radon-222	289
Propene	+ krypton (high pressure) 50, 51	
2-Propenoic acid	+ radon-222	297

R

Rabbit blood	+ krypton-85	127
choroid layer	+ krypton-85	127
erythrocytes	+ krypton-85	127
leg muscle (saline homogenate)	+ krypton	128
leg muscle (saline homogenate)	+ xenon	222

REGISTRY NUMBER INDEX

Underlined page numbers refer to evaluation text and those not underlined to compiled tables.

REGISTRY NUMBER INDEX

Underlined page numbers refer to evaluation text and those not underlined to compiled tables.

REGISTRY NUMBER INDEX

Underlined page numbers refer to evaluation text and those not underlined to compiled tables.

7440-63-3	134-136,137,148,149,150-161,162,163-192, 193-194,195,196,199-201,202-204,206-211, 216,219,220,222,224
7447-40-7	12-13,21,240,255,340
7447-41-8	12-13,21
7487-94-7	239,247
7601-54-9	12-13,23,149,154
7647-01-0	12-13,16
7647-14-5	12-13,20,21,126,128,132,133,149,151-153,240,251,252
7664-41-7	105
7664-93-9	12-13,17,18,340
7681-11-0	12-13,21,149,151
7697-37-2	12-13,16
7720-78-7	239,249
7722-64-7	240,254
7732-02-2	239,246
7732-18-5	1-3,4-8,12-13,14-24,121-126,128,132-133,134-136, 137-145,149,150-154,209-211,216-220,222,224, 227-229,230-234,238-241,242-258,340,343,344
7757-79-1	12-13,24
7757-82-6	12-13,22
7758-02-3	12-13,21
7758-98-7	12-13,17,18,239,244
7761-88-8	239,245
7789-20-0	146,235
9048-46-8	126
10022-31-8	239,250
10024-97-2	102,186
10099-74-8	239,248
10361-37-2	12-13,19
12125-02-9	12-13,14,239,242
13943-58-2	240,253
13983-27-2	7-8,17-18,98,99,112,113,117,120,121,124,125, 127,129-131,133
14835-02-0	339,340-342
14859-67-7	227-229,230-237,238-241,242-258,259-261, 262-264,265,266-272,273,274-290,291,292-293, 294,295-321,322,323-327,328,329-338
14932-42-4	181,182,197,198,205,207,208,212-215,217, 218,221,223,225,226
14995-63-1	207,208,221
22481-88-7	343,344-345
26140-60-3	69